Probability for Information Technology

Changho Suh

Probability for
Information Technology

 Springer

Changho Suh
Korea Advanced Institute of Science and Technology
Daejeon, Korea (Republic of)

ISBN 978-981-97-4031-4   ISBN 978-981-97-4032-1   (eBook)
https://doi.org/10.1007/978-981-97-4032-1

This Springer imprint is published by the registered company Springer Nature Singapore Pte Ltd.
The registered company address is: 152 Beach Road, #21-01/04 Gateway East, Singapore 189721,
Singapore

If disposing of this product, please recycle the paper.

To my family, Ian, Hyun Seung, and Yuni

Preface

Features of the Book

The writing of this book was inspired by research activities in information technology (IT) and the role of probability in this domain. This motivation shapes the key features of the book, facilitating three essential aspects.

The first feature entails the illustration of principles of probability in the context of IT applications. Probability, initially developed to design profitable gambling strategies in games such as card games, dice, and roulette wheels, has evolved into fundamental fields with foundational impacts across various scientific and engineering domains. Today, probability theory is indispensable in the design of communication systems, the internet, control systems, as well as the development of algorithms for machine learning and artificial intelligence. It plays a key role in understanding important theories and principles in science and engineering, covering applications in cloud storage, peer-to-peer file sharing, speech recognition, search engine ranking, network multiplexing, GPS positioning, DNA sequencing, and more. While numerous books on probability have been published over the decades, covering a wide range of contents (Bertsekas & Tsitsiklis, 2008; Yates & Goodman, 2014; Walrand, 2021; Jaynes, 2003; Ross, 2014; Feller, 1991; Kay, 2006; Pitman, 2012; Unpingco, 2016; DasGupta, 2011; Blitzstein & Hwang, 2019), this book focuses on a single field: *information technology (IT)*. Among the vast content, we emphasize probability concepts specifically related to IT applications. The applications we will focus on are: (1) communication; (2) community detection in social networks; (3) machine learning and deep learning; and (4) speech recognition.

Secondly, this book adopts a lecture style. This book aims to inspire readers interested in information technology and potentially deeply connected to other disciplines. We strive to create a compelling storyline that effectively illustrates the role of fundamentals in the field. To establish this motivating storyline, the book adopts a lecture-style organization. Each section functions as a note for a lecture spanning around 80 min, establishing an intimate connection across sections based on coherent themes and concepts. To facilitate a smooth transition between sections, two paragraphs are featured: (i) the "recap" paragraph that summarizes what has been covered; and (ii) the "look ahead" paragraph that introduces upcoming contents by connecting with past materials.

The final feature of this book is the inclusion of numerous programming exercises using two software languages: (i) Python; and (ii) TensorFlow. While C++ and MATLAB have been extensively used in many traditional fields, Python has emerged as a major software tool in recent areas such as the IT field. Given the coverage of numerous IT applications in this book, Python serves as the primary platform. For implementation of machine learning and deep learning algorithms, we employ TensorFlow, one of the most popular deep learning frameworks. TensorFlow provides a multitude of built-in functions that streamline many crucial procedures in deep learning. One notable advantage of TensorFlow is its integration with Keras, the most high-level library with a focus on fast user experimentation. Keras enables us to translate ideas into implementation with minimal steps.

Structure of the Book

This book is composed of course materials that we developed at KAIST, including: (i) EE210 Probability and Introductory Random Processes (offered in Spring 2021); (ii) EE623 Information Theory (offered in Fall 2012–2016, 2018, 2019); and (iii) EE321 Communication Engineering (Spring 2013–2015, 2022). The book is organized into three parts, each containing multiple sections. Each section covers the material presented in a single lecture that lasted approximately 80 min. Each problem set, which served as homework in the courses, is included in three or four sections. The detailed contents are summarized as follows.

I. *Probability basics (12 sections and 4 problem sets):* A brief introduction of probability and its role in IT applications; sample space; events; Monty Hall problem; conditional probability; independence; total probability law; Bayes' law; coupon collector problem; random variables; probability mass functions; expectation; BitTorrent and Python simulation; variance; Markov's inequality; Chebyshev's inequality; continuous random variables; probability density functions; Gaussian random variables.

II. *Random processes and key principles (8 sections and 2 problem sets):* Definition of random processes; the Bernoulli process; the Gaussian process; the Markov process; Maximum A Posteriori (MAP) estimation; error probability and Python simulation; Maximum Likelihood (ML) estimation; Mean Square Error (MSE) and Python simulation; Law of Large Numbers; Central Limit Theorem.

III. *IT applications (15 sections and 4 problem sets):* Communication and probability modeling of noise; the MAP decoder and Python implementation; community detection in social networks and probabilistic modeling; the ML estimate of community detection and its achievability proof; the spectral algorithm and Python implementation; machine learning and probabilistic modeling of data; the role of the ML principle in logistic regression and deep learning; gradient descent; TensorFlow implementation of a digit classifier; speech recognition and probabilistic modeling; the Markov process and graphical models; the MAP principle in speech recognition; the Viterbi algorithm for speech recognition and Python implementation.

Concerning IT applications, many sections are adapted from the author's previous books: (i) "Convex Optimization for Machine Learning" (Suh, 2022); (ii) "Communication Principles for Data Science" (Suh, 2023a); and (iii) "Information Theory for Data Science" (Suh, 2023b). The contents have been tailored to align with the theme of this book, focusing on the role of probabilistic concepts. Toward the end, two appendices provide brief tutorials on the employed programming languages (Python and TensorFlow). These tutorials are adapted from (Suh, 2022, 2023a, 2023b), with appropriate modifications to suit the focused topics. Additionally, a list of references related to the discussed contents is included. Detailed explanations are omitted, as the goal is not to comprehensively cover the extensive research literature.

How to Use This Book

This book is written as a textbook for a sophomore-level undergraduate course, although it is also suitable for a junior or senior-level undergraduate course. Readers are expected to have some mathematical maturity and exposure to programming.

For students and interested readers, we offer some guidelines:

1. *Study one section per day and two sections per week:* Given that each section is designed for a single lecture and most courses include two lectures per week, we recommend this way of reading.
2. *Go through all the contents in Parts I and II:* Some of the most crucial concepts in probability are the MAP and ML principles. If you are already familiar with these, feel free to skim through Parts I and II, and then delve into Part III. However, if you are not yet acquainted with these principles, we strongly recommend going through the contents of Parts I and II sequentially. A motivating storyline is woven across sections, and appropriate exercise problems are strategically placed in between. We believe that following this reading approach will enhance your motivation and understanding of the materials.
3. *Explore Part III in part depending on your interest:* Since Part III is dedicated to applications, you may choose to read them in parts. Nevertheless, we have crafted a logical storyline assuming that each section is read sequentially. One of the key features in Part III is the implementation using Python and TensorFlow. You should be able to implement all the covered algorithms by referring to the guidelines in the main body, along with the skeleton codes provided in problem sets and appendices.

4. *Solve four to five problems in each problem set:* Approximately 100 problems (comprising more than 260 subproblems) are included. Most of these elaborate on the concepts discussed in the main text. The exercises cover a spectrum, from basic principles of probability and random processes, relatively straightforward derivations of results in the main text, in-depth exploration of non-trivial concepts not fully explained in the main text, to programming implementation using Python or TensorFlow. All the problems are closely tied to the established storyline. Working on at least some of the problems is crucial for a thorough understanding of the materials.

In the course offerings at KAIST, we have typically covered most of the materials in Parts I and II, with only two to three applications in Part III. Depending on backgrounds and interests of students, as well as time availability, several alternative ways to structure a course are conceivable.

1. *Semester-based course (24–26 lectures):* Cover all the sections in Parts I and II, along with two to three applications in Part III, such as (i) communication and community detection, (ii) communication and speech recognition, or (iii) machine learning and speech recognition.
2. *Quarter-based course (18–20 lectures):* Cover nearly all the materials in Parts I and II, excluding some topics such as BitTorrent, error probability analysis, and MSE analysis. Explore two applications from Part III.
3. *A graduate course for students with mathematics basics:* Conduct a brief review of the contents in Parts I and II over approximately six to eight lectures. Cover a significant portion of the materials in Part III.

Programming exercises may be incorporated into homework assignments to streamline the learning process.

Daejeon, Korea (Republic of)   Changho Suh
November 2023

Acknowledgements We extend our sincere gratitude to Ian Suh and Hyun Seung Suh for their invaluable feedback and input on the book's structure, writing style, and accessibility to beginners. Their contributions have been immensely valuable, and we deeply appreciate their assistance.

Contents

Chapter 1
Probability Basics

If you find yourself uncertain about a
probability-related problem, it's crucial not to rely
solely on intuition. Instead, it is often best to go back
to basics and employ a more rigorous approach.

1.1 Overview of the Book

Outline
In this section, we will cover two basic topics. Firstly, we will discuss the logistics of the book, explaining how it is organized and is proceeded. Secondly, we will provide a brief summary of the book. At the outset, we will introduce the concept of probability and its significance in information technology (IT). Furthermore, we will showcase how probability plays a role as a critical tool in diverse applications across several fields. Finally, we will provide an overview of the specific topics to be covered in the book.

Prerequisite To benefit from this book, readers should possess a certain level of mathematical maturity and programming exposure. A high school-level understanding of probability concepts is sufficient. While some knowledge of random processes can be helpful, it is not mandatory. Familiarity with Python is recommended as it will be the primary programming language. However, if readers are not familiar

© The Author(s), under exclusive license to Springer Nature Singapore Pte Ltd. 2025
C. Suh, *Probability for Information Technology*,
https://doi.org/10.1007/978-981-97-4032-1_1

with Python, a basic understanding of programming concepts is adequate. For those interested in learning Python, we offer a tutorial in Appendix A for self study.

Problem sets Each problem set will be offered in three or four sections, resulting in a total of 10 problem sets. We encourage collaboration with peers when completing the problem sets. Discussion, teaching, and learning from others are all effective ways to maximize learning. Instructors will be the only ones with access to solutions upon request. Some problems may require the use of programming tools such as Python and TensorFlow. We will use Jupyter notebook. Installation instructions are available in Appendix A.1. Or you can consult with: https://jupyter.readthedocs. io/en/latest/install.html.

Tutorials for programming tools are also provided in appendices: (i) Appendix A for Python; and (ii) Appendix B for TensorFlow.

Probability and uncertainty We introduce the definition of probability and explore its applications in various fields. Additionally, we provide a comprehensive list of topics to be covered in this book.

Probability is defined as the ratio of the number of times an event of interest occurs to the total number of possible outcomes. For example, consider rolling a six-sided die. The probability of rolling 1 would be $\frac{1}{6}$, since there are six possible outcomes. Probability plays a crucial role in dealing with uncertainty, which is pervasive in many real-world scenarios. Probability offers a powerful tool to model and analyze the uncertain behavior of complex systems, making it an essential concept in numerous applications.

Applications Probability theory was developed primarily for the purpose of devising profitable gambling strategies for games such as card games, dice and roulette wheels. However, today's probability theory has become an indispensable tool for a wide range of applications across various fields. For example, it is crucial in the design of optimal communication systems, the internet, and control systems. It also plays a key role in the development of algorithms for artificial intelligence and machine learning. Furthermore, probability theory is essential in understanding many important theories and principles in science. Other applications include cloud storage, peer-to-peer file sharing, speech recognition, ranking in search engines, network multiplexing, GPS positioning, DNA sequencing, and more. In this book, we will focus on four applications to illustrate the role of probability: (1) communication; (2) community detection in social networks; (3) machine learning; and (4) speech recognition.

Application #1: Communication (Shannon 2001) Communication is the process of transmitting information from one end (called the transmitter) to the other end (called the receiver), often in the form of digital information represented by a sequence of 0/1 binary digits, known as bits. The transmitter sends the information through a physical medium (referred to as the channel) to the receiver. See Figure 1.1 for visualization. An example of a channel could be air or a copper wire.

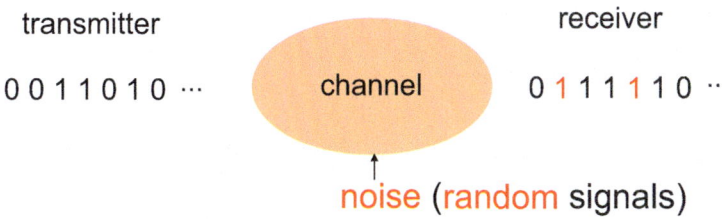

Fig. 1.1 In communication, information is transmitted from the transmitter to the receiver through a channel. However, the channel can introduce *random noise signals*, which can be modeled as a sequence of random variables, also known as a *random process*

The channel in communication highlights the connection with probability. By examining the behavior of the channel, it becomes clear why probability matters in communication. The channel is a system or function that takes a transmitted signal as input and produces a received signal as output. However, the channel is not a deterministic function, so it is not a one-to-one mapping. In other words, the input cannot be easily reconstructed from the output. This is because there is a random entity or noise that is added to the system. In most communication systems, the noise is additive, i.e., the received signal is the sum of the transmitted signal and the noise. Since the noise is random, we cannot predict its behavior. Random noise signals can be mathematically modeled as a sequence of random variables or a random process, which are key concepts in probability.

In fact, the purpose of communication goes beyond just transferring information over a channel. The purpose includes reconstruction of the transmitted signals $(0011010\cdots$ in the example in Figure 1.1) from the received signals that are corrupted by noise $(0111100\cdots$ in the example). Therefore, it is natural to aim at finding the best strategy for reconstruction. The optimal method is based on a key principle in probability called the Maximum A Posteriori probability (MAP) estimation. If you have not heard of the MAP estimation, don't worry. That is precisely why you are reading this book. We will delve deep into the MAP principle and its applications.

Application #2: Community detection in social networks (Abbe 2017; Fortunato 2010; Girvan and Newman 2002) Community detection is arguably one of the most prominent problems in data science and information technology, with significant ramifications across various domains such as social networks (Facebook, LinkedIn, Twitter, etc.) and biological networks. A community refers to a group of people sharing common interests and/or residing in the same locality. Therefore, the task of community detection involves identifying groups that exhibit similarity. Refer to Figure 1.2 for a visual representation. For instance, consider users represented as nodes in a graph (depicted on the left in Figure 1.2). Assume there are two communities, namely the blue and red communities. The objective of the problem is to determine to which community each user belongs, distinguishing between the blue and red communities (shown on the right in Figure 1.2).

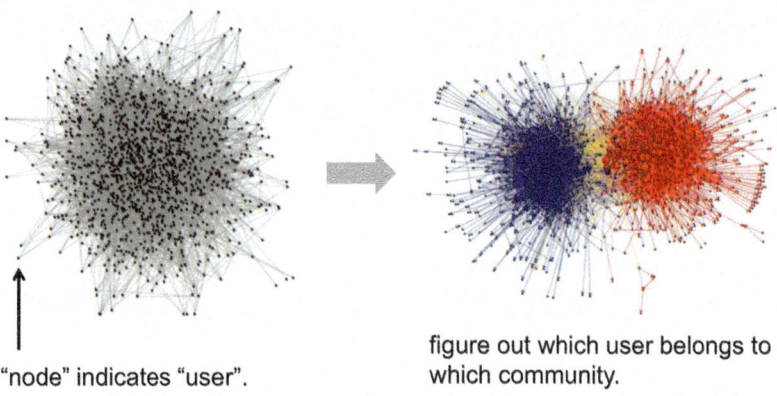

"node" indicates "user".

figure out which user belongs to
which community.

Fig. 1.2 Community detection is to identify groups that are alike

The connection with probability arises from the nature of information available for performing community detection. In many applications, a readily accessible source of information is *relationship information*, such as data on friendships in Meta's network (formerly Facebook). By aggregating more and more pairwise information, one can reliably identify the community memberships of users. For example, using the total number $\binom{n}{2}$ of all possible pairs (where n is the number of users) along with the community membership of a particular user, it becomes possible to deduce the memberships of all users. In practice, however, such similarity parities are provided in a *random* manner. This is where the concept of *probability* becomes crucial.

A common probabilistic assumption made in the literature is that the information for any pair of users is given with a probability, say p, independently across all other pairs. Under this probabilistic assumption, a fundamental task in social networks is to determine the fundamental limit on the probability p above which community detection becomes feasible. The Maximum Likelihood (ML) estimation, a key principle in probability, plays a foundational role in characterizing this limit. If you are unfamiliar with the ML estimation, don't worry. This book will thoroughly explore this principle.

Application #3: Machine learning (LeCun et al. 1998; Samuel 1967) Probability is essential in understanding machine learning which has received significant attention in recent years. Machine learning involves training machines to perform tasks like human beings. Figure 1.3 provides an overview of the machine learning process. Data is a critical component in machine learning, as it is used to train machines. Data is typically represented by a set of input-output pairs:

Fig. 1.3 Machine learning is a methodology used to train machines to perform tasks that require human-like intelligence. In many cases, the most effective training methodology relies on a key principle in probability known as the Maximum Likelihood estimation

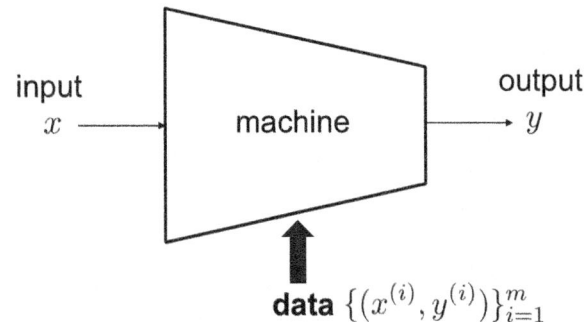

$$\{(x^{(i)}, y^{(i)})\}_{i=1}^{m} \tag{1.1}$$

where $(x^{(i)}, y^{(i)})$ is the ith training example (or sample) and m is the number of examples.

Probability plays a central role in the training process, especially in cases where the Maximum Likelihood (ML) estimation is the optimal training methodology. This is where the connection with probability becomes evident.

Application #4: Speech recognition (Vaseghi 2008) Probability is useful in the design of speech recognition systems such as Siri or Amazon Alexa. To see its relevance to probability, consider the goal of speech recognition, that is to convert voice signals, comprising spoken words picked up from a microphone, into a written command that can be represented in the form of text. See Figure 1.4 for illustration.

A key observation in speech recognition is that voice characteristics vary between speakers, introducing uncertainty into voice signals. As a result, voice signals can be modeled as a *random process*. Probability plays an important role in designing speech recognition systems, as the problem can be formulated as a communication

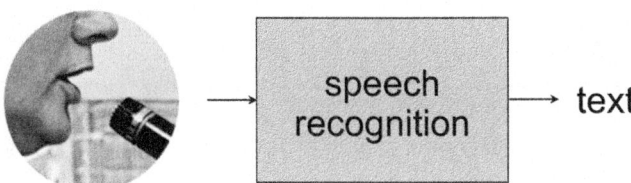

Fig. 1.4 The goal of speech recognition is to convert voice signals into written commands, while accounting for variations in voice characteristics across different speakers. This uncertainty in voice signals can be modeled as a random process, and probabilistic methods are commonly used to design speech recognition systems. One key principle in probability that underlies the design of speech recognition systems is the Maximum A Posteriori (MAP) estimation

problem, similar to the one discussed in Figure 1.1. The best way to design a speech recognition system is often based on the MAP estimation.

Book outline From the above applications, we observe several fundamental concepts and principles in probability, including (i) random variables, (ii) random processes, (iii) the MAP estimation, and (iv) the ML estimation. These are the primary topics this book aims to cover in depth. Comprehending these concepts necessitates familiarity with some more rudimentary probability concepts. Hence, we have structured this book into three parts.

In Part I, we will focus on the basic concepts of probability that you may have learned from high school. These include sample space, events, conditional probability, independence, total probability law, random variables, probability mass function, density, expectation and variance.

Moving onto Part II, we will delve into two key principles: (i) the MAP estimation; and (ii) the ML estimation. To discuss these principles, we will also need to explore the Law of Large Numbers and the Central Limit Theorem. Furthermore, we will cover popular inequality techniques such as Markov's inequality and Chebyshev's inequality. These are useful for proving the theorems concerning the principles.

Finally, in Part III, we will explore the four applications and emphasize the role of the concepts covered in Parts I and II. We will take a deeper dive into communication, social networks, machine learning, and speech recognition, highlighting how probability theory is instrumental in these applications. Overall, this book provides a comprehensive understanding of the essential concepts and principles, making it a valuable resource for those interested in information technology.

1.2 Sample Space and Events

Recap
In the previous section, we introduced the definition of probability: the fraction of occurrences of an interested event over all possible outcomes of an experiment. We emphasized the role of probability in several applications, including communication, social networks, machine learning, and speech recognition. We then discussed two concepts: (i) random variables; and (ii) random processes, along with two key principles: (i) the MAP (Maximum A Posteriori probability) estimation; and (ii) the ML (Maximum Likelihood) estimation. We mentioned that a deeper understanding of these requires knowledge of more basic concepts, such as sample space, events, conditional probability, independence, and total probability law. In the upcoming sections, we will delve into these.

Outline
In this section, we will explore sample space and events. Specifically, we will cover the following topics:

1. Definition of sample space;
2. Probability model;
3. Definition of an event;
4. Computing the probability of an event;
5. Five easy and one non-trivial examples.

First, we will explain what the sample sample is and build upon it to develop a probability model, which will then be used to connect the sample space to events. Next, we will define events and learn how to compute the probability of an interested event. Finally, we will practice these concepts via a suite of examples.

Sample space The sample space is a collection of all potential outcomes that occur from a given experiment. To illustrate this, consider a simple experiment where a coin is tossed four times. What are possible outcomes here? They are: HHHH, HHHT, HHTH, HHTT, all the way up to, TTTH and TTTT; see Figure 1.5. The sample space, commonly denoted as Ω, includes all possible outcomes of an experiment. For instance, in the case of flipping a coin four times, $\Omega = \{HHHT, HHHT, HHTH, HHTT, \ldots, TTTH, TTTT\}$. In this set, an element, typically denoted as $\omega \in \Omega$, represents each outcome. The concept of events is linked to the sample space via a *probability model*.

Probability model A probabilistic model is used to describe the likelihood of various outcomes and it comprises two entities. The first is the sample space Ω. The second is a *probability assignment*, denoted by $\mathbb{P}(\omega)$, for each $\omega \in \Omega$. This probability assignment, also referred to as *probability distribution*, must comply with the following two properties, called the non-negativity and the sum-up-to-one properties:

$$0 \le \mathbb{P}(\omega) \le 1 \quad \text{for all } \omega \in \Omega;$$
$$\sum_{\omega \in \Omega} \mathbb{P}(\omega) = 1. \tag{1.2}$$

The simplest way to assign probabilities to outcomes is by assigning each one an equal probability: $\mathbb{P}(\omega) = \frac{1}{|\Omega|} \ \forall \omega \in \Omega$. Here, the symbol "$\forall$" means "for all," and $|\Omega|$ indicates the number of elements in Ω, called the *cardinality* of Ω. This distribution is referred to as a *uniform distribution*. However, not all probability distributions are uniform. It is worth noting that while there are many potential outcomes in the sample space, the experiment results in only one specific outcome. The randomness lies in the fact that we cannot predict the exact outcome beforehand, and the only way to discuss the occurrence is in a *probabilistic* manner by appropriately modeling the probability distribution $\mathbb{P}(\omega)$.

Events An event refers to a specific outcome in an experiment. To understand this better, recall the coin-tossing example; see Figure 1.5. Suppose we are interested in the occurrence of "getting two heads" in a four-time coin toss. In this case, the corresponding outcomes would be HHTT, HTHT, HTTH, THHT, THTH, and TTHH. Here, an event E is simply a collection of all such outcomes. Formally, the event E is defined as the set of all outcomes corresponding to a specific occurrence; see Figure 1.6. The event E is a subset of the sample space, i.e., $E \subseteq \Omega$. This means that all outcomes in the event E must be contained within the sample space Ω.

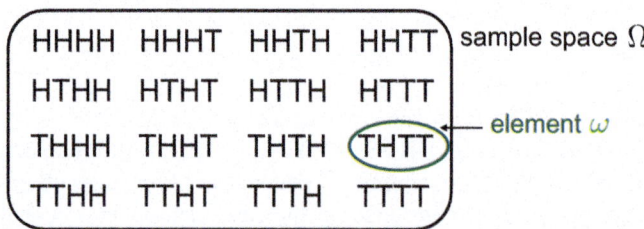

Fig. 1.5 When tossing a coin four times, the sample space is the set of all conceivable outcomes, represented by Ω. In this case, the letters H and T indicate "Head" and "Tail" respectively. Each possible outcome is an element in the set, represented by a lowercase letter ω

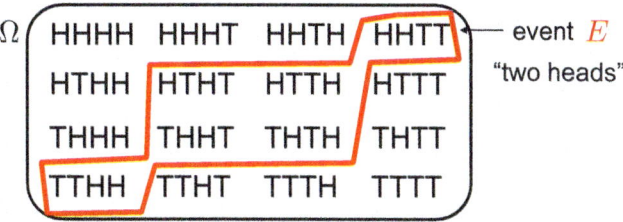

Fig. 1.6 An event E is defined as the set of outcomes that satisfy a specific condition of interest, such as the occurrence of "getting two heads out of four-times coin tossing" in our previous example. The event E consists of all outcomes that meet the given condition

One natural question that arises is: how to define the probability of an event E? The way should be to add up all the probabilities of the associated outcomes in E. For any event $E \subseteq \Omega$,

$$\mathbb{P}(E) := \sum_{\omega \in E} \mathbb{P}(\omega) \tag{1.3}$$

where the symbol ":=" means "is defined as". In the prior example in Figure 1.5, the probability of getting two heads in four coin tosses can be calculated as:

$$\mathbb{P}(E) = \binom{4}{2} \cdot \frac{1}{16} = \frac{6}{16}. \tag{1.4}$$

There are $|\Omega| = 16$ possible outcomes, and each outcome $\omega \in \Omega$ has the same probability $\frac{1}{16}$. There are six ("4-choose-2" $= \binom{4}{2}$) outcomes in E, yielding $\mathbb{P}(E) = \frac{6}{16}$. Here we can make the following observation.

> **Observation**
> For uniform distribution, the probability calculation of an event E boils down to *counting* the corresponding outcomes.

Easy examples We practice these concepts with several examples.

1. *Tossing a fair coin once*: The sample space is $\Omega = \{H, T\}$ where H represents the event of getting a head and T represents the event of getting a tail. Since the coin is fair, each outcome in Ω is equally likely, that is, $\mathbb{P}(\omega) = \frac{1}{2}$ for all $\omega \in \Omega$. One interested event is getting a head. Formally, $E = \{H\}$ and thus $\mathbb{P}(E) = \mathbb{P}(\{H\}) = \frac{1}{2}$. Similarly, the probability of getting a tail is also $\frac{1}{2}$.
2. *Tossing a fair coin three times*: The sample space would be:

$$\Omega = \{HHH, HHT, HTH, HTT, THH, THT, TTH, TTT\}. \tag{1.5}$$

Since the coin is still fair, all possible outcomes are equiprobable. Therefore, we can assign the same probability to each outcome as $\mathbb{P}(\omega) = \frac{1}{|\Omega|} = \frac{1}{8}$ for every $\omega \in \Omega$. Consider the event E of "getting two heads". We can then calculate the probability of E as:

$$\mathbb{P}(E) = \frac{\binom{3}{2}}{8} = \frac{3}{8}, \tag{1.6}$$

since there are 3 ($= \binom{3}{2}$) possible outcomes. A different event, such as "All three are same", has a probability of $\mathbb{P}(E) = \frac{1}{8} + \frac{1}{8}$, as there are only two outcomes: $E = \{HHH, TTT\}$.

3. *Tossing a biased coin once:* The sample space remains the same as that of the first example, $\Omega = \{H, T\}$, regardless of coin's characteristics. However, the bias of the coin would affect the probability distribution. Suppose that "Head" is twice as likely as "Tail". Then, the probability assignment would be:

$$\mathbb{P}(H) = \frac{2}{3}, \quad \mathbb{P}(T) = \frac{1}{3}.$$

Perhaps this serves as the simplest example for non-uniform probabilities.

4. *Tossing the biased coin twice:* The sample space reads:

$$\Omega = \{HH, HT, TH, TT\}.$$

However, assigning probabilities to these outcomes is not immediately obvious. This is because the bias of the coin tells us how to assign probabilities to the outcome of a *single* flip, not the outcome of *multiple* flips. The one thing we do know for sure is that the probabilities of the outcomes should not be uniform. While one may argue that $\mathbb{P}(HH) = \frac{2}{3} \times \frac{2}{3}$, this is based on intuition and prior knowledge of *independence*, which we have not yet covered. Since we did not learn about the independence, a rigorous discussion on independence will be deferred to a later section. Please bear with us.

5. *Rolling two fair dice:* In this case, the sample space reads:

$$\Omega = \{(1, 1), (1, 2), (1, 3), \ldots, (4, 6), (5, 6), (6, 6)\}.$$

All 36 outcomes are equally likely, with probability $\frac{1}{36}$ each. Consider the event E that the sum of the two dice is at least 10. Since the probability distribution is uniform, we compute the probability of E by counting the number of outcomes in Ω that satisfy this event:

$$(4, 6), (5, 5), (5, 6), (6, 4), (6, 5), (6, 6).$$

Hence, $\mathbb{P}(E) = \frac{6}{36}$.

A non-trivial example The examples we have looked at so far may have given you the impression that the concepts of sample space, probability assignment, and events are simple and easy to grasp. However, this is not always the case. In fact, there are numerous examples where it is challenging to come up with an appropriate sample space and calculate the probability of an event. Even experts in probability theory can easily make mistakes when dealing with these concepts. Nonetheless, there are many clever techniques that can be used to calculate probabilities efficiently. We will introduce one such technique via a non-trivial example.

The example is a well-known problem. Imagine a classroom with three students, each of whom has a unique birthday in a non-leap year having exactly 365 days. What would be the sample space Ω in this situation? It can be represented by a set of all possible triplets of birthdays, each ranging from the first to the 365th day of the year:

$$\Omega = \{(1, 1, 1), (1, 1, 2), (1, 1, 3), \ldots, (365, 365, 364), (365, 365, 365)\}.$$

The cardinality is $|\Omega| = 365^3$. Suppose that each of the 365 days is equally likely for each student's birthday, and the birthdays of the three students are independent of each other. Then, a natural probability assignment would be uniform: $\mathbb{P}(\omega) = \frac{1}{365^3} \ \forall \omega \in \Omega$.

Now consider the event E in which at least two students share the same birthday. Since the probability distribution is uniform, we can simply count the number of outcomes that satisfy this condition. There are several possibilities, such as two students sharing the same birthday or all three having the same birthday. If we extend the scenario to a more general case where there are n students in a classroom, determining the number of outcomes for such an event would be considerably more complex.

An interesting trick comes to the rescue. Whenever you encounter a scenario where the counting appears challenging (often indicated by phrases such as "at least" in the event description), consider its *complement*: $E^c := \{\omega : \omega \in \Omega, \omega \notin E\}$. Actually the above is the situation where counting $|E^c|$ is much simpler. To see this clearly, consider the complement of E: "No two students share the same birthday". How do we count the number of outcomes for this event? We utilize permutations. For the first student, there are 365 possible choices, followed by 364 choices for the second student (to ensure the birthday differs from the first student's), and 363 choices for the third student. Thus, the number of possible ways to select 3 choices out of 365 (where order matters) is equivalent to the number of outcomes for this event:

$$|E^c| = {}_{365}P_3 = \frac{365!}{(365 - 3)!} = 365 \times 364 \times 363 \tag{1.7}$$

where $n! := n \times (n - 1) \times \cdots \times 2 \times 1$. Using this, we can then compute:

$$\mathbb{P}(E) = \frac{|\Omega| - |E^c|}{|\Omega|} = 1 - \frac{{}_{365}P_3}{365^3}. \tag{1.8}$$

Also, one can readily extend this approach to a general scenario where there are n students. In such a case, the probability of the event E can be expressed as follows:

$$\mathbb{P}(E) = \frac{|\Omega| - |E^c|}{|\Omega|} = 1 - \frac{365 P_n}{365^n}. \tag{1.9}$$

Birthday paradox (Flajolet et al. 1992) This well-known example has an interesting naming. It is known as "Birthday Paradox". The naming comes from the fact that the actual value of $\mathbb{P}(E)$ contradicts many people's intuition. Initially, people believed that the likelihood of two students sharing the same birthday is small because there are many options for a birthday (365 days). However, the precise calculation reveals that for $n = 23$, $\mathbb{P}(E)$ exceeds 50%, and for $n = 60$, $\mathbb{P}(E)$ is over 99%.

Look ahead
As mentioned earlier, several challenging examples exist where computing the probability of an event is difficult. In the next section, we will examine another such example. Constructing an appropriate sample space can be intricate in the upcoming example, and hasty calculations based on intuition can lead to errors. This example emphasizes the importance of a systematic and rigorous approach based on the probability model, wherein we should first define a sample space and then calculate the probability of the event based on that sample space.

1.3 Monty Hall Problem and Python Simulation

Recap

In the previous section, we discussed sample space, probability model, and events. The sample space Ω refers to the set of all possible outcomes in an experiment. The probability model consists of the sample space Ω and the probability distribution $\mathbb{P}(\omega)$, which adheres to the non-negativity and sum-up-to-one constraints:

$$0 \leq \mathbb{P}(\omega) \leq 1 \quad \text{for all } \omega \in \Omega; \tag{1.10}$$

$$\sum_{\omega \in \Omega} \mathbb{P}(\omega) = 1. \tag{1.11}$$

An event E is a subset of the sample space Ω, consisting of specific outcomes associated with a particular occurrence of interest. The probability of an event is defined as:

$$\mathbb{P}(E) := \sum_{\omega \in E} \mathbb{P}(\omega). \tag{1.12}$$

We also went through several examples. When discussing the birthday paradox problem, we highlighted the existence of numerous non-trivial counter-intuitive examples where one can easily make mistakes.

Outline

In this section, we will delve into another counter-intuitive example. The section is divided into five parts. Firstly, we will explain the context in which the example arises. Next, we will introduce a question that was raised in the example. It turns out the introduced question can be misleadingly answered once we rely solely upon intuition. Thirdly, we will construct a suitable sample space and probability model to arrive at the correct answer. In the fourth part, we will highlight the importance of adopting a systematic approach based on the sample space and probability model. Finally, we will validate the correct answer numerically via Python simulation.

Context One interesting problem that we will discuss is the Monty Hall problem, which originated from a well-known American television game show named "Let's Make A Deal". This problem is named after the show's host, Monty Hall. During the show, a participant (referred to as the "trader") is selected from audience and makes a deal with Monty Hall. The Monty Hall problem emerged as one of the challenges presented during the game (Figure 1.7).

Monty Hall problem The problem is set up as follows. There are three doors, and behind one of them is the prize "car", while the other two conceal "goats". The trader does not know which door hides the car, while the host does. See Figure 1.8 for illustration of the problem.

Given the setting, the deal starts with the trader choosing one of the three doors, hoping to select the one hiding the prize "car". The host, who is aware of what lies behind each door, then opens one of the remaining doors, behind which is a goat (that we call the "goat door"). One scenario is illustrated in Figure 1.9. The trader happened to select door 1, which turned out to be the one hiding the car. The host opened door 3, revealing a goat. It is noteworthy that the host will always have a goat-door to open since there are two such doors in the setup. Therefore, even if the trader had initially picked a goat-door (in this case, door 2), there would still be one goat-door left for the host to reveal (door 3 in this scenario).

The trader is then presented with two options: (i) sticking with the original choice of door (in this case, door 1); or (ii) switching to the other unopened door (door 2 in this case). See Figure 1.10. To make a smart choice, the trader needs to figure out which strategy would be more beneficial. Therefore, it is essential to determine the probabilities of the following two events:

$$\mathbb{P}(\text{win w/ sticking}) \quad \text{vs} \quad \mathbb{P}(\text{win w/ switching}).$$

Fig. 1.7 (On the left) The logo of a well-known American game show; (On the right) A portrait of Monty Hall, the original host of the game show, after whom the "Monty Hall Problem" is named

Fig. 1.8 The Monty Hall problem is set up with three doors, behind one of which is the prize car, and behind the other two are goats. The trader does not know which door hides the car, while the host does

Fig. 1.9 Step 1 of the Monty Hall problem involves the trader being asked to select a door that she/he believes has the prize car behind it. In Step 2, the host, who knows which door hides the prize, opens a different door that has a goat behind it. Finally, in Step 3, the trader is given the option of either sticking with the initial door selection or switching to the other unopened door

Fig. 1.10 Sticking with the initial choice vs switching to another unopened door

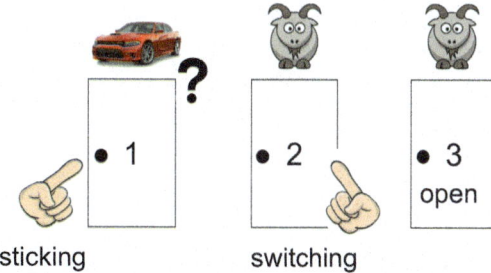

An immediate response Regarding the first probability, one can easily provide a response:

$$\mathbb{P}(\text{win w/ sticking}) = \frac{1}{3}.$$

With three doors available and the car hidden behind one of them, the likelihood of winning would remain at $\frac{1}{3}$ if no action is taken (i.e., if the trader chooses to stick with the initial door choice).

Regarding the second probability, one may believe that it might also be $\frac{1}{3}$ since there was no relocation of the car or goats, and thus no changes occurred.

$$(\text{Initial guess}): \mathbb{P}(\text{win w/ switching}) = \frac{1}{3}. \tag{1.13}$$

When the author of this book first encountered this problem in a probability course at UC Berkeley, the initial guess was the same as the one mentioned earlier. Interestingly, even Paul Erdős, one of the most famous mathematicians in history, also held this belief; see Wikipedia.

Switching is more beneficial   Remember that this is a non-trivial and counter-intuitive example. So it is easy to imagine that the second probability is not $\frac{1}{3}$. If you're wondering what the actual value of the second probability is, you need to consider what has changed between the initial timing and the later timing when the trader is given the option to stick or switch. The critical difference is that the host opens one of the goat-doors in the later timing, which removes that door as a viable choice. Naturally, this may lead you to speculate that the second probability is:

$$\text{(Second guess): } \mathbb{P}(\text{win w/ switching}) = \frac{1}{2}, \qquad (1.14)$$

since there are only two doors left and one is a goat-door while the other is undoubt-edly the car-door. As it turns out, this guess is also incorrect. You may be wondering what went wrong. The correct answer is that the probability of winning by switching is twice as much as the probability of winning by sticking:

$$\text{(Correct answer): } \mathbb{P}(\text{win w/ switching}) = \frac{2}{3}. \qquad (1.15)$$

In the sequel, we will present a rigorous proof of (1.15).

Sample space   To provide a rigorous proof, we should rely on the concept of a *sample space* instead of relying on potentially unreliable intuition. To construct a sample space, we first need to consider where uncertainty arises. In the problem setting, there are three sources of uncertainty: (i) the location of the car; (ii) the trader's initial choice; and (iii) the host's choice. Therefore, we can consider the following triplet as a starting point:

$$\text{(car's location, trader's choice, host's choice).} \qquad (1.16)$$

The next question is: what are the possible triplets? To make it easier to identify them, let's consider two cases:

$$\text{(Case I): car's location } = \text{ trader's choice;}$$
$$\text{(Case II): car's location } \neq \text{ trader's choice.}$$

In Case I (for example, when the car's location and the trader's initial choice are both door 1 as shown in Figure 1.11(Left)), there are two goat-doors remaining. Therefore, the host's choice would be either door 2 or door 3.   The following six triplets belong to this category:

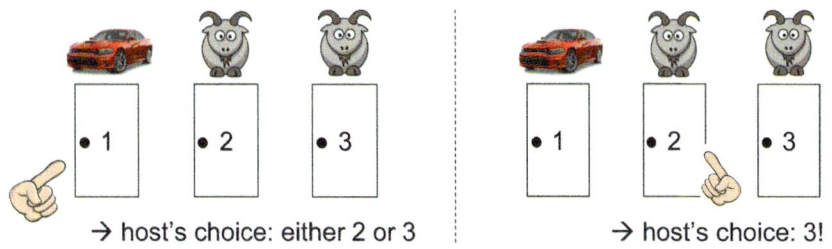

→ host's choice: either 2 or 3   → host's choice: 3!

Fig. 1.11 (Left) Case I: car's location = trader's choice; (Right) Case II: car's location ≠ trader's choice

$$(1, 1, 2), (1, 1, 3), (2, 2, 3), (2, 2, 1), (3, 3, 1), (3, 3, 2).$$

In Case II (for example, where the car's location is door 1 and the trader's choice is door 2 as shown in Figure 1.11(Right)), there is only one goat-door remaining. Therefore, the host must choose door 3. There are six triplets of this type because $_3P_2 = 3 \times 2 = 6$. Thus, the following triplets belong to this category:

$$(1, 2, 3), (2, 1, 3), (1, 3, 2), (3, 1, 2), (2, 3, 1), (3, 2, 1).$$

We can construct a sample space by aggregating all of the possible triplets we mentioned above:

$$\Omega = \{(1, 1, 2), (1, 1, 3), (2, 2, 3), (2, 2, 1), (3, 3, 1), (3, 3, 2) \\ (1, 2, 3), (2, 1, 3), (1, 3, 2), (3, 1, 2), (2, 3, 1), (3, 2, 1)\}. \tag{1.17}$$

Probability distribution Next, we need to determine the probability distribution $\mathbb{P}(\omega)$ for all $\omega \in \Omega$. To do so, we consider all possible configurations that depend on the car's location and the trader's choice. Since there are three possible choices for each, there are a total of 9 possible cases. Since these choices are random, we assume that all 9 cases are equally likely. This motivates us to construct the probability distribution as follows:

$$\mathbb{P}((1, 2, 3)) = \mathbb{P}((2, 1, 3)) = \mathbb{P}((1, 3, 2))$$

$$= \mathbb{P}((3, 1, 2)) = \mathbb{P}((2, 3, 1)) = \mathbb{P}((3, 2, 1)) = \frac{1}{9}; \tag{1.18}$$

$$\mathbb{P}((1, 1, 2)) + \mathbb{P}((1, 1, 3)) = \mathbb{P}((2, 2, 3)) + \mathbb{P}((2, 2, 1))$$

$$= \mathbb{P}((3, 3, 1)) + \mathbb{P}((3, 3, 2)) = \frac{1}{9}. \tag{1.19}$$

We include two probabilities when the car's location is the same as the trader's choice, e.g., $\mathbb{P}((1, 1, 2)) + \mathbb{P}((1, 1, 3)) = \frac{1}{9}$. The reason is that for such cases, there are two sub-cases. This is also demonstrated in Figure 1.12.

Assuming that if the car's location is the same as the trader's choice, the host's choice is random between the two remaining goat-doors, we can assign equal probabilities to the two sub-cases:

$$\mathbb{P}((1, 1, 2)) = \mathbb{P}((1, 1, 3)) = \mathbb{P}((2, 2, 3)) = \mathbb{P}((2, 2, 1))$$

$$= \mathbb{P}((3, 3, 1)) = \mathbb{P}((3, 3, 2)) = \frac{1}{18}. \qquad (1.20)$$

Computation of $\mathbb{P}$(win w/ switching) Using the sample space Ω we constructed earlier, we can now prove (1.15). Winning with switching corresponds to all the triplets where car's location differs from trader's choice. Therefore, the probability of winning with switching is:

$$\mathbb{P}(\text{win w/ switching}) = \mathbb{P}((1, 2, 3)) + \mathbb{P}((2, 1, 3)) + \mathbb{P}((1, 3, 2))$$

$$+ \mathbb{P}((3, 1, 2)) + \mathbb{P}((2, 3, 1)) + \mathbb{P}((3, 2, 1)) = \frac{6}{9} = \frac{2}{3}.$$

We can also verify the probability of winning by sticking, which we guessed as $\frac{1}{3}$ earlier. The event of winning with sticking corresponds to all the triplets where car's location is the same as the trader's initial choice. Thus, we have:

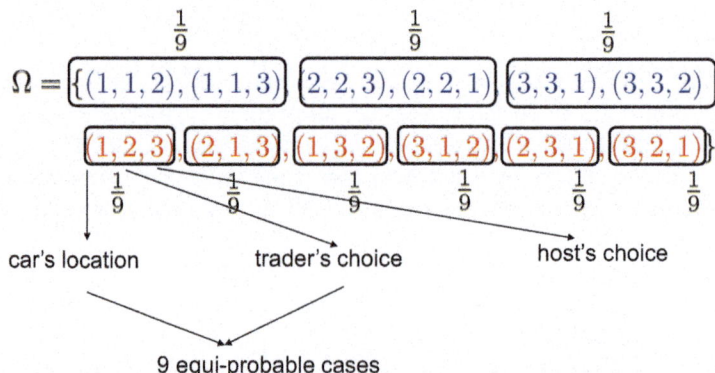

Fig. 1.12 To construct the probability distribution, consider the 9 equiprobable cases that arise from the random choices of car's location and trader's choice. Therefore, we assign a probability of $\frac{1}{9}$ to each of these cases. For example, $\mathbb{P}((1, 2, 3)) = \frac{1}{9}$. However, when the car's location is the same as the trader's choice, there are two sub-cases. Therefore, we assign a combined probability of $\frac{1}{9}$ to these sub-cases. Specifically, we have $\mathbb{P}((1, 1, 2)) + \mathbb{P}((1, 1, 3)) = \frac{1}{9}$

$$\mathbb{P}(\text{win w/ sticking}) = \mathbb{P}((1, 1, 2)) + \mathbb{P}((1, 1, 3)) + \mathbb{P}((2, 2, 3))$$
$$+ \mathbb{P}((2, 2, 1)) + \mathbb{P}((3, 3, 1)) + \mathbb{P}((3, 3, 2)) = \frac{6}{18} = \frac{1}{3}.$$

Lesson We would like to highlight one lesson that we can learn from the Monty Hall problem:

> **Lesson**
> If you feel unsure about how to solve a problem, it's important not to rely solely on intuition. Instead, it's often best to go back to the basics and rely on a more rigorous approach.

Our initial and second guesses ((1.13) and (1.14)) were incorrect because we relied solely on intuition without verifying the probability based on the probability model. This illustrates the importance of not relying on intuition alone and instead using a probability model to obtain a reliable answer.

To confirm the validity of the proof, we can also conduct a computer simulation. This step can be helpful in convincing those who may not trust a proof, even if it is rigorous. The ability to program and run simulations is a valuable skill in such situations. In fact, the renowned mathematician Paul Erdős did not accept the proof of (1.15) until he was presented with a confirming computer simulation. Below, we will use Python to conduct a simulation and confirm the result.

Python simulation We will conduct simulations to empirically validate what we proved in the Monty Hall problem:

$$\mathbb{P}(\text{win w/ sticking}) = \frac{1}{3}; \tag{1.21}$$

$$\mathbb{P}(\text{win w/ switching}) = \frac{2}{3}. \tag{1.22}$$

Initially, consider the scenario in which the trader employs the *sticking* strategy. We will develop a function that outputs 1 for a win and 0 for a loss in each game. Subsequently, we will conduct this game independently for N trials and calculate the empirical winning rate, defined as the ratio of the number of winnings to the total number of trials, as a function of N. This analysis will enable us to compare the empirical result with the theoretical probability of $\frac{1}{3}$. The verification of convergence in this process relies on the *Monte Carlo simulation* approach (Kroese et al. 2014), grounded in the *Law of Large Numbers (LLN)* (Bernoulli 1713; Bertsekas and Tsitsiklis 2008), a prominent theory that will be extensively explored in Sect. 2.5. The subsequent code implementation illustrates this methodology.

```
import numpy as np
import matplotlib.pyplot as plt

Nexp=np.arange(1,20,1) # Nexp=[1,2,3,...,19]
N=2**Nexp # N=[2**1, 2**2, 2**3, ..., 2**19]
emp_rate = np.zeros(len(N)) # initialization
theory=np.zeros(len(N))+1/3 # theoretic prob

for k in range(len(N)):
    car=np.random.randint(3,size=N[k])+1
    pick= np.random.randint(3,size=N[k])+1
    emp_rate[k]=sum(car==pick)/N[k]

plt.figure(figsize=(4,4),dpi=150)
plt.plot(N,emp_rate,color='blue',
         label='empirical rate')
plt.plot(N,theory,color='red',
         label='theoretic probability=1/3')
plt.xlabel('N (number of trials)')
plt.ylabel('empirical rate')
plt.title('empirical rate for sticking')
plt.xscale('log')
plt.legend()
```

In Figure 1.13, the empirical success rate for the sticking strategy is plotted, alongside a comparison to the theoretical probability. For visualization purpose, a log scale is applied to the x-axis. As the number of trials N increases, the empirical rate converges towards the theoretical probability of $\frac{1}{3}$.

Next, examine the scenario in which the trader adopts the *switching* strategy. A crucial observation in this case is that whenever car's location differs from the initial choice, the switched choice aligns with car's location. Leveraging this insight, we can streamline the code as shown below.

```
import numpy as np
import matplotlib.pyplot as plt

Nexp=np.arange(1,20,1) # Nexp=[1,2,3,...,19]
N=2**Nexp # N=[2**1, 2**2, 2**3, ..., 2**19]
emp_rate = np.zeros(len(N)) # initialization
theory=np.zeros(len(N))+2/3 # theoretic prob

for k in range(len(N)):
    car=np.random.randint(3,size=N[k])+1
    pick= np.random.randint(3,size=N[k])+1
    # key observation: whenver the car's location
    # is different from the first choice,
    # the switched choice matches the car.
    emp_rate[k]=sum(car!=pick)/N[k]

plt.figure(figsize=(4,4),dpi=150)
plt.plot(N,emp_rate,color='blue',
```

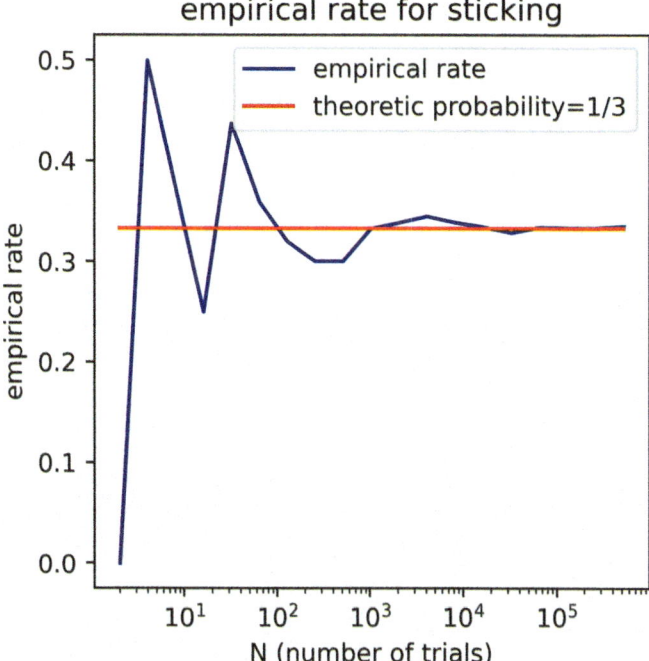

Fig. 1.13 An empirical rate for the sticking strategy versus theoretic probability $\frac{1}{3}$

```
            label='empirical rate')
plt.plot(N,theory,color='red',
            label='theoretic probability=2/3')
plt.xlabel('N (number of trials)')
plt.ylabel('empirical rate')
plt.title('empirical rate for switching')
plt.xscale('log')
plt.legend()
```

Figure 1.14 demonstrates the empirical success rate for the switching strategy with a comparison to the theoretical probability. We see that as the number of trials N increases, the empirical rate approaches the theoretic probability $\frac{2}{3}$.

Look ahead

We have covered fundamental concepts like sample space, probability models, and events through various examples. In the upcoming section, we will delve deeper into probability theory, exploring more concepts such as conditional probability, total probability law, and Bayes' law.

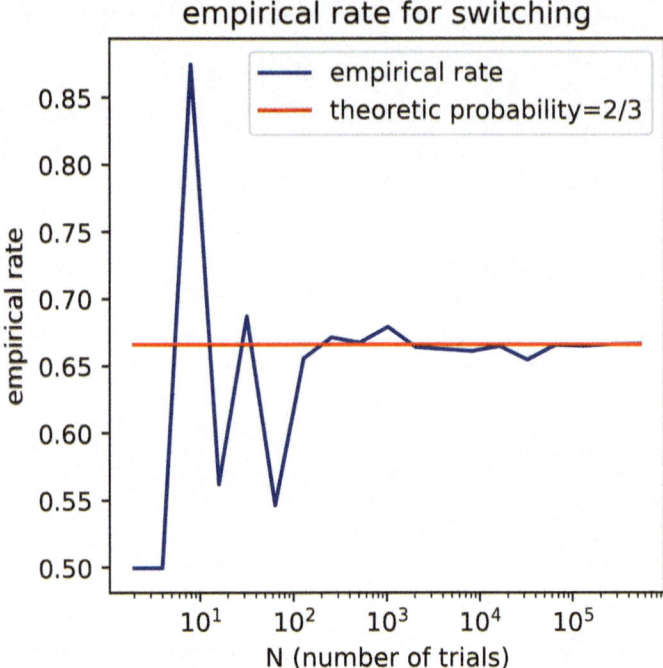

Fig. 1.14 An empirical rate for the switching strategy versus theoretic probability $\frac{2}{3}$

Problem Set 1

Problem 1.1 (*Sample space and events*) Consider an experiment of tossing a coin n times where $n \geq 3$.

(a) Construct a sample space Ω. What is $|\Omega|$?
(b) Let A be the event that the first flip is "Head". What is $|A|$?
(c) Let B be the event that the last (nth) flip is "Head". What is $|B|$?
(d) Let C be the event that the first flip and the last flip are both "Head"s. What is $|C|$?
(e) Let D be the event that the first flip or the last flip is "Head". What is $|D|$?
(f) Are the events A and B disjoint? Express the event C in terms of A and B, using the set operations like "$\cap$" and "$\cup$". Express the event D in terms of A and B.

Problem 1.2 (*Probability model: Exercise #1*) Consider an experiment of flipping a fair coin until a cumulative count of three instances of obtaining a "head" is reached.

(a) Construct a sample space Ω.
(b) Construct the probability distribution $\mathbb{P}(\omega)$ for all $\omega \in \Omega$.
(c) Using part (b) or otherwise, argue that

$$\sum_{i=3}^{\infty} \frac{\binom{i-1}{2}}{2^i} = 1. \tag{1.23}$$

(d) Empirically confirm (1.23) using a Python simulation. It may be helpful to generate a plot with n on the x-axis and $\sum_{i=3}^{n} \frac{\binom{i-1}{2}}{2^i}$ on the y-axis for a suitable range of n, such as $3 \leq n \leq 30$.
Hint: Consider utilizing built-in functions such as `scipy.special.comb` in your implementation.

Problem 1.3 (*Probability model: Exercise #2*) Pick 3 balls without replacement from an urn with 15 balls that are identical except that 10 are red and 5 are blue.

(a) Construct a sample space Ω.
(b) Construct the probability distribution $\mathbb{P}(\omega)$ for all $\omega \in \Omega$.

Problem 1.4 (*Counting: Exercise #1*) Consider a standard deck of 52 cards that consists of four suits (clubs, diamonds, hearts, and spades), each containing 13 ranks (Ace, 2, 3, ..., 10, Jack, Queen, King). Suppose that there are four players and a dealer, and the dealer distributes the 52 cards evenly amongst the players such that each player receives 13 cards.

(a) In how many ways can 52 cards be distributed evenly among four players?
Hint: First, consider the total number of ways to arrange 52 cards, which is 52!.
(b) Suppose A represents the event in which each player receives exactly one "Ace" in a given deal. What is the cardinality of A?

(*c*) Consider the scenario in which this distribution of 52 cards amongst four players is repeated three times. Let *B* denote the event in which event *A* occurs at least once in these three deals. What is the cardinality of *B*?

Problem 1.5 (*Counting: Exercise #2*) Select a number at random from the range of 1 to 999999, with each number having an equal probability. What is the probability that the sum of its digits will be 23? For instance, the number 7646 falls within the range of 1 to 999999 and has a digit sum of 23 (i.e., $7 + 6 + 4 + 6$).

Problem 1.6 (*Counting: Exercise #3*) What is the count of three-digit positive integers that have the same digit repeated in all three positions?

Problem 1.7 (*Birthday*) Assuming that the birthdays of *n* students, numbered from 1 to *n*, are uniformly distributed across 365 days, what is the probability that students 1, 2 and 3 share the same birthday?

Problem 1.8 (*4 sided die & 6 sided die*) You possess a die with four sides labelled 1 through 4 and another die with six sides labelled 1 through 6.

(*a*) You roll one of the dice with an equal probability of $1/2$ for each die. If the result of the roll is *x*, compute the probability that the 4-sided die was selected. Express the answer as a function of *x*.
(*b*) You select one of the two dice with an equal probability of $1/2$, then roll it twice. If the result of the first roll is *x*, compute the probability that the second roll will be *y*. Express the answer as a function of both *x* and *y*.

Problem 1.9 (*A betting strategy*) Suppose a player possesses two $1 dollar bills and is presented with a table consisting of 11 squares as depicted in Figure 1.15.

Prior to the commencement of the game, the player is instructed to place the two $1 dollar bills on either a single square (like the one displayed in Figure 1.15(Top)) or

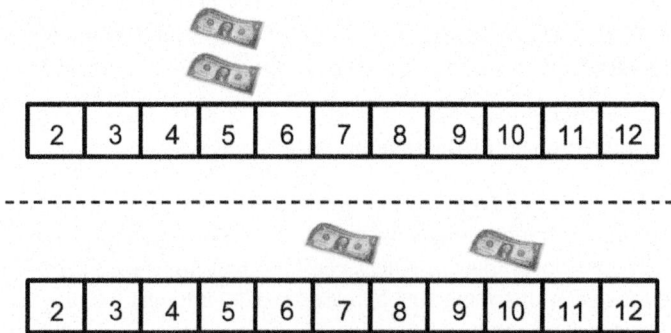

Fig. 1.15 (Top) The top row consists of 11 squares, labeled from 2 to 12. The value displayed in each square represents the sum of the two numbers that appear on the dice when they are rolled. The player must place two $1 dollar bills on either the same square or on two different squares. In this example, both bills are placed on the square labeled 5; (Bottom) The bottom figure illustrates another scenario in which the two bills are positioned on separate squares labeled 7 and 10

on two distinct squares (similar to Figure 1.15(Bottom)). The choice is entirely up to the player. Next, two fair dice are rolled, and if the sum of the numbers on the dice matches the value displayed on a square containing one or two bills, then a single bill is removed from the respective square. This process is then repeated, and if the sum of the newly rolled dice matches the value displayed on a square containing the remaining bill, then the player will remove that bill as well. If the player successfully removes both bills, she/he wins; otherwise, loses.

(a) Let us define a quadruplet (i, j, k, ℓ) where (i, j) denote the two numbers that appear on the dice during the first roll, and (k, ℓ) denote the numbers displayed during the second roll. We can use this quadruplet to construct a sample space Ω and its corresponding probability distribution $\mathbb{P}(\omega)$ where $\omega \in \Omega$.

(b) If the player decides to put two bills on the square labeled 7, what is the probability that the player will win the game?

(c) Which strategy should the player use to maximize the probability of winning the game? In other words, where should the player place two bills? After determining the optimal placement, what is the probability of winning given that placement?

Problem 1.10 (*Monty Hall problem: Variation #1*) Suppose there are four doors, one of which has a prize car behind it, while the other three have goats behind them. The trader selects one door, unaware of what is behind it. Then, the host opens one of the other three doors to reveal a goat. At this point, the trader is given two choices: (i) keep the original pick or (ii) switch to another unopened door. The host will then open a second goat-door (different from the current pick). Finally, the trader must choose between (i) sticking with the door he/she picked on the previous decision, or (ii) switching to the only other remaining door.

(a) How many ways can the trader make decisions in this game?

(b) Calculate the probability of winning for each of the possible strategies by systematically defining a probability model, as we did in Sect. 1.3. Determine which strategy has the highest probability of winning and is therefore the best strategy.

Problem 1.11 (*Monty Hall variation #1:* **Python** *lab*) This problem asks you to conduct **Python** simulations to confirm empirically the results obtained in Problem 1.10.

(a) Suppose the trader takes one strategy, say *sticking-sticking* strategy. Implement a **Python** function that returns 1 (for winning) or 0 (for losing) for one game.

(b) Plot the empirical winning rate as a function of the number of games played N using the function from part (a). The empirical winning rate is defined as $\frac{\text{number of winnings}}{N}$. Determine where the winning rate converges as N increases.

(c) Repeat the process in parts (a) and (b) for all other possible strategies, and compare the empirical results to the theoretical results obtained in Problem 1.10 (b).

Problem 1.12 (*Monty Hall problem: Variation #2*) Suppose there are 100 doors, one of which has a prize car behind it, while the other 99 have goats behind them. The trader selects one door, unaware of what is behind it. Then, the host opens one of the other 98 doors to reveal a goat. The trader is given two choices: (i) stick with the original pick or (ii) switch to another unopened door.

(*a*) Find the probability of winning for both strategies: (i) keeping the original pick, and (ii) switching to another unopened door. Then, determine which strategy yields the highest probability of winning. Does your conclusion align with your intuition?

(*b*) Use a Python simulation to empirically verify your answer in part (*a*).

Problem 1.13 (*True or False?*)

(*a*) The sample space Ω is finite.

(*b*) A man and a woman each have five coins consisting of two double-headed coins, one double-tailed coin, and two normal coins. The man randomly chooses a coin and flips it once, observing that it shows "Head" on the upper face. The woman also randomly chooses a coin and flips it twice, observing that it shows "Head" on both tosses. The probability of the woman having chosen a double-headed coin is greater than that of the man.

(*c*) Consider an experiment in which a coin is flipped four times. We define event A as the occurrence of "Head" on the first flip; event B as the occurrence of "Head" on the second flip, and event C as the occurrence of "Head" on the third flip. Then, events (A, B, C) are disjoint.

1.4 Conditional Probability and Total Probability Law

Recap
In the preceding section, we explored a well-known but counter-intuitive problem called the Monty Hall problem, which led even experts to make mistakes. One takeaway from this problem is the importance of not relying solely on intuition. Instead, if you feel uncertain, it is best to go back to the basics. Specifically, when dealing with probability, the lesson is to first establish a sample space, then determine the corresponding probability distribution, and finally calculate the probability of the desired event based on the probability model.

Outline
In this section, we will delve into another set of fundamental concepts: conditional probability, total probability law, and Bayes' law. The section consists of five parts. Firstly, we will provide the definition of conditional probability and explore the rationale behind this definition. Secondly, we will apply this concept to a well-known instance called the "disease testing" problem. Through this exercise, we will introduce the total probability law. Next, we will mention another important theorem that is relevant to the disease testing problem, namely Bayes' law. Finally, we will investigate applications and significance of Bayes' law.

Definition of conditional probability Conditional probability is a concept that pertains to multiple events. To simplify matters, consider a scenario where we are concerned with two events, say A and B. The probability of event A given that event B has occurred is represented by $\mathbb{P}(A|B)$ and is defined as follows:

$$\mathbb{P}(A|B) := \frac{\mathbb{P}(A \cap B)}{\mathbb{P}(B)}. \tag{1.24}$$

One important point to note is that the notation $\mathbb{P}(\cdot)$ on the right-hand side of the equation is distinct from $\mathbb{P}(\cdot|\cdot)$ on the left-hand side, despite employing the same symbol $\mathbb{P}$. The two notations can be easily distinguished since they take a different number of arguments: $\mathbb{P}(\cdot)$ takes one argument, whereas $\mathbb{P}(\cdot|\cdot)$ takes two. To avoid confusion, one may use the following notations:

$$\mathbb{P}_B(A) = \mathbb{P}(A|B), \quad \mathbb{P}_\Omega(A) = \mathbb{P}(A). \tag{1.25}$$

In this notation, the sample space is indicated by the subscript of $\mathbb{P}$. Specifically, for $\mathbb{P}(\cdot)$, the sample space is denoted as Ω, whereas for $\mathbb{P}(\cdot|\cdot)$, the sample space is restricted to the specific event B as it has already occurred (hence the term "conditioned on"). However, it is customary not to use subscript notation in this manner. As such, we will follow the convention of using the same symbol $\mathbb{P}$ without any subscripts.

$$\mathbb{P}(A|B), \quad \mathbb{P}(A). \tag{1.26}$$

As mentioned before, the two notations can be distinguished by the number of arguments they take.

Rationale behind the definition $\mathbb{P}(A|B) := \frac{\mathbb{P}(A \cap B)}{\mathbb{P}(B)}$ You might be curious about the motivation behind defining conditional probability as shown in (1.24). Every definition has a rationale behind it, and in this case, it can be readily understood from a Venn diagram, as depicted in Figure 1.16. An important observation is that when we condition on event B, it becomes the new sample space. As B has already occurred, any outcome that falls outside of B is no longer a possibility.

After establishing the new sample space, the next step is to determine the corresponding probability distribution. Let $\mathbb{P}(\omega|B)$ denote the probability distribution with respect to the new sample space. Notice that the probability assignment must satisfy the constraint that the probabilities sum up to one:

$$\sum_{\omega \in B} \mathbb{P}(\omega|B) = 1. \tag{1.27}$$

Now, the question arises as to how to define $\mathbb{P}(\omega|B)$? The following observation provides a clue. The probability of event B is given by:

$$\sum_{\omega \in B} \mathbb{P}(\omega) = \mathbb{P}(B). \tag{1.28}$$

Fig. 1.16 When we condition on event B, B becomes the new sample space, since it has already occurred. Therefore, we do not need to consider any outcomes outside of B (denoted as B^c)

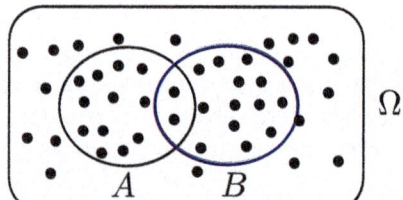

Dividing both sides by $\mathbb{P}(B)$, we get:

$$\sum_{\omega \in B} \frac{\mathbb{P}(\omega)}{\mathbb{P}(B)} = 1. \tag{1.29}$$

Looking at (1.27) and (1.29), one natural definition for $\mathbb{P}(\omega|B)$ is:

$$\mathbb{P}(\omega|B) := \frac{\mathbb{P}(\omega)}{\mathbb{P}(B)}. \tag{1.30}$$

Only considering (1.27) and (1.29), there could be other ways to define $\mathbb{P}(\omega|B)$, as the equality for summation does not necessarily imply the equality for each individual participating in the summation. However, normalizing $\mathbb{P}(\omega)$ by $\mathbb{P}(B)$ as shown in (1.30) also preserves the probability behavior with respect to ω. Therefore, this approach is commonly accepted as the definition.

In the new sample space B, the probability of event A is computed as follows:

$$\mathbb{P}(A|B) = \sum_{\omega \in A \cap B} \mathbb{P}(\omega|B). \tag{1.31}$$

This is because the event A conditioned on B is the intersection of events A and B, denoted as $A \cap B$; this can be visualized in Figure 1.17. Applying the definition (1.30) to (1.31), we get:

$$\begin{aligned} \mathbb{P}(A|B) &= \sum_{\omega \in A \cap B} \mathbb{P}(\omega|B) \\ &:= \sum_{\omega \in A \cap B} \frac{\mathbb{P}(\omega)}{\mathbb{P}(B)} \\ &= \frac{\mathbb{P}(A \cap B)}{\mathbb{P}(B)} \end{aligned} \tag{1.32}$$

where the last equality comes from the definition of $\mathbb{P}(A \cap B)$. We can observe that the above equation aligns with the definition of conditional probability given in (1.24).

Fig. 1.17 Given B has occurred, A occurring can be interpreted as the intersection of events A and B

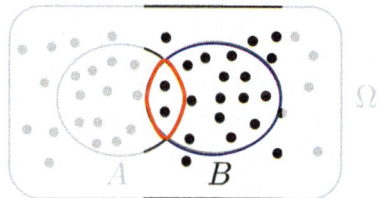

The disease testing problem We will examine one prominent problem that relies on conditional probability. This problem is known as the disease testing problem, and we will specifically focus on cancer testing.

If a person tests positive for cancer, the person is likely to question the accuracy of the test and would want to confirm its correctness. To do so, the person may be interested in computing the probability of actually having cancer given a positive test result. Conditional probability comes into play here, and we can interpret the probability of interest as $\mathbb{P}(A|B)$ where A represents the event that the person has cancer and B represents the event that the test result is positive.

Now how to compute $\mathbb{P}(A|B)$? Recall the definition:

$$\mathbb{P}(A|B) := \frac{\mathbb{P}(A \cap B)}{\mathbb{P}(B)}. \tag{1.33}$$

Clearly we need to calculate two quantities to determine $\mathbb{P}(A|B)$: (i) $\mathbb{P}(A \cap B)$; and (ii) $\mathbb{P}(B)$.

Computation of $\mathbb{P}(A \cap B)$ Let us first tackle the computation of the numerator. We face a challenge here: the exact value of $\mathbb{P}(A \cap B)$ is unknown. But we have good news. We can infer this quantity from statistical data obtained through clinical trials. To understand this, refer to Figure 1.18. In clinical trials, statistical data can be obtained by testing a certain population and gathering test results. For example, in the case where all tested individuals are from the cancer population as shown in Figure 1.18, if we conduct many tests on a large number of such individuals and find that approximately 95% of them test positive, we can use this information to estimate $\mathbb{P}(B|A)$. The accuracy of this estimate improves with an increase in the number of tested individuals. In practice, such tests are often conducted to measure the accuracy of the test itself, with a higher fraction indicating a more accurate test. In this scenario, the 95% accuracy is referred to as the true positive rate (TPR), while the 5% misdetection rate is known as the false negative rate (FNR) (Yerushalmy 1947).

Are we finished with computing $\mathbb{P}(A \cap B)$ once we have a good estimate of $\mathbb{P}(B|A)$? Not quite. To understand why, let us rewrite $\mathbb{P}(A \cap B)$ using the definition of conditional probability:

$$\mathbb{P}(A \cap B) = \mathbb{P}(A)\mathbb{P}(B|A).$$

Here, the challenge is to determine $\mathbb{P}(A)$, which can also be addressed using statistical data. The prevalence of the disease in the population, which can be obtained from statistical data, can be used to estimate $\mathbb{P}(A)$. For example, if statistics show that 10% of the entire population has cancer, we can assume $\mathbb{P}(A) = 0.1$. By combining this estimate with the above estimate of $\mathbb{P}(B|A)$, we can compute:

$$\mathbb{P}(A \cap B) = \mathbb{P}(A)\mathbb{P}(B|A) = 0.1 \times 0.95. \tag{1.34}$$

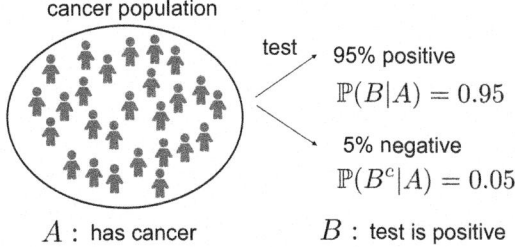

A : has cancer B : test is positive

Fig. 1.18 In clinical trials, test results can be obtained for a specific population, such as the cancer population in this example. If approximately 95% of the tests are positive, this positive rate can be used as an estimate for the conditional probability $\mathbb{P}(B|A)$. Similarly, the negative rate of 0.05 can be used as an estimate for the flipped conditional probability $\mathbb{P}(B^c|A)$

Computation of $\mathbb{P}(B)$ Moving on to the computation of the denominator in (1.33), $\mathbb{P}(B)$, it is a bit challenging. We can estimate $\mathbb{P}(B|A^c)$ by using clinical data from testing on the normal population A^c. Suppose the fraction of positive test results in the normal population is 0.2, which is called the false positive rate (FPR). Then, we can use this number as an estimate of $\mathbb{P}(B|A^c)$. In practice, we can estimate both $\mathbb{P}(B|A)$ and $\mathbb{P}(B|A^c)$ through testing. Although the interested quantity $\mathbb{P}(B)$ is not directly available, we can still compute it indirectly using *total probability law*.

Total probability law Let us now explain what the total probability law is. We begin by manipulating the event B as follows:

$$B = (A \cap B) \cup (A^c \cap B).$$

This is immediate from the Venn diagram in Figure 1.19. An important observation to make is that the events $A \cap B$ and $A^c \cap B$ are disjoint. This is clear because A and A^c are disjoint. Therefore, we have:

$$\mathbb{P}(B) = \mathbb{P}(A \cap B) + \mathbb{P}(A^c \cap B). \tag{1.35}$$

Why? Consider the definition of the probability of an event. Equation (1.35) is precisely what the *total probability law* states.

With the help of the total probability law (1.35), and using the estimated values of $\mathbb{P}(B|A)$, $\mathbb{P}(B|A^c)$, and $\mathbb{P}(A)$ obtained earlier, we can now compute the probability of interest as follows:

$$\begin{aligned}
\mathbb{P}(B) &= \mathbb{P}(A \cap B) + \mathbb{P}(A^c \cap B) \\
&= \mathbb{P}(A)\mathbb{P}(B|A) + \mathbb{P}(A^c)\mathbb{P}(B|A^c) \\
&= 0.1 \times 0.95 + 0.9 \times 0.2
\end{aligned} \tag{1.36}$$

Fig. 1.19 The total probability law states that an event B can be expressed as the union of two disjoint subsets: $A \cap B$ and $A^c \cap B$. Therefore, the probability of B can be obtained by adding the probabilities of these two subsets: $\mathbb{P}(B) = \mathbb{P}(A \cap B) + \mathbb{P}(A^c \cap B)$

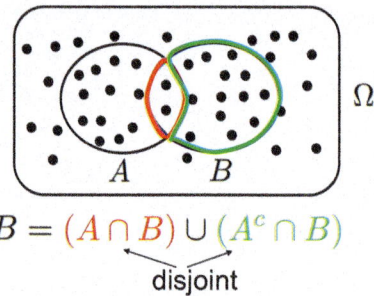

$$B = (A \cap B) \cup (A^c \cap B)$$
$$\underbrace{\qquad\qquad\qquad}_{\text{disjoint}}$$

where the second equality is due to the definition of conditional probability and the last comes from the estimates.

Computation of $\mathbb{P}(A|B)$ After plugging in the values of $\mathbb{P}(B|A)$, $\mathbb{P}(B|A^c)$, and $\mathbb{P}(A)$ from the previous steps into (1.34) and (1.36) and then substituting the results into (1.33), we obtain the following expression for conditional probability $\mathbb{P}(A|B)$:

$$
\mathbb{P}(A|B) := \frac{\mathbb{P}(A \cap B)}{\mathbb{P}(B)} = \frac{\mathbb{P}(A)\mathbb{P}(B|A)}{\mathbb{P}(A)\mathbb{P}(B|A) + \mathbb{P}(A^c)\mathbb{P}(B|A^c)}
$$
$$
= \frac{0.1 \times 0.95}{0.1 \times 0.95 + 0.9 \times 0.2} \approx 0.3455. \tag{1.37}
$$

Discussion on $\mathbb{P}(A|B)$ Observe that the probability given in (1.37) is approximately 35%. This could potentially ease the person's concerns. Why? This is because despite the positive test result, the probability of having cancer, i.e., $\mathbb{P}(A|B)$, is less than 50%. In fact, this scenario is often encountered in reality, where even with a positive test result, the chance of being normal is higher. Hence, many individuals who receive positive cancer test results often opt for a second test from a different medical facility.

You may be wondering why, in reality, we obtain a small value for $\mathbb{P}(A|B)$. This is due to the fact that the cancer population A is typically quite small, such as 10% in the previous example. As the size of cancer population decreases, the conditional probability $\mathbb{P}(A|B)$ also decreases. This is consistent with the definition of conditional probability:

$$
\mathbb{P}(A|B) := \frac{\mathbb{P}(A \cap B)}{\mathbb{P}(B)} = \mathbb{P}(A) \cdot \frac{\mathbb{P}(B|A)}{\mathbb{P}(B)} \tag{1.38}
$$

where the second equality comes from $\mathbb{P}(B|A) := \frac{\mathbb{P}(A \cap B)}{\mathbb{P}(A)}$. The definition of conditional probability reveals that $\mathbb{P}(A)$ is multiplied in front of the conditional probability, and it also affects the denominator $\mathbb{P}(B)$ through the total probability law. Remember $\mathbb{P}(B) = \mathbb{P}(A)\mathbb{P}(B|A) + \mathbb{P}(A^c)\mathbb{P}(B|A^c)$. However, reducing $\mathbb{P}(A)$ does not directly decrease $\mathbb{P}(B)$ because of the term $\mathbb{P}(A^c)\mathbb{P}(B|A^c)$. Instead, the value of $\mathbb{P}(A)$ plays a more significant role in determining $\mathbb{P}(A|B)$. As a result, a smaller value of $\mathbb{P}(A)$ leads to a smaller value of $\mathbb{P}(A|B)$. This example demonstrates the interesting interpretation provided by the concept of conditional probability.

Bayes' law (Bayes 1958) Finally, we want to highlight a fundamental law that emerges while computing $\mathbb{P}(A|B)$. That is, Bayes' law. Despite its simplicity, this law is very powerful. It is a direct consequence of applying the definition of conditional probability twice:

$$\mathbb{P}(A|B) := \frac{\mathbb{P}(A \cap B)}{\mathbb{P}(B)} = \frac{\mathbb{P}(A)\mathbb{P}(B|A)}{\mathbb{P}(B)} \qquad (1.39)$$

where the second equality comes from $\mathbb{P}(B|A) := \frac{\mathbb{P}(A \cap B)}{\mathbb{P}(A)}$. The Bayes' law may look trivial, but it has wide applications. There are many scenarios where we need to compute $\mathbb{P}(A|B)$ but are only given $\mathbb{P}(B|A)$, the flipped version of the conditional probability. Such examples will be provided in Problem Set 2.

Look ahead
The primary topic of this section is the notion of conditional probability. There is a related concept that naturally follows, namely, independence. We will delve into the idea of independence in the next section.

1.5   Independence

Recap

In the previous section, we have studied one important concept: *conditional probability* defined as below:

$$\mathbb{P}(A|B) := \frac{\mathbb{P}(A \cap B)}{\mathbb{P}(B)}. \qquad (1.40)$$

After explaining the rationale behind the definition, which involves using the new sample space B and the corresponding probability distribution $\mathbb{P}(\omega|B)$ for $\omega \in B$, we explored the application of conditional probability in the context of the cancer testing problem. In this context, we highlighted two important laws that arise when computing the desired probabilities. One of them is the total probability law, which simplifies the computation of the denominator in the conditional probability formula (1.40):

$$\mathbb{P}(B) = \mathbb{P}(A \cap B) + \mathbb{P}(A^c \cap B). \qquad (1.41)$$

The proof of this is straightforward due to: (i) the events $A \cap B$ and $A^c \cap B$ are disjoint; and (ii) $\mathbb{P}(B)$ is calculated by summing up over all $\omega \in B = (A \cap B) \cup (A^c \cap B)$.

Another important law that we emphasized is the Bayes' law:

$$\mathbb{P}(A|B) = \frac{\mathbb{P}(A)\mathbb{P}(B|A)}{\mathbb{P}(B)}. \qquad (1.42)$$

The proof is also straightforward. It follows directly from applying the definition of conditional probability twice. As previously mentioned, Bayes' law is particularly useful in cases where we want to compute $\mathbb{P}(A|B)$ but are only given $\mathbb{P}(B|A)$.

Outline

In this section, our focus will be on another crucial concept that follows from conditional probability: *independence*. We will cover four main topics. First, we will introduce the definition of independence for a simple scenario where only two events are considered. Then, we will provide the reasoning behind the definition. Next, we will extend the definition to cases involving more than two events. Finally, as we have done in previous sections, we will reinforce our understanding of the concept through examples.

Definition of independence for two events Let's begin by examining a simple scenario involving two events, say A and B. If the probability of the intersection of events A and B is equal to the multiplication of their individual probabilities, we say that events A and B are *independent*:

$$\mathbb{P}(A \cap B) = \mathbb{P}(A)\mathbb{P}(B). \tag{1.43}$$

There is a reason behind defining independence as mentioned above.

Rationale behind the definition (1.43) The reason for defining independence in the aforementioned way is closely tied to the intuitive notion of independence. To put it naturally, we would describe A and B as independent if the occurrence of event B has no impact on the probability of event A, meaning that $\mathbb{P}(A)$ remains constant regardless of whether B is conditioned upon or not:

$$\mathbb{P}(A|B) = \mathbb{P}(A). \tag{1.44}$$

Using the definition of conditional probability, we can then rewrite the natural condition (1.44) as:

$$\frac{\mathbb{P}(A \cap B)}{\mathbb{P}(B)} = \mathbb{P}(A). \tag{1.45}$$

Clearly, this aligns with the initial definition (1.43): $\mathbb{P}(A \cap B) = \mathbb{P}(A)\mathbb{P}(B)$.

You may wonder why we do not utilize the more intuitive condition (1.45) instead of (1.43). The reason is that if we adopt the former, we would require an additional condition:

$$\mathbb{P}(B) > 0,$$

as otherwise $\mathbb{P}(A|B)$ is not definable. Alternatively, we would necessitate $\mathbb{P}(A) > 0$ if we were to use the inverse version of the condition: $\mathbb{P}(B|A) = \mathbb{P}(B)$. However, in the initial definition (1.43), we are not bound by any such prerequisites, which is evidently preferable.

Definition for the three events case Let us now consider a scenario involving three events, say A, B, and C. In this case, we say that (A, B, C) are *mutually independent* if the intersection of any pair of events is equal to the product of their individual probabilities, and the intersection of all three events, $A \cap B \cap C$, follows a similar pattern:

$$\mathbb{P}(A \cap B) = \mathbb{P}(A)\mathbb{P}(B);$$
$$\mathbb{P}(B \cap C) = \mathbb{P}(B)\mathbb{P}(C);$$
$$\mathbb{P}(C \cap A) = \mathbb{P}(C)\mathbb{P}(A); \qquad (1.46)$$
$$\mathbb{P}(A \cap B \cap C) = \mathbb{P}(A)\mathbb{P}(B)\mathbb{P}(C).$$

The reasoning behind this definition is identical to that of the previous one. This is derived from the conditions that appear intuitive and are expressed using both conditional probability and its unconditional counterpart, as shown in Figure 1.20. The conditions that appear natural are intended to convey that the occurrence of any number of events does not provide any information about the remaining events. In Figure 1.20, we have listed *only a portion* of such conditions. Notably, the combination of $\mathbb{P}(A|B \cap C) = \mathbb{P}(A)$ and $\mathbb{P}(B|C) = \mathbb{P}(B)$ results in $\mathbb{P}(A \cap B \cap C) = \mathbb{P}(A)\mathbb{P}(B)\mathbb{P}(C)$. However, there are numerous other ways of formulating these conditions. For instance, $\mathbb{P}(A \cap B|C) = \mathbb{P}(A \cap B)$ in combination with $\mathbb{P}(A|B) = \mathbb{P}(B)$ also leads to $\mathbb{P}(A \cap B \cap C) = \mathbb{P}(A)\mathbb{P}(B)\mathbb{P}(C)$. This underscores another benefit of the original definition (1.46): it is more concise than the one based on the conditions that appear natural.

It is worth noting that we use the term "mutual independence" for the independence of three or more events, rather than just "independence". The reason is that there exists another definition of independence, known as "pairwise independence". In the case of three events (A, B, C), we say that they are "pairwise independent" if only the pairwise intersections satisfy the product rule:

$$\mathbb{P}(A \cap B) = \mathbb{P}(A)\mathbb{P}(B);$$
$$\mathbb{P}(B \cap C) = \mathbb{P}(B)\mathbb{P}(C); \qquad (1.47)$$
$$\mathbb{P}(C \cap A) = \mathbb{P}(C)\mathbb{P}(A).$$

There are instances where events are pairwise independent but not mutually independent. An example illustrating this will be presented in the sequel.

On a different note, some curious readers may wonder whether the following condition

$$\mathbb{P}(A \cap B \cap C) = \mathbb{P}(A)\mathbb{P}(B)\mathbb{P}(C) \text{ (that we call } \textit{three-way independence)} \quad (1.48)$$

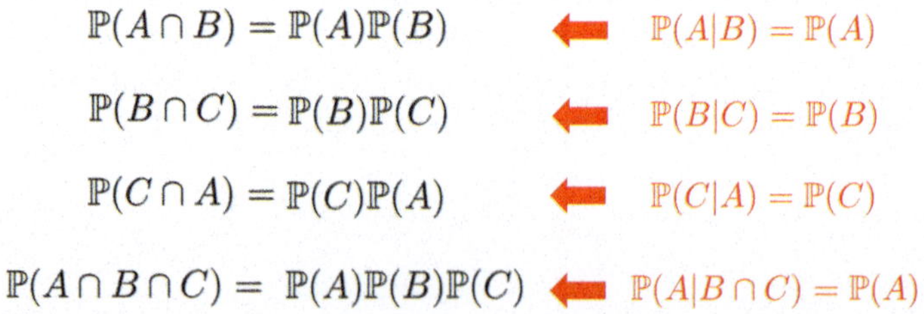

$$\mathbb{P}(A \cap B) = \mathbb{P}(A)\mathbb{P}(B) \quad \Longleftarrow \quad \mathbb{P}(A|B) = \mathbb{P}(A)$$

$$\mathbb{P}(B \cap C) = \mathbb{P}(B)\mathbb{P}(C) \quad \Longleftarrow \quad \mathbb{P}(B|C) = \mathbb{P}(B)$$

$$\mathbb{P}(C \cap A) = \mathbb{P}(C)\mathbb{P}(A) \quad \Longleftarrow \quad \mathbb{P}(C|A) = \mathbb{P}(C)$$

$$\mathbb{P}(A \cap B \cap C) = \mathbb{P}(A)\mathbb{P}(B)\mathbb{P}(C) \quad \Longleftarrow \quad \mathbb{P}(A|B \cap C) = \mathbb{P}(A)$$

Fig. 1.20 The motivation for defining independence is rooted in the natural conditions (highlighted in red) that are expressed using both conditional probability and its unconditional counterpart

implies pairwise independence of any events. It turns out that this is not the case, and there are many counter-examples. We will shortly provide one such example.

You may find the concept of independence for three or more events confusing. It is indeed be a challenging concept to grasp. To avoid confusion, you may want to focus on the *original definition* (1.46) and check whether all the conditions are met. You need not worry about other conditions that may seem natural or intuitive. Therefore, memorizing the definition is crucial in this case.

Example: Pairwise independent but not mutually independent Consider the experiment of flipping a fair coin twice. Let A and B denote the events that the first and second flips show "Head" and "Tail", respectively. Let C denote the event that the first and second flips are different. Intuitively, we can see that the occurrence of the event C is determined by the events A and B, which means that these events are not mutually independent. Let's verify this using the definition (1.46). First, let's compute:

$$\mathbb{P}(A \cap B \cap C) = \mathbb{P}(A \cap B)\mathbb{P}(C|A \cap B)$$
$$= \mathbb{P}(A \cap B) \qquad (1.49)$$
$$= \frac{1}{4}$$

where the first equality is due the definition of conditional probability; the second equality comes from the fact that given A and B, the event C must occur, i.e., $\mathbb{P}(C|A \cap B) = 1$; and the last is because HT is one of the four events $\{HH, HT, TH, TT\}$ equally likely. On the other hand,

$$\mathbb{P}(A)\mathbb{P}(B)\mathbb{P}(C) = \frac{1}{2} \times \frac{1}{2} \times \frac{1}{2} = \frac{1}{8} \neq \frac{1}{4}. \qquad (1.50)$$

We can observe that the occurrence of event C is associated with the outcomes $\{HT, TH\}$ out of four possible outcomes, and therefore $\mathbb{P}(C) = \frac{2}{4} = \frac{1}{2}$. This indicates that the events A, B, and C are not mutually independent.

But these are *pairwise* independent. To see this, we first check:

$$\mathbb{P}(A \cap B) = \frac{1}{4} = \frac{1}{2} \times \frac{1}{2} = \mathbb{P}(A)\mathbb{P}(B). \qquad (1.51)$$

What about A and C? We first compute:

$$\mathbb{P}(A \cap C) = \mathbb{P}(A)\mathbb{P}(C|A)$$
$$= \frac{1}{2}\mathbb{P}(B|A)$$
$$= \frac{1}{2}\mathbb{P}(B) \qquad (1.52)$$
$$= \frac{1}{2} \times \frac{1}{2}$$

where the second equality comes from the fact that conditioned on A (first flip is "Head"), the event C (two are different) is equivalent to B (second flip is "Tail"); and the third is because of the independence of A and B. As we computed earlier, $\mathbb{P}(C) = \frac{1}{2}$. This together with $\mathbb{P}(A) = \frac{1}{2}$ and (1.52) proves the independence of A and C. By symmetry, one can also prove the independence of B and C. Hence, (A, B, C) are *pairwise* independent.

Example: Three-way independent but not pairwise independent Let's examine a counter-intuitive example that is highly contrived, where the three-way independence does not necessarily entail the pairwise independence. Consider an experiment of rolling two fair dice and the following three events are:

$$A = \{\text{first shows } 1, 2 \text{ or } 3\};$$
$$B = \{\text{first shows } 3, 4 \text{ or } 5\}; \qquad (1.53)$$
$$C = \{\text{the sum of the two dice is } 9\}.$$

The occurrence of the event $A \cap B \cap C$ corresponds to $(3, 6)$ out of 36 equally likely events, so its probability is $\mathbb{P}(A \cap B \cap C) = \frac{1}{36}$. We can also compute $\mathbb{P}(A) = \frac{3}{6} = \frac{1}{2}$, $\mathbb{P}(B) = \frac{3}{6} = \frac{1}{2}$, and $\mathbb{P}(C) = \frac{4}{36} = \frac{1}{9}$, which corresponds to the events $(3, 6), (4, 5), (5, 4), (6, 3)$. Therefore, we have:

$$\mathbb{P}(A \cap B \cap C) = \frac{1}{36} = \frac{1}{2} \times \frac{1}{2} \times \frac{1}{9} = \mathbb{P}(A)\mathbb{P}(B)\mathbb{P}(C). \qquad (1.54)$$

On the other hand, these are not pairwise independent. For instance,

$$\mathbb{P}(A \cap B) = \frac{1}{6} \neq \frac{1}{2} \times \frac{1}{2} = \mathbb{P}(A)\mathbb{P}(B).$$

Mutual independence among n events Let's now consider a general scenario where we have an arbitrary number of events, denoted as A_1 through A_n. We say that these events $(A_1, \ldots, A_n)$ are considered *mutually independent* if, for any subset $I \subseteq \{1, 2, \ldots, n\}$, the probability of the intersection of the events associated with I equals the product of their respective probabilities:

$$\mathbb{P}\left(\bigcap_{i \in I} A_i\right) = \prod_{i \in I} \mathbb{P}(A_i) \tag{1.55}$$

where $\bigcap_{i \in I} A_i := A_{i_1} \cap \cdots \cap A_{i_{|I|}}$, $\prod_{i \in I} \mathbb{P}(A_i) := \mathbb{P}(A_{i_1}) \cdots \mathbb{P}(A_{i_{|I|}})$ and $(i_1, \ldots, i_{|I|})$ are the elements of I. The justification for this definition is identical to the three-events case. Thus, we recommend that you just memorize this definition and utilize it when needed.

We have completed the definition of independence. As mentioned earlier, let's reinforce our understanding of independence through a few examples. We will explore a total of four examples.

Example #1: Tossing the $\frac{2}{3}$-biased coin twice The first example is one we previously discussed in Sect. 1.2 where we used intuition rather than a rigorous approach. The experiment involves flipping a biased coin (with a $\frac{2}{3}$ probability of showing "Head") twice. The corresponding sample space is:

$$\Omega = \{HH, HT, TH, TT\}.$$

Recall that we were interested in computing the probability $\mathbb{P}(HH)$. We can now construct the probability distribution in a more formal way. Let A and B be the events that the first and second flips show "Head" respectively. Then, we can express $\mathbb{P}(HH)$ as:

$$\begin{aligned}
\mathbb{P}(HH) &= \mathbb{P}(A \cap B) \\
&= \mathbb{P}(A)\mathbb{P}(B) \\
&= \frac{2}{3} \times \frac{2}{3}
\end{aligned}$$

where the second equality is due to the reasonable assumption that two flips are *independent*.

Example #2: Monty Hall problem The second example is the Monty Hall problem, which we previously discussed in Sect. 1.3. Refer to Figure 1.21. In Sect. 1.3, we examined the sample space where each outcome consists of a triplet: (car's location, trader's choice, host's choice). Here, we can construct an alternative sample space that only takes into account the two uncertainties of the car's location and the trader's choice.

$$\Omega = \{(1, 1), (1, 2), (1, 3), (2, 1), (2, 2), (2, 3), (3, 1), (3, 2), (3, 3)\}.$$

What is the probability distribution in this case? It is clearly uniform since there are 9 equally likely cases due to symmetry. We can also prove this rigorously using the concept of independence. Let A be the event that the car is located behind door i

Fig. 1.21 The scenario of the Monty Hall problem involves a door behind which a prize car is hidden, with the other two doors hiding goats. The trader does not know which door has the car, while the host is aware of the setup. The trader is asked to select a door for the prize car

and B the event that the trader's choice is door j. Then, we can express $\mathbb{P}((i, j))$ as follows:

$$\begin{aligned} \mathbb{P}((i, j)) &= \mathbb{P}(A \cap B) \\ &= \mathbb{P}(A)\mathbb{P}(B) \\ &= \frac{1}{3} \times \frac{1}{3} \end{aligned} \qquad (1.56)$$

where the second equality follows from the independence of the car's location and the trader's choice.

Example #3: Balls & Bins The third is a well-known problem called the *Balls-&-Bins* problem, which we have not yet discussed. Refer to Figure 1.22. k balls are to be thrown into n bins independently at random, where each ball is equally likely to land in any bin.

Let's say we want to calculate the probability that the first bin remains empty after throwing all k balls. To do so, we define the event A_i as the event that the ith ball is not placed in bin 1. Using this notation, we can express the probability as follows:

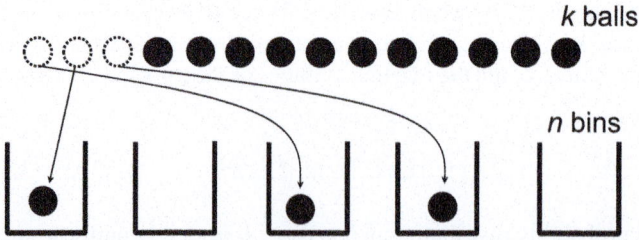

Fig. 1.22 Balls-&-Bins problem: Multiple balls are thrown independently and uniformly at random into multiple bins

$$\mathbb{P}(\text{first bin empty}) = \mathbb{P}(A_1 \cap A_2 \cap \cdots \cap A_k)$$
$$= \mathbb{P}(A_1) \cap \mathbb{P}(A_2) \cap \cdots \cap \mathbb{P}(A_k)$$
$$= \mathbb{P}(A_1)\mathbb{P}(A_2) \cdots \mathbb{P}(A_k) \qquad (1.57)$$
$$= \left(\frac{n-1}{n}\right)^k$$

where the second equality is due to the mutual independence of $(A_1, \ldots, A_k)$; and the last equality is due to $\mathbb{P}(A_i) = \frac{n-1}{n}$ $\forall i \in \{1, 2, \ldots, k\}$.

Example #4: Fair versus biased coins The final example is a bit tricky. Refer to Figure 1.23. There are two coins: one is fair and the other is biased with probability p of showing "Head". Two experiments are considered. In Experiment 1, we randomly choose one of the two coins and flip it once. We repeat this procedure in an independent manner. The question of interest is to determine $\mathbb{P}(HH)$. To do so, we define A_i as the event that the ith flip is "Head" and B_i as the event that the fair coin is selected for the ith flip. We can then express $\mathbb{P}(HH)$ as:

$$\mathbb{P}(HH) = \mathbb{P}(A_1 \cap A_2)$$
$$= \mathbb{P}(A_1)\mathbb{P}(A_2) \qquad (1.58)$$

where the last equality holds because the two trials in Experiment 1 are independent. We focus on one probability:

$$\mathbb{P}(A_1) = \mathbb{P}(B_1 \cap A_1) + \mathbb{P}(B_1^c \cap A_1)$$
$$= \mathbb{P}(B_1)\mathbb{P}(A_1|B_1) + \mathbb{P}(B_1^c)\mathbb{P}(A_1|B_1^c)$$
$$= \frac{1}{2}\left(\frac{1}{2} + p\right)$$

Experiment #1:   Experiment #2:

Choose a coin btw the two.   Choose a coin btw the two.
Flip the coin.   Flip the coin twice.

Repeat this independently

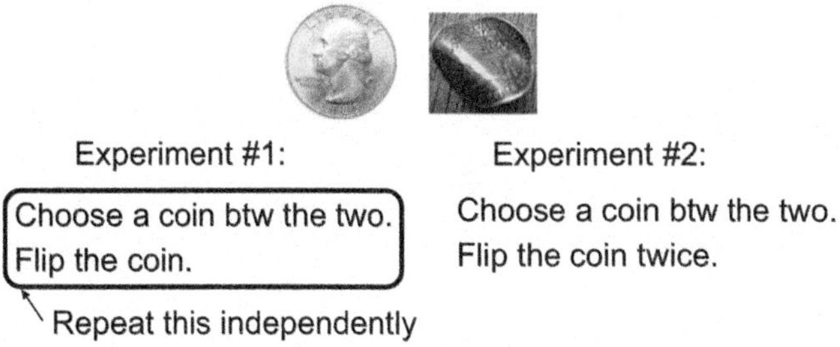

Fig. 1.23 Fair-versus-biased coins: There are two coins available. One is fair and the other is biased, with the probability of showing "Head" being p. In Experiment 1, we randomly select one of the coins and flip it. This whole process is repeated independently. In Experiment 2, we randomly choose one of the coins and flip it twice

where the first equality is due to the total probability law and the second follows from the definition of conditional probability. Plugging this into (1.58) and applying the symmetry argument w.r.t. $\mathbb{P}(A_2)$, we get:

$$\mathbb{P}(HH) = \mathbb{P}(A_1 \cap A_2) = \frac{1}{4} \left(\frac{1}{2} + p \right)^2. \qquad (1.59)$$

Let us now examine Experiment 2, which differs from Experiment 1 in that we randomly select a coin between the two and then flip the same coin twice. Again we want to compute the probability of obtaining two consecutive "Head" outcomes:

$$\mathbb{P}(HH) = \mathbb{P}(A_1 \cap A_2). \qquad (1.60)$$

One question that naturally arises is whether the events A_i are independent. It turns out they are not. Therefore, we need to compute the desired probability using a different method. To do so, we introduce the key event B, which represents the event that the initially chosen coin is fair. Using this event together with the total probability law, we can obtain:

$$\begin{aligned}
\mathbb{P}(HH) &= \mathbb{P}(A_1 \cap A_2) \\
&= \mathbb{P}(B \cap A_1 \cap A_2) + \mathbb{P}(B^c \cap A_1 \cap A_2) \\
&= \mathbb{P}(B)\mathbb{P}(A_1 \cap A_2|B) + \mathbb{P}(B^c)\mathbb{P}(A_1 \cap A_2|B^c) \qquad (1.61) \\
&= \frac{1}{2} \left(\frac{1}{4} + p^2 \right)
\end{aligned}$$

where the second and third equalities follow from the total probability law and the definition of conditional probability, respectively. The last step is crucial. After a coin is chosen, given the event B (the initially chosen coin is fair), flipping the coin in the first trial is independent of the second trial. Therefore, given B, A_1 and A_2 are independent:

$$\mathbb{P}(A_1 \cap A_2|B) = \mathbb{P}(A_1|B)\mathbb{P}(A_2|B) = \frac{1}{2} \times \frac{1}{2} = \frac{1}{4}.$$

Similarly,

$$\mathbb{P}(A_1 \cap A_2|B^c) = \mathbb{P}(A_1|B^c)\mathbb{P}(A_2|B^c) = p \times p = p^2.$$

So we can obtain the last equality in (1.61). One observation is that two dependent events can be conditionally independent. The formal definition of conditional independence is: We say events A and B are *conditionally independent* with respect to C if

$$\mathbb{P}(A \cap B|C) = \mathbb{P}(A|C)\mathbb{P}(B|C). \qquad (1.62)$$

There are also scenarios where two events that are independent can become conditionally dependent. An example of this is the counter-example we examined earlier, where the events are pairwise independent but not mutually independent. Why? Think about it.

Finally let us verify the claim that we made earlier: the events A_1 and A_2 are dependent. We can do this by computing:

$$\mathbb{P}(A_1) = \mathbb{P}(B)\mathbb{P}(A_1|B) + \mathbb{P}(B^c)\mathbb{P}(A_1|B^c)$$
$$= \frac{1}{2}\left(\frac{1}{2} + p\right) \tag{1.63}$$

where the first equality comes again from the total probability law and the definition of conditional probability. Hence, we verify the dependence via:

$$\mathbb{P}(A_1 \cap A_2) = \frac{1}{2}\left(\frac{1}{4} + p^2\right) \neq \frac{1}{4}\left(\frac{1}{2} + p\right)^2 = \mathbb{P}(A_1)\mathbb{P}(A_2).$$

Look ahead
Thus far, we have worked through relatively straightforward examples to understand the concept of independence. As a result, you may feel confident in your grasp of the concept. However, the concept of independence is actually quite deep and nuanced. There are many complex examples where answering interesting questions requires clever manipulation of the concept. In the next section, we will explore one such example: the *coupon collector problem*.

1.6   Coupon Collector Problem and **Python** Simulation

Recap

In the preceding section, we familiarized ourselves with a significant concept: independence. When considering two relevant events in a simple scenario, the definition was clear and intuitive, especially with the aid of the conditional probability definition. We define two events A and B as independent if they satisfy the following condition:

$$\mathbb{P}(A \cap B) = \mathbb{P}(A)\mathbb{P}(B). \tag{1.64}$$

The situation becomes complex when dealing with an arbitrary number of events. Different types of independence, such as mutual independence, pairwise independence, and k-way independence, exist. There are numerous cases in which one form of independence holds while the other is not. To handle this confusion, we advised just memorizing the definition of mutual independence. We say that events $(A_1, \ldots, A_n)$ are mutually independent if

$$\mathbb{P}\left(\bigcap_{i \in I} A_i\right) = \prod_{i \in I} \mathbb{P}(A_i) \quad \forall I \subseteq \{1, 2, \ldots, n\}. \tag{1.65}$$

During our practice with the concept of independence, we encountered several examples that were not overly challenging. However, we mentioned that many non-trivial instances exist in which independence plays a significant role, yet applying it is not straightforward.

Outline

In this section, we will explore one such example, namely the "coupon collector problem". The section is composed of four parts. First, we will introduce the problem and present an intriguing question that arises in the problem context. As it turns out, it is challenging to address the question directly. Therefore, in the second part, we will investigate a simplified version of the problem that can offer valuable insights, with the help of the independence concept. The third part will examine an essential technique that enables us to address the original challenging question, in combination with the solution to the simpler version. Finally, we will apply the technique to come up with a solution to the original question.

k snacks

n different coupons

Fig. 1.24 The coupon collector problem involves n unique coupons, where one coupon is randomly chosen (from n) to be placed in a snack. If we purchase k snacks, a popular question that arises is: What is the probability of collecting all n coupons from the k snacks?

Coupon collector problem (Flajolet et al. 1992) The problem we will explore is well-known and referred to as the "coupon collector problem". The problem setup is as follows. Imagine a snack that includes a coupon inside, such as "Cheetos". Cheetos coupon's appearance is depicted in Figure 1.24.

Suppose there are n distinct coupons, and each snack has one coupon randomly chosen from the n possibilities, independently of the other snacks. An interesting question that arises in this problem context is:

> **? Questions**
> What is the minimum number of snacks that need to be purchased to have a 90% probability of collecting all the coupons?

Formally, this question can be expressed as follows. Let k denote the number of snacks purchased. What is the value of k such that

$$\mathbb{P}(\text{obtain every coupon from } k \text{ snacks}) \geq 0.9? \qquad (1.66)$$

As previously mentioned, directly addressing the question is difficult. Therefore, we will introduce a simpler version of the problem that provides insight.

A simpler version of the question (1.66) The simplified version of the problem focuses on obtaining only one particular coupon, say Coupon 1. Thus, the relevant question becomes: What is the value of k such that

$$\mathbb{P}(\text{obtain "Coupon 1" from } k \text{ snacks}) \geq 0.9? \qquad (1.67)$$

The independence concept can be used to answer this question. The crucial step is to introduce multiple events in a smart way such that they are independent. One can be inspired by the fact that each snack containing a coupon is independent of the other snacks and introduce the following events. Let A_i be the event that the ith snack contains "Coupon 1". These events are independent because of our underlying assumption. Using this notation, one can express the probability on the left-hand side of (1.67) as:

$$\mathbb{P}(\text{obtain ``Coupon 1'' from } k \text{ snacks})$$
$$= \mathbb{P}(A_1 \cup A_2 \cup \cdots \cup A_k). \tag{1.68}$$

However, an issue arises because the interested event is expressed as the *union* of the events A_i: $\bigcup_{i=1}^{k} A_i$. This prevents us from exploiting the independence property: $\mathbb{P}(A_1 \cap A_2 \cap \cdots \cap A_k)$. Fortunately, there is a trick that allows us to resolve this issue: the complement trick.

We can use the complement trick together with an important law that you may have learned from high school, De Morgan's law. By applying De Morgan's law $(A_1 \cup A_2)^c = A_1^c \cap A_2^c$ to (1.68) $k - 1$ times, we obtain:

$$\mathbb{P}(\text{obtain ``Coupon 1'' from } k \text{ snacks})$$
$$= \mathbb{P}(A_1 \cup A_2 \cup \cdots \cup A_k) \tag{1.69}$$
$$= 1 - \mathbb{P}(A_1^c \cap A_2^c \cap \cdots \cap A_k^c).$$

One crucial observation here is that the events $(A_1^c, \ldots, A_k^c)$ are also mutually independent. This is due to our assumption that each snack contains a coupon chosen uniformly at random, independently of the other snacks. By applying the independence property, we obtain:

$$\mathbb{P}(\text{obtain ``Coupon 1'' from } k \text{ snacks})$$
$$= 1 - \mathbb{P}(A_1^c \cap A_2^c \cap \cdots \cap A_k^c)$$
$$= 1 - \mathbb{P}(A_1^c) \cdots \mathbb{P}(A_k^c) \tag{1.70}$$
$$= 1 - \left(1 - \frac{1}{n}\right)^k$$

where the last equality follows from $\mathbb{P}(A_i^c) = \frac{n-1}{n} \; \forall i \in \{1, 2, \ldots, k\}$. By applying this into the left-hand side of (1.67) and then simplifying, we obtain:

$$k \log\left(1 - \frac{1}{n}\right) \le \log 0.1.$$

Since $\log\left(1 - \frac{1}{n}\right) < 0$, diving both sides by $\log\left(1 - \frac{1}{n}\right)$ gives:

$$k \ge \frac{\log 0.1}{\log\left(1 - \frac{1}{n}\right)}. \tag{1.71}$$

To make the expression (1.71) more intuitive, let's focus on an extreme case where n is very large. To simplify the expression, we can use an approximation based on Taylor's series. We define $f(x) = \log(1 + x)$, where log is the natural logarithm.

Then, for small values of x, we have:

$$\log(1 + x) \approx f(0) + \frac{f'(0)}{1!} x = x, \tag{1.72}$$

as $f'(0) = \frac{1}{1+x}\big|_{x=0} = 1$. Applying this into $\log\left(1 - \frac{1}{n}\right)$ for a very large n, we can map x to $-\frac{1}{n}$, thus yielding:

$$\log\left(1 - \frac{1}{n}\right) \approx -\frac{1}{n}.$$

Applying this approximation into (1.71), we get:

$$k \geq \frac{\log 0.1}{\log\left(1 - \frac{1}{n}\right)} \approx n \log 10.$$

This result implies that the number of snacks we need to buy to obtain one particular coupon increases *linearly* with n.

Go back to the original question (1.66) Let us now revisit the original problem, which involves finding the probability of collecting all n coupons after purchasing k snacks:

$$\underbrace{\mathbb{P}(\text{obtain every coupon from } k \text{ snacks})}_{=:E}.$$

To compute $\mathbb{P}(E)$, we can define E_i as the event of obtaining "Coupon i" from k snacks. Then, the interested event E can be expressed as the intersection of all E_i's: $E = E_1 \cap E_2 \cap \cdots \cap E_n$. Therefore, we have:

$$\mathbb{P}(E) = \mathbb{P}(E_1 \cap E_2 \cap \cdots \cap E_n). \tag{1.73}$$

In (1.70), $\mathbb{P}(E_1) = 1 - \left(1 - \frac{1}{n}\right)^k$. By symmetry, $\mathbb{P}(E_1) = \cdots = \mathbb{P}(E_n)$. This then gives:

$$\mathbb{P}(E_i) = 1 - \left(1 - \frac{1}{n}\right)^k, \quad \forall i \in \{1, 2, \ldots, n\}. \tag{1.74}$$

Looking at (1.73) and (1.74), one might be tempted to believe that the events $(E_1, \ldots, E_n)$ are mutually independent. Unfortunately, this is not the case. Here is a counterexample. For instance, in a simple setting where $(n, k) = (2, 2)$, the probability of each event is:

$$P(E_i) = 1 - \left(1 - \frac{1}{2}\right)^2 = \frac{3}{4} \quad \longrightarrow \quad P(E_1)P(E_2) = \frac{9}{16}. \tag{1.75}$$

On the other hand, the probability of the intersected event is:

$$
\begin{aligned}
P(E_1 \cap E_2) &= P(\text{obtain every coupon from two snacks}) \\
&= P(\text{two coupons in the two snacks are different}) \\
&= \frac{2}{4} = \frac{1}{2}
\end{aligned}
\tag{1.76}
$$

where the second-to-last equality comes from the fact that there are two coupon configurations for the desired event out of 4: (Coupon 1, Coupon 1), (Coupon 1, Coupon 2), (Coupon 2, Coupon 1), and (Coupon 2, Coupon 2). With (1.75) and (1.76), we see that (E_1, E_2) are dependent:

$$P(E_1 \cap E_2) = \frac{1}{2} \neq \frac{9}{16} = P(E_1)P(E_2). \tag{1.77}$$

Relax the goal What can we do in this situation where $(E_1, \ldots, E_n)$ are dependent in (1.73)? There is a useful approach. The idea is to relax the problem by deriving a lower or upper bound on the interested probability instead of finding the exact probability. To apply this approach to our problem, let's review the original question. What is k such that

$$P(\text{obtain every coupon from } k \text{ snacks}) = P(E) \geq 0.9? \tag{1.78}$$

Suppose we have a lower bound $\mathcal{L}_{\text{bound}}$ on $P(E)$: $P(E) \geq \mathcal{L}_{\text{bound}}$. One important observation is that if the lower bound is greater than 0.9, then the exact probability is also greater than 0.9.

$$\mathcal{L}_{\text{bound}} \geq 0.9 \implies P(E) \geq 0.9. \tag{1.79}$$

This observation can lead us to set up the following relaxed goal: Finding k such that

$$P(E) \geq \mathcal{L}_{\text{bound}} \geq 0.9. \tag{1.80}$$

Under this relaxed goal, we are then interested in computing $\mathcal{L}_{\text{bound}}$ instead. It turns out there is an important bounding technique that enables us to compute $\mathcal{L}_{\text{bound}}$: the *union bound*.

Union bound The statement of the union bound is straightforward. Given two events (A, B), it suggests:

$$P(A \cup B) \leq P(A) + P(B). \tag{1.81}$$

The proof is straightforward. Using the definition of an event,

$$\mathbb{P}(A \cup B) = \mathbb{P}(A) + \mathbb{P}(B) - \mathbb{P}(A \cap B). \tag{1.82}$$

Why? Think about a Venn diagram. Since $\mathbb{P}(A \cap B) \geq 0$, we get (1.81).
  Now how to apply the union bound into the interested probability:

$$\mathbb{P}(E) = \mathbb{P}(E_1 \cap E_2 \cap \cdots \cap E_n)? \tag{1.83}$$

To this end, using De Morgan's law, we first obtain:

$$\mathbb{P}(E^c) = \mathbb{P}(E_1^c \cup E_2^c \cup \cdots \cup E_n^c). \tag{1.84}$$

Applying the union bound (1.81) into the above multiple times (precisely $n - 1$ times), we get:

$$
\begin{aligned}
\mathbb{P}(E^c) &= \mathbb{P}(E_1^c \cup E_2^c \cup \cdots \cup E_n^c) \\
&\leq \mathbb{P}(E_1^c) + \mathbb{P}(E_2^c \cup \cdots \cup E_n^c) \\
&\vdots \\
&\leq \mathbb{P}(E_1^c) + \mathbb{P}(E_2^c) + \cdots + \mathbb{P}(E_n^c) \\
&= n \left(1 - \frac{1}{n} \right)^k
\end{aligned}
\tag{1.85}
$$

where the last equality follows from $\mathbb{P}(E_i^c) = \left(1 - \frac{1}{n}\right)^k \ \forall i \in \{1, \ldots, n\}$; see (1.74). Hence, we get:

$$\mathbb{P}(E) = 1 - \mathbb{P}(E^c) \geq 1 - n \left(1 - \frac{1}{n} \right)^k =: \mathcal{L}_{\text{bound}}. \tag{1.86}$$

How many snacks required to buy under the relaxed goal? Let's now determine the number of snacks needed to collect all the coupons under the relaxed goal:

$$\mathbb{P}(E) \geq \mathcal{L}_{\text{bound}} \geq 0.9. \tag{1.87}$$

Putting (1.86) into the above, we get:

$$k \log \left(1 - \frac{1}{n} \right) \leq \log \frac{0.1}{n}.$$

Dividing $\log \left(1 - \frac{1}{n} \right) < 0$ on both sides,

$$k \geq \frac{\log \frac{0.1}{n}}{\log \left(1 - \frac{1}{n}\right)}. \tag{1.88}$$

Again, focusing on one extreme case in which n is very large, we obtain:

$$\log \left(1 - \frac{1}{n}\right) \approx -\frac{1}{n}$$

using Taylor's series approximation. Applying this into (1.88), we get:

$$k \geq \frac{\log \frac{0.1}{n}}{\log \left(1 - \frac{1}{n}\right)} \approx n \log(10n). \tag{1.89}$$

The result implies that we need to purchase a large number of snacks to collect all n coupons, and the number of snacks required grows *super-linearly* in n.

Python simulation The lower bound result (1.89) asserts that, when the quantity of snacks to be purchased exceeds $n \log(10n)$, a 90% success rate in coupon collection is guaranteed. We will empirically validate this result using Python. For this purpose, we will designate the number k of snacks to be purchased as $\lceil n \log(10n) \rceil$, calculate the empirical success rate under this configuration, and then compare it with the targeted probability of 90%.

We start by constructing a Python function that returns 1 if all the coupons are successfully collected under the specified conditions, and 0 otherwise.

```python
import numpy as np

def coupon_success(k,n):
    # k: # snacks purchased
    # n: # coupons
    coupons=set(np.arange(1,n+1,1))
    count=k # count initialization
    while count!=0:
        # coupon number picked
        pick=np.random.randint(n)+1
        count-=1 # count decreased
        if pick in coupons:
            coupons.remove(pick)
            if len(coupons)==0: return 1
    return 0

print(coupon_success(100,10))
print(coupon_success(15,10))
```

1
0

Note that when k significantly exceeds $n \log(10n)$ (as illustrated in the example with $(k, n) = (100, 10)$), coupon collection is successful. Conversely, when k falls well below the lower bound (as demonstrated in the second example with $(k, n) = (10, 10)$), coupon collection fails.

Expanding on the `coupon_success` function, we proceed to develop another function that calculates the empirical success rate of coupon completion. For an accurate empirical rate, we employ a large number N of independent trials, such as setting $N = 1000$.

```python
def avg_coupon_success(k,n,N):
    # k: # snacks purchased
    # n: # coupons
    # N: # independent trials
    tot_success=0
    for i in range(N):
        tot_success += coupon_success(k,n)
    return tot_success/N

print(avg_coupon_success(60,10,1000))
```

0.989

Notice that when k surpasses $n \log(10n)$ (as demonstrated in the example with $60 > 10 \log(10 \cdot 10) \approx 46.05$), the empirical success rate is 98.9%, surpassing the expected target rate of 90%.

Lastly, we will graphically represent the empirical success rate and compare it with the targeted probability of 0.9 across a broad range of n, say $2 \le n \le 1024$.

```python
import matplotlib.pyplot as plt

N=1000
nexp=np.arange(1,11,1) # [1,2,3,...,10]
n=2**nexp # [2**1, 2**2,..., 2**10]
emp_rate = np.zeros(len(n))
compare = np.zeros(len(n)) +  0.9

for i in range(len(n)):
    k=int(n[i]*np.log(10*n[i]))+1
    emp_rate[i]=avg_coupon_success(k,n[i],N)

plt.figure(figsize=(4,4),dpi=150)
plt.plot(n,emp_rate,color='blue',
         label='empirical success rate')
plt.plot(n,compare,color='red',
         label='0.9')
plt.xlabel('$n$ (number of coupons)')
plt.ylabel('empirical success rate')
plt.legend()
plt.xscale('log')
plt.show()
```

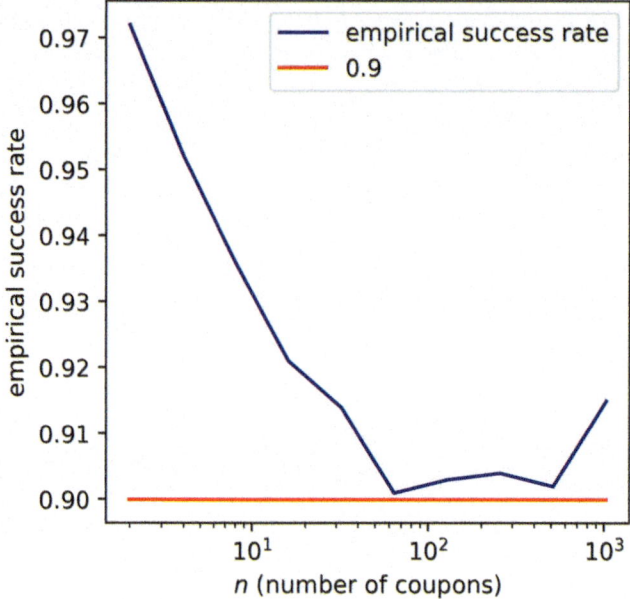

Fig. 1.25 An empirical success rate for coupon completion

From Figure 1.25, it is evident that the empirical success rate consistently exceeds 0.9 across the entire range of n as guaranteed by probability theory.

Look ahead

So far, we have covered many crucial concepts, laws, and techniques such as sample space, probability model, probability distribution, events, conditional probability, total probability law, Bayes' law, independence, and union bound. However, there are still more concepts that we need to explore to comprehend the principles of MAP and ML estimation that we intend to learn. In the next session, we will delve into one of the remaining concepts. That is, *random variables*.

Problem Set 2

Problem 2.1 (*Conditional probability: Exercise # 1*) Suppose there are two coins: one is fair; and the other is biased with $\mathbb{P}(H) = 0.6$ and $\mathbb{P}(T) = 0.4$. You are given one of the two coins with an equal chance of getting either coin. After flipping the chosen coin four times, you observe that three of the outcomes are heads. What is the probability that the coin you have is the fair one?

Problem 2.2 (*Conditional probability: Exercise # 2*) Suppose I possess a bag that comprises of either a $1 or a $5 bill (each with probability $\frac{1}{2}$). I subsequently include another $1 bill to the bag, making it a total of two bills. After shaking the bag, you randomly draw one bill from it without peeking. If it happens to be a $1 bill, what is the likelihood that a second person drawing the remaining bill from the bag will also draw a $1 bill?

Problem 2.3 (*Probability model construction*) You have a "Go" tournament ahead of you, and to win, you must win two games in a row out of three. Winning the first and the third game but losing the second results in losing the tournament. You are given two options: (i) playing against Changho Suh, then AlphaGo, then Changho Suh (SAS), or (ii) playing against AlphaGo, then Changho Suh, then AlphaGo (ASA). Changho Suh is a novice Go player and can be easily defeated with probability $p > \frac{1}{2}$. However, AlphaGo is a formidable opponent, and your probability of winning is $q < \frac{1}{2}$.

(*a*) Construct an appropriate sample space for the SAS option. What is the probability of winning the tournament?
(*b*) Given the ASA option, repeat part (*a*).
(*c*) Which option will maximize your chances of winning the tournament? Does your choice feel intuitive? If it does, explain why. If it doesn't, explain why the solution seems counter-intuitive.

Problem 2.4 (*Conditional probability: Exercise # 3*) In her pocket, a woman has 5 coins. Among these coins, two have heads on both sides, one has tails on both sides, and the remaining two are normal coins, with one head and one tail. The woman can only determine which coin is which by looking at them.

(*a*) The woman shuts her eye, chooses a coin at random, and tosses it. What is the probability that the lower face of the coin is "Head"?
(*b*) After taking the action in part (*a*) (i.e., choosing a coin at random and then tossing it), she opens her eyes and sees that the upper face of the coin is "Head". What is the probability that the lower face is "Head"?
(*c*) Given the situation in part (*b*) (i.e., choosing a coin at random, tossing it and then seeing that the upper face of the coin is "Head"), she shuts her eyes again, picks up the same coin, and tosses it again. What is the probability that the lower face is "Head"?

(*d*) Given the situation in part (*c*) (i.e., choosing a coin at random, tossing it, seeing that the upper face of the coin is "Head", picking up the same coin, then tossing it again while shutting her eyes), she opens her eyes and sees that the upper face is "Head". What is the probability that the lower face is "Head"?

Problem 2.5 (*Rationale behind the definition of mutual independence*) Consider three events (A_1, A_2, A_3). Suppose that:

$$\mathbb{P}\left(\bigcap_{i \in I} A_i\right) \neq 0 \quad \forall I \subseteq \{1, 2, 3\}.$$

One can readily conceive of the following condition for independence:

$$\mathbb{P}\left(\bigcap_{i \in I} A_i \Big| \bigcap_{i \in I^c} A_i\right) = \mathbb{P}\left(\bigcap_{i \in I} A_i\right) \quad \forall I \subseteq \{1, 2, 3\}. \tag{1.90}$$

Also, consider the criterion that constitutes the definition of mutual independence:

$$\mathbb{P}\left(\bigcap_{i \in I} A_i\right) = \prod_{i \in I} \mathbb{P}(A_i) \quad \forall I \subseteq \{1, 2, 3\}. \tag{1.91}$$

(*a*) Show that (1.90) implies (1.91).
(*b*) Show that (1.91) implies (1.90).

Problem 2.6 (*Pairwise vs. mutual independence*) Consider a sample space of 8 equiprobable outcomes:

$$\Omega = \{\omega_1, \omega_2, \ldots, \omega_8\};$$
$$\mathbb{P}(\omega_i) = \frac{1}{8}, \quad \forall \omega_i \in \Omega.$$

Come up with three events (A_1, A_2, A_3) such that:

$$\mathbb{P}(A_1) = \mathbb{P}(A_2) = \mathbb{P}(A_3) = 0.5; \tag{1.92}$$

$$\mathbb{P}(A_1 \cap A_2) = \mathbb{P}(A_1 \cap A_3) = \frac{1}{4}, \quad \mathbb{P}(A_2 \cap A_3) = \frac{1}{8}; \tag{1.93}$$

$$\mathbb{P}(A_1 \cap A_2 \cap A_3) = \mathbb{P}(A_1)\mathbb{P}(A_2)\mathbb{P}(A_3). \tag{1.94}$$

Note: The answer is not unique.

Remark: Observe in your example that A_2 and A_3 are not independent, as $\mathbb{P}(A_2 \cap A_3) = \frac{1}{8} \neq \frac{1}{4} = \mathbb{P}(A_2)\mathbb{P}(A_3)$. Note that the definition of independence would be very strange if it allowed A_1, A_2, A_3 to be independent while A_2 and A_3 are dependent. This illustrates why the definition of mutual independence requires

$$\mathbb{P}\left(\bigcap_{i\in I} A_i\right) = \prod_{i\in I} \mathbb{P}(A_i), \quad \forall I \subseteq \{1, 2, 3\},$$ (1.95)

rather than just

$$\mathbb{P}\left(\bigcap_{i=1}^{3} A_i\right) = \prod_{i=1}^{3} \mathbb{P}(A_i).$$ (1.96)

Problem 2.7 (*Independence*)

(a) Let A and B be independent events. Show that A^c and B are independent.
(b) Let A and B be two events. If the occurrence of event B makes A more likely, then does the occurrence of the event A make B more likely? Justify your answer.
(c) If event A is independent of itself, show that $\mathbb{P}(A)$ is 1 or 0.
(d) If $\mathbb{P}(A)$ is 1 or 0, show that A is independent of all events B.

Problem 2.8 (*Random variables in Balls-&-Bins*) Consider the setup of the Balls-&-Bins problem that we discussed in Sect. 1.5, where there are k balls and n bins, and we randomly throw each ball into a bin with uniform probability, independently from other balls. Define

X_i : the index of the bin where the ith ball lands, $i \in \{1, 2, \ldots, k\}$;

Y_j : number of balls in bin j, $j \in \{1, 2, \ldots, n\}$.

Assume that $n \geq 3$.

(a) Is the information regarding Y_j's equally present in all X_i's? In simpler terms, can we determine Y_j's using X_i's, and vice versa? If not, which set of random variables holds more information, and could you provide an instance of an event that can be expressed using one set of random variables but not the other?
(b) Are X_i's are mutually independent?
(c) Compute $\mathbb{P}(Y_j = 0)$ for $j \in \{1, 2, \ldots, n\}$.
(d) Compute $\mathbb{P}(Y_i = 0, Y_j = 0)$ for $i \neq j$.
(e) Are Y_i and Y_j are independent?
(f) When n is significantly large, what can be inferred about the independence of Y_i and Y_j? Could you provide some insight into your response?

Problem 2.9 (*Serve on the jury*) During the murder trial of OJ Simpson, he was accused of killing his ex-wife, Nicole Simpson. The prosecution presented evidence indicating that OJ had previously subjected Nicole to abuse. In defense of OJ Simpson, one of his lawyers, Alan Dershowitz, made the following argument. Dershowitz contended that the fact that OJ Simpson had previously abused his wife was irrelevant and should be dismissed because only 1 in 1,000 women abused by their partners are ultimately killed by their abuser. For the purpose of this question, let us assume that Dershowitz's statistic of 1 in 1,000 is accurate.

(a) Can we infer that there is only a probability of $\frac{1}{1000}$ that OJ Simpson committed the murder of Nicole based on this information? What are the reasons supporting or opposing this conclusion?

(b) Consider the scenario where we randomly choose a woman who has been subjected to spousal abuse. We define the following events: M represents the event where the woman is murdered at some stage of her life, and G denotes the event where the woman is murdered by her abuser at some point in her life. Based on a plausible estimation, 0.2% of women who are abused by their partner will be murdered by someone other than their abuser at some point in their life. Determine the probability that the woman selected is killed by her abuser, given that she is murdered.

(c) Given your response to part (b), what is your stance on Dershowitz's argument? Considering your computation, would you perceive it significant that OJ Simpson had previously subjected Nicole to abuse? Which number do you believe is more precise to utilize: the 1 in 1,000 number or the number you computed in part (b)? Explain why.

Problem 2.10 (*True or False?*)

(a) Suppose two events A and B are independent. Then, their complementary events A^c and B^c are independent as well.

(b) Let X, Y, Z be three random variables defined on the same probability space. Suppose X and Y are pairwise independent, Y and Z are pairwise independent, and X and Z are pairwise independent. Then, X, Y, Z are mutually independent.

1.7 Random Variables

Recap
We have thus far covered various concepts, laws, and techniques related to probability: sample space, probability model, probability distribution, events, complement counting trick, conditional probability, total probability law, Bayes' law, independence, De Morgan's law, and union bound. Through several challenging examples such as the birthday paradox problem, the Monty Hall problem, the disease testing problem, and the coupon collector problem, we hope you have gained a solid understanding of these concepts. Remember that our primary objective was to comprehend the two critical principles: the MAP and ML estimation. We mentioned that to achieve this goal, we need to explore two more concepts: (i) *random variables*; and (ii) *random processes*.

Outline
In this section, we will delve into one of them: random variables. The section is divided into five parts. Firstly, we will figure out what a random variable is and its associated probability model. Second, we will examine the concept of independence between random variables. Third, we will introduce the definition of independence for random variables. In the fourth part, we will investigate functions of random variables, particularly the summation function, and its corresponding probability model. Finally, as usual, we will apply the learned concepts through examples.

Definition of a random variable To put it simply, a random variable is a function that transforms outcomes (which are not necessarily numerical values) into numbers. It has a domain, which is the outcome in a sample space Ω, and a range, which is a real-valued number. To illustrate this concept, let's take the example of rolling two fair dice, which we studied earlier. The sample space for this experiment consists of pairs (ω_1, ω_2) where ω_1 and ω_2 denote the first and second dice numbers, respectively. This sample space has 6×6 elements, which can be visualized as dots in a two-dimensional picture. See the left figure in Figure 1.26.

In this example, we consider a random variable that takes the first dice number ω_1. This random variable is denoted by a capital letter, say X. It can be thought of as a mapping that yields the first dice number from an outcome pair (ω_1, ω_2). We can denote the random variable as $X(\omega)$ to emphasize that it is a function of the outcome ω, but many people prefer to use a simpler expression X. The cyan-blue arrows in Figure 1.26 illustrate the mapping of the random variable X, which, for example, yields the value 1 for six outcomes: $(1, 1), (1, 2), \ldots, (1, 6)$.

As a random variable is essentially a function, it is not hard to imagine that there could be a multitude of random variables. Indeed, there are numerous possibilities

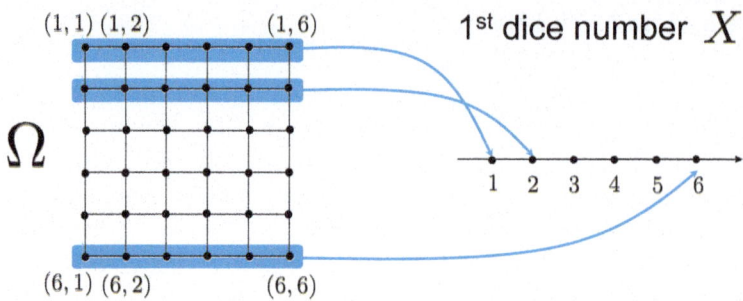

Fig. 1.26 An example of a random variable is the first dice number in an experiment of rolling two fair dice with a sample space of $\Omega = (1, 1), (1, 2), \ldots, (5, 6), (6, 6)$. The random variable, denoted by X, is a function that assigns a *real-valued* number to each outcome in Ω. For example, $X = 1$ corresponds to 6 outcomes in which the first dice number is 1: $(1, 1), (1, 2), \ldots, (1, 6)$

to construct them. For instance, one straightforward example is to define another random variable, denoted as Y, that outputs the second dice number from an outcome. Alternatively, we could also define a more intricate random variable, denoted as S, which returns the sum of the two dice numbers. This example is visualized in Figure 1.27. In this case, if we take a specific value of S, say $S = 4$ ($S(\omega) = 4$), it is associated with three outcomes: $(1, 3), (2, 2), (3, 1)$, which are indicated by a green diagonal rounded-square.

Rationale behind the definition One may question why we define random variables. In other words, why are we concerned with such mappings from outcomes to numbers? The reason is that we are often interested in numerical values with respect to outcomes. Rolling two dice is just one of many experiments where numerical

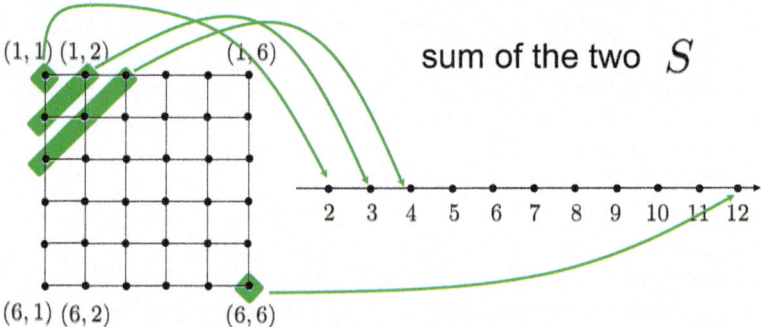

Fig. 1.27 Consider another random variable, denoted by S, which maps an outcome in Ω to the sum of the two dice numbers. For a given value s, the event $S = s$ is associated with multiple outcomes in Ω satisfying the condition $\omega_1 + \omega_2 = s$, where $(\omega_1, \omega_2) \in \Omega$. For example, the event $S = 4$ is mapped to three outcomes: $(1, 3), (2, 2)$, and $(3, 1)$

values are involved. Consider other experiments, such as: (i) counting the number of "Heads" in a coin-flipping experiment performed n times; (ii) determining the number of students who share the same birthday in the birthday paradox problem; or (iii) counting the number of collected coupons from purchasing k snacks in the coupon collector problem. These are merely a few instances, and there are countless others that come to mind.

Probability model of a random variable In Sect. 1.2, we introduced a crucial concept—the probability model comprising a sample space and its corresponding probability distribution. When it comes to random variables, the situation is similar. We also have a probability model related to a random variable. As with the previous case, this probability model consists of two components. The first is the set of values that a given random variable, denoted by X, can take on. This set is often represented by the caligraphic version of the letter X, denoted as $\mathcal{X}$:

$$\mathcal{X} = \{x_1, x_2, \ldots, x_n\}$$

where x_i indicates a numerical value that X can take on as one particular realization. By convention, we use a small letter to indicate a certain realization. The set $\mathcal{X}$ is called the *range*.

The second component is the probability distribution, which is denoted slightly differently from the probability distribution regarding the sample space $\mathbb{P}(\omega)$. We use the notation $\mathbb{P}_X(x)$, $\forall x \in \mathcal{X}$ to distinguish it from $\mathbb{P}(\omega)$. The random variable X is placed in the subscript to indicate that we are referring to the probability distribution associated with this particular random variable. The probability distribution is also known as the probability mass function, abbreviated as "pmf". The computation of the pmf is similar to that of an event, as $X = x$ can be interpreted as a specific event:

$$\mathbb{P}_X(x) = \mathbb{P}(X = x) = \sum_{\omega \in \Omega : \omega \xrightarrow{X} x} \mathbb{P}(\omega). \tag{1.97}$$

Here "$\omega \in \Omega : \omega \xrightarrow{X} x$" placed below in the summation means "over all ω's such that ω yields x via the mapping X. Here the symbol colon ":" means "such that" or "subject to".

In the rolling-two-dice experiment described above, we can easily establish the probability model. The set of possible values for the first dice number X is $\mathcal{X} = \{1, 2, \ldots, 6\}$, and its corresponding probability distribution is given by:

$$\mathbb{P}_X(x) := \sum_{\omega = (\omega_1, \omega_2) \in \Omega : \omega_1 = x} \mathbb{P}(\omega) = \frac{6}{36} = \frac{1}{6}.$$

Again, "$\omega = (\omega_1, \omega_2) \in \Omega : \omega_1 = x$" placed below in the summation means "over all ω's subject to $\omega_1 = x$". For another random variable S, on the other hand, the range reads: $S = \{2, 3, \ldots, 12\}$. The corresponding probability distribution can be computed as follows: for instance, when $s = 4$,

$$\mathbb{P}_S(4) := \sum_{\omega=(\omega_1,\omega_2)\in\Omega:\omega_1+\omega_2=4} \mathbb{P}(\omega) = \frac{3}{36}.$$

One key property of the probability distribution There is a crucial property that any probability distribution must satisfy. The property comes from two observations applicable to all random variables. For the sake of clarity, we will explain these observations using the random variable S. Firstly, if we consider any two events, such as $S = s_1$ and $S = s_2$, they are always *disjoint* whenever $s_1 \neq s_2$. This is because the occurrence of one event excludes the possibility of the other. It is a fundamental property of functions that random variables belong to. A single input cannot produce two distinct outputs. Secondly, the union of all disjoint events must cover the entire sample space Ω. For every outcome $\omega \in \Omega$, there should be a corresponding event, and all possible events must encompass the entire elements of Ω. These two observations are demonstrated in Figure 1.28.

These two observations then yield:

$$\sum_{s\in S} \mathbb{P}_S(s) = \sum_{s\in S} \mathbb{P}(S = s)$$

$$= \sum_{s\in S} \sum_{\omega\in\Omega:S(\omega)=s} \mathbb{P}(\omega) \qquad (1.98)$$

$$= \sum_{\omega\in\Omega} \mathbb{P}(\omega) = 1$$

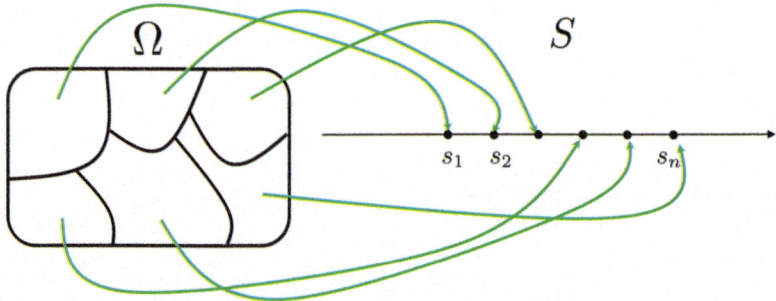

Fig. 1.28 Two key observations that apply to all random variables: (i) events defined by $S = s$ are always *disjoint* for different values of s; (ii) for every outcome $\omega \in \Omega$, there is a corresponding s, and as a result, all possible events should cover all the elements in Ω

where the last equality follows from the fact that the events $S = s$ are disjoint for different s's (first observation) and the union of all the disjoint events spans Ω (second observation). This can be perceived as a similar constraint to the sum-up-to-one property of the sample space, but in the context of random variables.

Histogram is a common visualization tool for probability distributions. As an example, the probability distribution $\mathbb{P}_S(s)$ can be illustrated by a histogram, as shown in Figure 1.29, where the height of each bar represents its corresponding probability. It is evident from this histogram that the sum of $\mathbb{P}_S(s)$ for all possible values of s equals 1. We can confirm this by computing $\frac{1}{36}(1 + 2 + \cdots + 5 + 6 + 5 + \cdots + 2 + 1) = \frac{21+15}{36} = 1$.

Independence of random variables The concept of independence applies to random variables similar to events. Two random variables, X and Y, are independent if any two events $X = x$ and $Y = y$ are independent $\forall x \in \mathcal{X}$ and $\forall y \in \mathcal{Y}$. For multiple random variables, $(X_1, \ldots, X_n)$ are mutually independent if all events $(X_1 = x_1, \ldots, X_n = x_n)$ are mutually independent $\forall x_1 \in \mathcal{X}_1, \ldots, \forall x_n \in \mathcal{X}_n$.

A function of random variables A random variable can be viewed as a function, and it is possible to form a composite function using it. This composite function can be thought of as a function of basic random variables and is also a random variable itself. One commonly used composite function is the summation. Here we will delve into a topic related to the summation of random variables.

To make this investigation more specific, we will use the previous example of the random variable S (the sum of two dice numbers in the rolling-two-dice experiment). It is possible to express S in terms of more basic random variables, namely X and Y, which represent the numbers on the first and second dice, respectively:

$$S = X + Y.$$

Fig. 1.29 The histogram represents the probability distribution of the random variable S, which denotes the sum of the two numbers obtained by rolling two fair dice

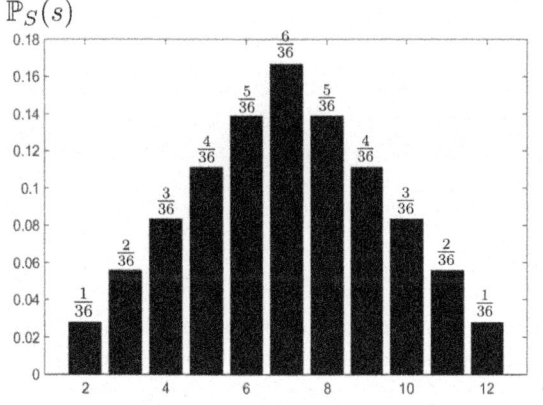

It is an ideal example that pertains to the summation of random variables.

An important point to emphasize is a different method to calculate $\mathbb{P}_S(s)$. In the previous section, we computed $\mathbb{P}_S(s)$ by counting the number of cases where $w_1 + w_2 = s$. However, there is an alternative way that involves using the probability distributions of the component random variables $\mathbb{P}_X(x)$ and $\mathbb{P}_Y(y)$. Here is how it works:

$$\mathbb{P}_S(s) = \mathbb{P}(X + Y = s).$$

This is because the event $S = X + Y = s$ can be considered as an event composed of two sources of uncertainty: one is related to X, and the other is related to Y. Due to this multiple uncertainty, there are several subcases that yield the event $S = s$. To handle such scenarios, we can make use of an important law we learned, called the *total probability law*. Using this law, we can write:

$$\mathbb{P}_S(s) = \mathbb{P}(X + Y = s)$$
$$= \sum_{x=1}^{s-1} \mathbb{P}(\{X = x\} \cap \{Y = s - x\}).$$

To simplify the expression that includes "$\cap$", people often use the following alternative expression: $\mathbb{P}(\{X = x\} \cap \{Y = s - x\}) = \mathbb{P}(X = x, Y = s - x)$, where the comma "," symbol means "and". By using this alternative expression, we can rewrite the previous equation as follows:

$$\mathbb{P}_S(s) = \mathbb{P}(X + Y = s)$$
$$= \sum_{x=1}^{s-1} \mathbb{P}(X = x, Y = s - x)$$
$$= \sum_{x=1}^{s-1} \underbrace{\mathbb{P}(X = x)}_{\mathbb{P}_X(x)} \underbrace{\mathbb{P}(Y = s - x)}_{\mathbb{P}_Y(s-x)}$$

where the last equality comes from the fact that the events $X = x$ and $Y = s - x$ are independent (i.e., X and Y are independent). By recognizing $\mathbb{P}(X = x)$ and $\mathbb{P}(Y = s - x)$ as $\mathbb{P}_X(x)$ and $\mathbb{P}_Y(s - x)$ respectively, we arrive at:

$$\mathbb{P}_S(s) = \sum_{x=1}^{s-1} \mathbb{P}_X(x)\mathbb{P}_Y(s - x). \tag{1.99}$$

The right-hand side expression appears complicated. Actually it is a well-known operation called the *convolution* (Oppenheim et al. 1997), denoted by $(\mathbb{P}_X * \mathbb{P}_Y)(s)$.

The convolution is a powerful operation that arises in a wide variety of engineering and science fields. It is also used in probability and statistics. The formal definition of convolution is given by:

$$(\mathbb{P}_X * \mathbb{P}_Y)(s) := \sum_{x=0}^{s} \mathbb{P}_X(x)\mathbb{P}_Y(s-x)$$

where the starting point is 0 and the end point is s (the interested point). The expression in (1.99) is not an exact match for the formal definition of convolution. However, they are equivalent. This is because the two endpoints $x = 0$ and $x = s$ do not contribute to the summation as $\mathbb{P}_X(0) = 0$ and $\mathbb{P}_Y(0) = 0$. Therefore, we have:

$$
\begin{aligned}
\mathbb{P}_S(s) &= \sum_{x=1}^{s-1} \mathbb{P}_X(x)\mathbb{P}_Y(s-x) \\
&= \sum_{x=0}^{s} \mathbb{P}_X(x)\mathbb{P}_Y(s-x) \\
&=: (\mathbb{P}_X * \mathbb{P}_Y)(s).
\end{aligned}
\tag{1.100}
$$

The convolution operation is typically used when X and Y are independent. However, if they are *dependent*, then (1.100) is not applicable. In such cases, alternative methods for computing $\mathbb{P}_S(s)$ may be necessary, which can vary depending on interested scenarios.

Example #1: Tossing a p-biased coin n times Consider two examples that are related to the sum of random variables. The first is a simple case where we flip a p-biased coin n times. Here, "p-biased" means that the probability of getting "Head" is p.

Suppose we want to figure out the total number of "Head"s. We can represent this number as the sum of component random variables X_i:

$$X_i = \begin{cases} 1, & \text{if the } i \text{ th flips shows "Head";} \\ 0, & \text{otherwise.} \end{cases}$$

Someone may prefer a shorthand notation: $X_i = \mathbf{1}\{i\text{th flips shows "Head"}\}$ where $\mathbf{1}\{\cdot\}$ is the indicator function that returns 1 when $(\cdot)$ is true while returning 0 otherwise. Using these, we write:

$$S = X_1 + X_2 + \cdots + X_n$$

where the range is $S = \{0, 1, \ldots, n\}$.

Let us explore a way to compute the probability distribution $\mathbb{P}_S(s)$. Instead of using the convolution operation (1.100), we can exploit the property of *symmetry*. When considering an event $S = s$, there are multiple configurations that yield $S = s$, but their corresponding probabilities are equal due to symmetry. Therefore, we have:

$$\mathbb{P}_S(s) = \text{(number of flip patterns yielding } s)$$
$$\times \mathbb{P}(X_1 = 1, \ldots, X_s = 1, X_{s+1} = 0, \ldots, X_n = 0).$$

Here the number of flip patterns yielding s is $\binom{n}{s}$. The second probability can readily be computed using the independence of $(X_1, \ldots, X_n)$:

$$\mathbb{P}(X_1 = 1, \ldots, X_s = 1, X_{s+1} = 0, \ldots, X_n = 0)$$
$$= \mathbb{P}(X_1 = 1) \cdots \mathbb{P}(X_s = 1)\mathbb{P}(X_{s+1} = 0) \cdots \mathbb{P}(X_n = 0)$$
$$= p^s(1 - p)^{n-s}.$$

This together with the counting number $\binom{n}{s}$ then gives:

$$\mathbb{P}_S(s) = \binom{n}{s} p^s(1 - p)^{n-s}, \qquad s \in \mathcal{S} = \{0, 1, \ldots, n\}.$$

In fact, this distribution is a well-known probability distribution called the *binomial* distribution, often denoted as $S \sim \mathsf{Bin}(n, p)$, where the symbol "$\sim$" means "is distributed according to". Figure 1.30 provides some examples of probability density functions that illustrate the binomial distribution.

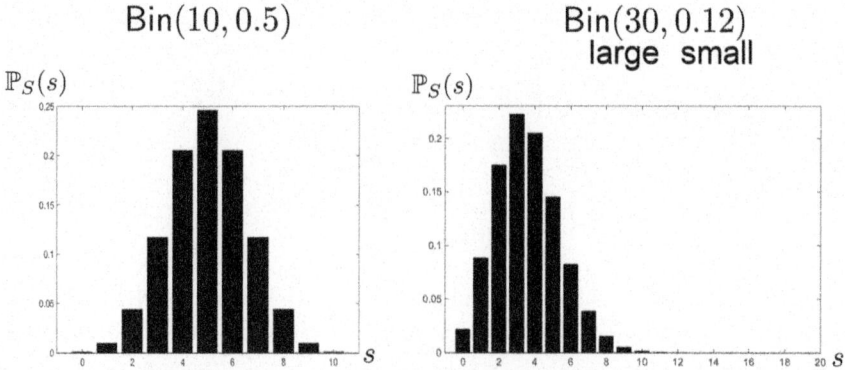

Fig. 1.30 The binomial distribution can be visualized using histograms. (Left): An example with parameters $(n, p) = (10, 0.5)$; (Right) An example with $(n, p) = (30, 0.12)$, where n is large and p is small

Example #2: Homework matching Consider another example where the computation of the probability distribution of a random variable is not as simple, even though it can also be expressed in terms of component random variables. See Figure 1.31. The homeworks of n students are collected, shuffled randomly, and then redistributed. As a result, each student may not receive her own homework.

Suppose we are interested in the number of students who receive their own homeworks. This number can be expressed as the sum of the following component random variables X_i's:

$$X_i = \begin{cases} 1, & \text{if the } i\text{th student receives her own homework;} \\ 0, & \text{otherwise.} \end{cases}$$

$$S = X_1 + X_2 + \cdots + X_n.$$

Similar to the prior example, we can use the *symmetry* property to obtain:

$$\mathbb{P}_S(s) = (\text{\# of matching patterns yielding } s)$$
$$\times \mathbb{P}(X_1 = 1, \ldots, X_s = 1, X_{s+1} = 0, \ldots, X_n = 0).$$

However, there are two challenges that make the computation of the above probability difficult. First, computing the number of matching patterns yielding s is complicated. Second, the random variables X_i's are *dependent*, which makes it difficult to compute the second probability quantity. For example, $\mathbb{P}(X_1 = 1) = \frac{1}{n}$, but the conditional probability $\mathbb{P}(X_1 = 1 | X_2 = 1) = \frac{1}{n-1}$, as the second student receiving her

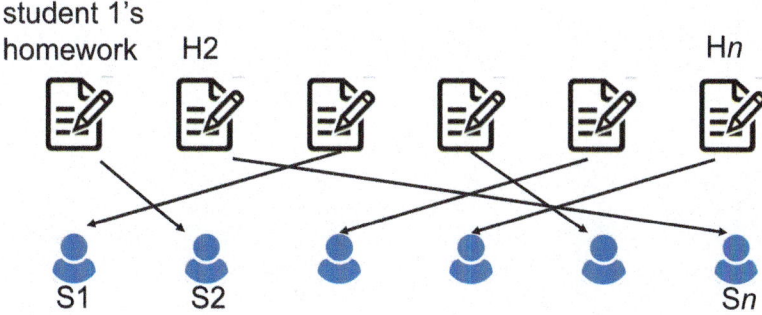

Fig. 1.31 A homework matching problem where n students submit their homework, which is then randomly shuffled and returned to the students. Each student receives only one homework and it is a one-to-one matching, but there is no guarantee that the returned homework belongs to the respective student

own homework means there is only one possibility for matching out of $n - 1$ candidates. Thus, $\mathbb{P}(X_1 = 1) = \frac{1}{n} \neq \frac{1}{n-1} = \mathbb{P}(X_1 = 1|X_2 = 1)$. These challenges make it difficult to compute $\mathbb{P}_S(s)$, and therefore we will not attempt to compute it.

Look ahead
This section has dealt with the probability model of a single random variable. However, there exists a probability model for multiple random variables, which we will discuss in the next section. Additionally, we will explore another quantity that can be used to represent a random variable with uncertainty. That is, *expectation*.

1.8   Joint Distribution and Expectation

Recap

In the previous section, we discussed the notion of a *random variable*. A random variable is simply a function that maps outcomes from a sample space to real values. Its probability model is defined by the range $\mathcal{X} = \{x_1, x_2, \ldots, x_n\}$ and the corresponding probability distribution $\mathbb{P}_X(x) \ \forall x \in \mathcal{X}$. We also examined the *summation* function as a special function of random variables, e.g., $S = X + Y$. In the case where (X, Y) are *independent*, we demonstrated that the probability distribution $\mathbb{P}_S(s)$ can be represented as the *convolution* of individual probability distributions:

$$\mathbb{P}_S(s) = (\mathbb{P}_X * \mathbb{P}_Y)(s).$$

However, the computation of $\mathbb{P}_S(s)$ can become quite complex when (X, Y) are dependent. While we have studied the probability model for a single random variable, it is worth noting that there is also a probability model for multiple random variables.

Outline

In this section, we will delve into the probability model associated with multiple random variables. The section is divided into five parts. First of all, we will extensively discuss the two components that comprise the probability model: the range and the probability distribution. We will establish that the probability distribution is closely tied to individual random variable distributions. Secondly, we will introduce a crucial concept that can serve as a deterministic and representative quantity of a random variable. That is, expectation. We will explain its definition and the rationale behind it. In fact, there are two key properties of expectation that play a role to simplify complicated-looking computations. The third part will investigate the first property, known as the function invariance property. Then we will investigate the second property, called the linearity of expectation. Finally, we will demonstrate the power of these properties with three examples.

Probability model of multiple random variables Consider a simple scenario where there are two random variables, X and Y. As usual, the probability model associated with these random variables comprises two elements. The first element is the range, which in this case, represents the collection of *pairs* that (X, Y) can take on. To see this clearly, let us assume that $\mathcal{X} = \{x_1, \ldots, x_n\}$ and $\mathcal{Y} = \{y_1, \ldots, y_k\}$. Then, we can define the range as:

$$
\begin{aligned}
\mathcal{X} \times \mathcal{Y} := \{ &(x_1, y_1), (x_1, y_2), \ldots, (x_1, y_k), \\
&(x_2, y_1), (x_2, y_2), \ldots, (x_2, y_k), \\
&\quad\vdots \qquad \vdots \qquad\qquad \vdots \\
&(x_n, y_1), (x_n, y_2), \ldots, (x_n, y_k) \}.
\end{aligned} \tag{1.101}
$$

Observe that the size of the range is equal to the product of the sizes of the individual sets, that is, $|\mathcal{X} \times \mathcal{Y}| = |\mathcal{X}| \cdot |\mathcal{Y}|$.

The second entity is the probability distribution, denoted by:

$$
\mathbb{P}_{X,Y}(x, y) \; \forall (x, y) \in \mathcal{X} \times \mathcal{Y}. \tag{1.102}
$$

Its computation is similar to that of a single random variable:

$$
\mathbb{P}_{X,Y}(x, y) = \mathbb{P}(X = x, Y = y) = \sum_{\omega \in \Omega: \omega \xrightarrow{X} x, \omega \xrightarrow{Y} y} \mathbb{P}(\omega) \tag{1.103}
$$

where Ω is the sample space and the colon ":" (placed below the summation) means "such that". To distinguish it from the probability distribution of a single random variable, the probability distribution associated with multiple random variables is referred to as the "joint" probability distribution or joint pmf. Alternatively, it can be called the joint distribution. Since events like $(X, Y) = (x_1, y_1)$ and $(X, Y) = (x_2, y_2)$ are disjoint for $(x_1, y_1) \neq (x_2, y_2)$, and the union of all such events spans Ω, the sum-up-to-one constraint still holds:

$$
\sum_{(x,y) \in \mathcal{X} \times \mathcal{Y}} \mathbb{P}_{X,Y}(x, y) = \sum_{\omega \in \Omega} \mathbb{P}(\omega) = 1. \tag{1.104}
$$

Relationship between $\mathbb{P}_{X,Y}(x, y)$ and $(\mathbb{P}_X(x), \mathbb{P}_Y(y))$ As previously stated, there is a strong relationship between $\mathbb{P}_{X,Y}(x, y)$ and $(\mathbb{P}_X(x), \mathbb{P}_Y(y))$. Specifically, this relationship is expressed as:

$$
\begin{aligned}
\sum_{y \in \mathcal{Y}} \mathbb{P}_{X,Y}(x, y) &= \sum_{y \in \mathcal{Y}} \mathbb{P}(X = x, Y = y) \\
&= \mathbb{P}_X(x)
\end{aligned}
$$

where the second equality is due to the total probability law. Similarly, we get:

$$
\sum_{x \in \mathcal{X}} \mathbb{P}_{X,Y}(x, y) = \mathbb{P}_Y(y).
$$

This is nothing but a consequence of applying the total probability law. The process of summing over one specific random variable is known as *marginalization*. The resulting components $(\mathbb{P}_X(x), \mathbb{P}_Y(y))$ are referred to as the *marginal* distributions.

Expectation   A random variable is inherently uncertain, and can only be expressed in terms of its probability distribution, which indicates the likelihood of various values occurring. However, people often seek a single, deterministic value that can represent this random quantity. This is where the concept of expectation kicks in. The expectation of a random variable X is denoted by $\mathbb{E}[X]$ and is defined as follows:

$$\mathbb{E}[X] := \sum_{x \in \mathcal{X}} x \cdot \mathbb{P}_X(x). \tag{1.105}$$

The reason why the expectation is defined as above is apparent. It is because, in this way, the expectation can be seen as a weighted average of all the feasible values. The weight, $\mathbb{P}_X(x)$ in (1.105), represents the frequency of the occurrence of x, making it reasonable to view it as a representative average.

We present two examples for computing expectation. In the first example, we consider the experiment of rolling a dice, and let X denote the dice number. Computing the expectation in this case is straightforward.

$$\mathbb{E}[X] = \frac{1}{6}(1 + 2 + \cdots + 6) = \frac{21}{6} = 3.5.$$

Now let's consider another experiment where two fair dice are rolled, and let S denote the sum of the two dice numbers. In Sect. 1.7, we already computed the probability distribution of S, which is shown in Figure 1.32. This together with a tedious calculation yields:

$$\mathbb{E}[S] = \frac{1}{36}(2 \cdot 1 + 3 \cdot 2 + \cdots + 6 \cdot 5 + 7 \cdot 6 + 8 \cdot 5 + \cdots + 12 \cdot 1) = 7.$$

Property #1: Function invariance of expectation   As mentioned in the beginning, there are two important properties w.r.t. expectation. The first property is related to

s	2	3	4	5	6	7	8	9	10	11	12
$\mathbb{P}_S(s)$	$\frac{1}{36}$	$\frac{2}{36}$	$\frac{3}{36}$	$\frac{4}{36}$	$\frac{5}{36}$	$\frac{6}{36}$	$\frac{5}{36}$	$\frac{4}{36}$	$\frac{3}{36}$	$\frac{2}{36}$	$\frac{1}{36}$

Fig. 1.32 The probability distribution of the random variable S: the sum of the two dice number in the rolling-two-dice experiment

a function of a random variable, known as the function invariance of expectation. It states that for a random variable X, if we have a function of X as $Y = g(X)$, then:

$$\mathbb{E}[Y] = \sum_{x \in \mathcal{X}} g(x)\mathbb{P}_X(x). \tag{1.106}$$

Notice that we calculate the expected value of Y using $\mathbb{P}_X(x)$, rather than $\mathbb{P}_Y(y)$. This has the advantage of avoiding the need to derive $\mathbb{P}_Y(y)$, which can be cumbersome. With the information about the function $g(\cdot)$ and $\mathbb{P}_X(x)$, we can easily compute $\mathbb{E}[Y]$.

Here is how we prove (1.106). Beginning with the definition of $\mathbb{E}[Y]$, we have:

$$\begin{aligned}
\mathbb{E}[Y] &= \sum_{y \in \mathcal{Y}} y\mathbb{P}_Y(y) \\
&= \sum_{y \in \mathcal{Y}} y\mathbb{P}(g(X) = y) \\
&= \sum_{y \in \mathcal{Y}} y \sum_{x \in \mathcal{X}: g(x) = y} \mathbb{P}_X(x)
\end{aligned} \tag{1.107}$$

where the second equality comes from the fact that the event $Y = y$ is equivalent to $g(X) = y$; and the last equality follows the definition of the event $g(X) = y$.

Let us examine the summation order in the last line of (1.107). Initially, we select a fixed y value from $\mathcal{Y}$ and subsequently choose all the x values that satisfy $g(x) = y$. We then gather the $\mathbb{P}_X(x)$ probabilities for each of these x values. This process is repeated for all other y values in $\mathcal{Y}$. Figure 1.33 (Left) depicts a particular instance of this process to provide a better understanding of how it is computed. Here (x_1, x_2) yield $g(x_1) = g(x_2) = y_1$; for others, we have $g(x_3) = g(x_4) = y_2$ and $g(x_5) = g(x_6) = y_3$. In this case, we have:

$$\begin{aligned}
&\sum_{y \in \mathcal{Y}} y \sum_{x \in \mathcal{X}: g(x) = y} \mathbb{P}_X(x) \\
&= y_1(\mathbb{P}_X(x_1) + \mathbb{P}_X(x_2)) + y_2(\mathbb{P}_X(x_3) + \mathbb{P}_X(x_4)) + y_3(\mathbb{P}_X(x_5) + \mathbb{P}_X(x_6)).
\end{aligned} \tag{1.108}$$

Let us now examine the *opposite* order of the summation, where we initially select a fixed x value from $\mathcal{X}$ and subsequently identify the corresponding $y = g(x)$. We then gather all such pairs for all x values. This process can be represented as shown in Figure 1.33 (Right), for a specific instance:

$$\begin{aligned}
&y_1(\mathbb{P}_X(x_1) + \mathbb{P}_X(x_2)) + y_2(\mathbb{P}_X(x_3) + \mathbb{P}_X(x_4)) + y_3(\mathbb{P}_X(x_5) + \mathbb{P}_X(x_6)) \\
&= g(x_1)\mathbb{P}_X(x_1) + g(x_2)\mathbb{P}_X(x_2) + g(x_3)\mathbb{P}_X(x_3) \\
&+ g(x_4)\mathbb{P}_X(x_4) + g(x_5)\mathbb{P}_X(x_5) + g(x_6)\mathbb{P}_X(x_6).
\end{aligned} \tag{1.109}$$

This is because $g(x_1) = g(x_2) = y_1, g(x_3) = g(x_4) = y_2$ and $g(x_5) = g(x_6) = y_3$. A succinct way to represent the above is:

$$g(x_1)\mathbb{P}_X(x_1) + g(x_2)\mathbb{P}_X(x_2) + g(x_3)\mathbb{P}_X(x_3)$$
$$+ g(x_4)\mathbb{P}_X(x_4) + g(x_5)\mathbb{P}_X(x_5) + g(x_6)\mathbb{P}_X(x_6) \tag{1.110}$$
$$= \sum_{x\in\mathcal{X}} g(x)\mathbb{P}_X(x).$$

This together with (1.109), (1.108) and (1.107) gives the claimed result:

$$\mathbb{E}[Y] = \sum_{y\in\mathcal{Y}} y \sum_{x\in\mathcal{X}:g(x)=y} \mathbb{P}_X(x) = \sum_{x\in\mathcal{X}} g(x)\mathbb{P}_X(x). \tag{1.111}$$

One may wonder whether this applies solely to the specific instance illustrated in Figure 1.33. However, this holds true for general scenarios as well. The rationale behind this is comparable to the one we provided in Sect. 1.7 when discussing two observations regarding the sum-up-to-one constraint of a random variable. This is because: (i) all possible y values cover the entire domain $\mathcal{X}$, meaning that there is always a specific y value for each $x \in \mathcal{X}$; and (ii) one-to-many mappings for a function are unacceptable, implying that each $x \in \mathcal{X}$ corresponds to a *unique* output y.

Property #2: Linearity of expectation The second property, known as the *linearity of expectation*, is a powerful property that comprises two sub-properties. The first sub-property is *additivity*:

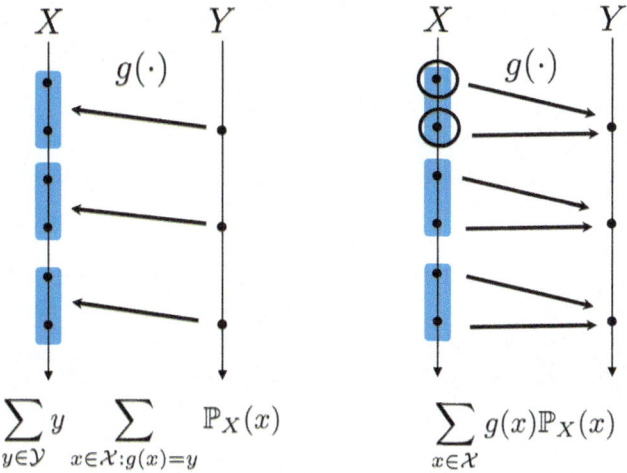

Fig. 1.33 (Left): Illustration of the computation of $\sum_{y\in\mathcal{Y}} y \sum_{x\in\mathcal{X}:g(x)=y} \mathbb{P}_X(x)$. We first fix a particular value of y. Next we find all of x's such that $g(x) = y$. We then aggregate all of them over all y's; (Right): Illustration of the computation of $\sum_{x\in\mathcal{X}} g(x)\mathbb{P}_X(x)$. We first fix a particular value of x. Next we find a corresponding $y = g(x)$. We then aggregate all of them over all x's

$$\mathbb{E}[X + Y] = \mathbb{E}[X] + \mathbb{E}[Y]. \tag{1.112}$$

Here the expectation in the left-hand-side is w.r.t. $\mathbb{P}_{X,Y}(x, y)$. On the other hand, the expectations in the right-hand-side are w.r.t. $\mathbb{P}_X(x)$ and $\mathbb{P}_Y(y)$, respectively. The second is *homogeneity* property:

$$\mathbb{E}[cX] = c\mathbb{E}[X] \text{ for any constant } c. \tag{1.113}$$

The proofs are straightforward. First we obtain:

$$\begin{aligned}
\mathbb{E}[X + Y] &= \sum_{x \in \mathcal{X}} \sum_{y \in \mathcal{Y}} (x + y)\mathbb{P}_{X,Y}(x, y) \\
&= \sum_{x \in \mathcal{X}} x \sum_{y \in \mathcal{Y}} \mathbb{P}_{X,Y}(x, y) + \sum_{y \in \mathcal{Y}} y \sum_{x \in \mathcal{Y}} \mathbb{P}_{X,Y}(x, y) \\
&= \sum_{x \in \mathcal{X}} x\mathbb{P}_X(x) + \sum_{y \in \mathcal{Y}} y\mathbb{P}_Y(y) \\
&= \mathbb{E}[X] + \mathbb{E}[Y]
\end{aligned}$$

where the first and last equalities are due to the definition of expectation; and the third follows from the total probability law. For the proof of homogeneity, we get:

$$\begin{aligned}
\mathbb{E}[cX] &= \sum_{x \in \mathcal{X}} (cx)\mathbb{P}_X(x) \\
&= c \sum_{x \in \mathcal{X}} x\mathbb{P}_X(x) = c\mathbb{E}[X]
\end{aligned}$$

where the first equality is due to the *function invariance property* (1.106).

The linearity property, as demonstrated in (1.112) and (1.113), is incredibly useful in simplifying complex computations. We will examine three instances where this property proves advantageous.

Example #1: Rolling two dice Recall the example we examined earlier: rolling two fair dice. Consider $S = X + Y$ where X and Y represent the values of the first and second dice, respectively. Earlier, we computed $\mathbb{E}[S]$ using $\mathbb{P}_S(s)$. However, utilizing the linearity property (1.112) allows us to simplify the computation as follows:

$$\mathbb{E}[S] = \mathbb{E}[X] + \mathbb{E}[Y] = 3.5 + 3.5 = 7.$$

Example #2: Tossing a p-biased coin n times The second example is the one we investigated in Sect. 1.7: tossing a p-biased coin n times. Consider the total number of "Head"s: S. Our focus will be on $\mathbb{E}[S]$. Recall that we represented S as the sum of the following component random variables:

$$X_i = \begin{cases} 1, & \text{if the } i\text{th flips shows "Head";} \\ 0, & \text{otherwise;} \end{cases}$$

$$S = X_1 + X_2 + \cdots + X_n.$$

We also computed the probability distribution of S:

$$\mathbb{P}_S(s) = \binom{n}{s} p^s (1-p)^{n-s}, \qquad \forall s \in S = \{0, 1, \ldots, n\}.$$

One naive way to compute $\mathbb{E}[S]$ is to simply apply the definition of expectation with $\mathbb{P}_S(s)$:

$$\mathbb{E}[S] = \sum_{s=0}^{n} s \cdot \binom{n}{s} p^s (1-p)^{n-s}. \tag{1.114}$$

What are your thoughts on (1.114)? Are you willing to give it a shot? Instead, we can use the linearity of expectation to simplify the computation significantly, as follows:

$$\mathbb{E}[S] = \mathbb{E}[X_1] + \cdots + \mathbb{E}[X_n] = np \tag{1.115}$$

where the second equality follows from $\mathbb{E}[X_i] = p \cdot 1 + (1-p) \cdot 0 = p$, $\forall i \in \{1, 2, \ldots, n\}$.

Example #3: Homework matching The final example pertains to the homework matching problem, as depicted in Figure 1.34, for which we gave up computing the probability distribution. Recall that we represented the number of students S who receive their own homeworks as the sum of the following individual random variables:

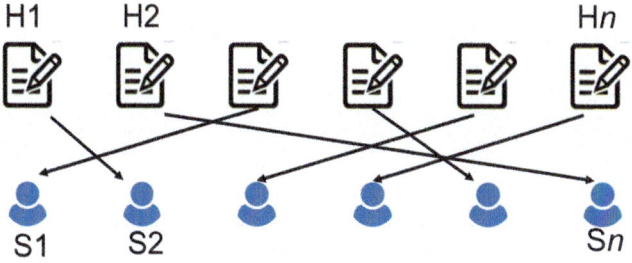

Fig. 1.34 Homework matching problem: The homeworks collected from n students are shuffled and returned back to the students. Each student receives only one homework (one-to-one matching), yet the returned homework is not necessarily her own homework

$$X_i = \begin{cases} 1, & \text{if the } i\text{th student receives her own homework;} \\ 0, & \text{otherwise;} \end{cases}$$

$$S = X_1 + X_2 + \cdots + X_n.$$

In Sect. 1.7, it was mentioned that the computation of $\mathbb{P}_S(s)$ becomes intractable due to the dependence of X_i's. Here the linearity of expectation comes to the rescue in this case:

$$\mathbb{E}[S] = \mathbb{E}[X_1] + \cdots + \mathbb{E}[X_n] = \frac{1}{n} \cdot n = 1$$

where the second equality follows from $\mathbb{E}[X_i] = \frac{1}{n} \, \forall i \in \{1, 2, \ldots, n\}$.

Look ahead
We argued that the expectation can serve as a representative quantity of a random variable. Also we have learned that one can easily compute the expectation using two key properties associated: (i) function invariance of expectation; and (ii) linearity of expectation. In the next section, we will investigate one prominent example that provides an opportunity to engage in the computation of expectation.

1.9   BitTorrent and **Python** Simulation

Recap
In the previous section, we introduced the concept of expectation and learned about two important properties which ease the computation of expectation. We have also investigated a couple of examples where we can exercise ourselves for the use of the properties. In fact, there are numerous interesting and non-trivial examples where the computation of expectation is very difficult when directly applying the definition, yet becomes much easier when employing the properties.

Outline
In this section, we will explore one such example. It consists of four parts. First, we will explain the context of the example, which turns out to be intimately related to the renowned problem that we previously examined: the coupon collector problem. Following that, we will pose a compelling question related to expectations and articulate it in mathematical terms with appropriate notations. Next, we will tackle the question by leveraging the two properties learned from the previous section. Finally, we will conduct a **Python** simulation to empirically validate the obtained result.

BitTorrent (Pouwelse et al. 2005) The example we will examine is regarding a prominent peer-to-peer file sharing system, named *BitTorrent*. BitTorrent offers entertainment services such as movies and TV shows. Since the file size of such contents is typically enormous, efficient data management is required and BitTorrent employs a smart idea to enable an efficient service. Here is how the idea works.

It does not follow a traditional method where a single chunk of the entire file is transferred from a server to a user. Instead, a movie file of a large size, e.g., ~ 10 GB is split into many chunks, say n chunks: $C_1, C_2, \ldots, C_n$. Many copies of these n chunks are then distributed over numerous servers in a network. See Figure 1.35.

A user wishes to download the entire movie file by requesting chunks to many servers which are located preferably nearby. Depending on how chunks are spread, the number of servers that the user needs to access can vary significantly. Therefore, the required number can be viewed as a random variable. Regarding a configuration of chunks distribution, here is the one that the real system employs. Each server stores only one chuck, chosen from the n possibilities uniformly at random. The chunk configuration of each server is independent of the others.

An interesting question One aspect that the system wants to equip with is data transfer efficiency. Hence, one intriguing question that one can ask regarding the efficiency arises. How many servers on average are needed to be accessed in order

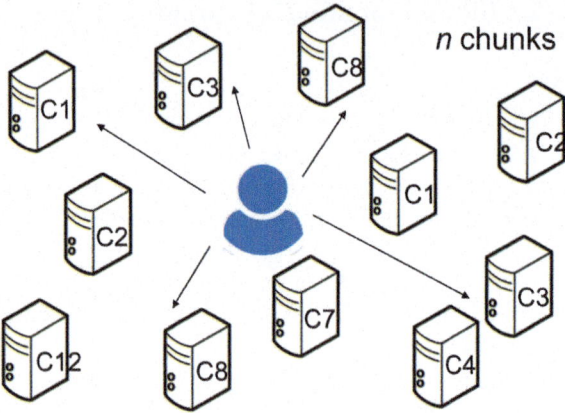

Fig. 1.35 BitTorrent: A peer-to-peer file sharing system. A user requests chunks to many servers to download the entire movie file

for the user to download the entire file successfully? In other words, what is the *expectation* of the number of servers that need to be accessed for reconstructing the entire file?

To be specific, the question can be phrased with a random variable, say K, which indicates the number of servers requested to download all of the n different chunks. The question then reads: What is $\mathbb{E}[K]$?

The answer One naive way to compute $\mathbb{E}[K]$ is to simply rely upon the definition of expectation. This way, however, requires knowledge of the probability distribution of K, which is highly non-trivial to be derived. To make a progress, we exploit the second property of expectation: *linearity*.

The idea starts with expressing K as follows:

$$K = X_1 + X_2 + \cdots + X_n.$$

Here the ith component random variable X_i denotes the number of servers requested in order for a user to obtain a new chunk, given that the user has already collected $(i-1)$ chunks where $i \in \{1, 2, \ldots, n\}$. For instance, X_1 is always 1, as no chunk has been collected and therefore any chunk obtained is new. From X_2 onward, an obtained chunk will be new or old depending on situations; hence, it would be a random variable with uncertainty.

Using the linear of expectation, we then obtain:

$$\mathbb{E}[K] = \mathbb{E}[X_1] + \mathbb{E}[X_2] + \cdots + \mathbb{E}[X_n]. \tag{1.116}$$

In this way, it suffices to figure out the probability distribution of each component random variable X_i. Notice that

$$\mathbb{P}(X_i = m)$$
$$= \mathbb{P}(\text{old chunk up to } m - 1 \text{ accesses}) \times \mathbb{P}(\text{new chunk in the } m\text{th access})$$
$$= \left(\frac{i-1}{n}\right)^{m-1} \times \left(\frac{n-i+1}{n}\right).$$

Given that $(i - 1)$ chunks are already collected, the probability of obtaining a new chunk is $\frac{n-i+1}{n}$, while that of obtaining an existing chunk is $\frac{i-1}{n}$. Also chunk requests are independent of each other.

In fact, this is a well-known distribution, called *geometric distribution*. It is often denoted by $X_i \sim \text{Geom}(p)$. Here the parameter p indicates the probability of a successful event, corresponding to obtaining a new chunk in the discussed example: $p = \frac{n-i+1}{n}$. It is widely acknowledged that the expectation of a geometric random variable is $\frac{1}{p}$. For the sake of practice, let us derive this expectation. Consider:

$$\mathbb{E}[X_i] = \sum_{m=1}^{\infty} m(1-p)^{m-1}p. \tag{1.117}$$

Here, the trick is to introduce the following function:

$$f(q) := \sum_{m=0}^{\infty} q^m \tag{1.118}$$

where $q := 1 - p$; and then make a connection with $\mathbb{E}[X_i]$. A key observation is:

$$\mathbb{E}[X_i] = p \cdot \frac{df(q)}{dq}. \tag{1.119}$$

Since $f(q)$ is the sum of a well-known geometric series, it is easy to compute:

$$f(q) = \frac{1}{1-q}. \tag{1.120}$$

Taking a derivative of $f(q)$ w.r.t. q and then applying it to (1.119), we get:

$$\mathbb{E}[X_i] = p \cdot \frac{1}{(1-q)^2} = \frac{1}{p} = \frac{n}{n-i+1}. \tag{1.121}$$

Finally, plugging this into (1.116), we obtain:

$$\mathbb{E}[K] = \frac{n}{n} + \frac{n}{n-1} + \cdots + \frac{n}{n-n+1}$$
$$= n\left(1 + \frac{1}{2} + \frac{1}{3} \cdots + \frac{1}{n-1} + \frac{1}{n}\right). \tag{1.122}$$

For a simpler expression, we may want to use Euler-Maclaurin formula (check wikipedia): in the limit of n,

$$\sum_{m=1}^{n} \frac{1}{m} \approx \log n.$$

Applying this yields:

$$\mathbb{E}[K] \approx n \log n. \tag{1.123}$$

Python simulation We will verify that the *average* number of snacks required for completion closely aligns with the approximated number $n \log n$ as n increases. To be more precise, let k_i denote the number of snacks purchased for completion in the ith trial, where $i \in \{1, 2, \ldots, N\}$. Conducting N independent trials, we will calculate the empirical average of k_i's ($\frac{k_1 + \cdots + k_N}{N}$) and compare it with $n \log n$, especially for large values of n.

Firstly, we implement a Python function below that returns k_i based on n.

```python
import numpy as np

def num_snacks_for_completion(n):
    # coupon set={1,2,...,n}
    coupons=set(np.arange(1,n+1,1))
    # number of snacks purchased
    count=0  #
    while len(coupons)!=0:
        # coupon number picked
        pick=np.random.randint(n)+1
        count+=1
        if pick in coupons: coupons.remove(pick)
    return count

print(num_snacks_for_completion(10))
```

22

In the example provided where $n = 10$, a total of 22 snacks were purchased to achieve completion.

Utilizing the num_snacks_for_completion built as above, we implement another function that calculates the empirical average of the snacks bought for completion.

```
def avg_snacks_for_completion(n,N):
    # n=number of coupons
    # N=number of independent trials
    tot_snacks=0
    for i in range(N):
        tot_snacks += num_snacks_for_completion(n)
    return tot_snacks/N

print(avg_snacks_for_completion(10,100))
```

29.44

In the given example with $n = 10$ and the number of trials N set at 100, the empirical average of snacks purchased is 29.44, resembling $n \log n = 10 \cdot \log 10 \approx 23.026$.

Lastly, we plot the average number of snacks purchased for completion and compare it with $n \log n$, for a wide range of n.

```
import matplotlib.pyplot as plt

N=100
nexp=np.arange(1,11,1) # [1,2,3,...,10]
n=2**nexp # [2**1, 2**2,..., 2**10]

avg_snacks = np.zeros(len(n))
compare = n*np.log(n)

for i in range(len(n)):
    avg_snacks[i]=avg_snacks_for_completion(n[i],N)

plt.figure(figsize=(4,4),dpi=150)
plt.plot(n,avg_snacks,color='blue',
         label='avg # snacks for completion')
plt.plot(n,compare,color='red',
         label='$n*\log n$')
plt.xlabel('$n$ (number of coupons)')
plt.ylabel('avg # snacks for completion')
plt.legend()
plt.show()
```

Observing Figure 1.36, it is evident that the empirical average of the number of snacks purchased closely aligns with the approximated value $n \log n$ across a broad range of n.

Look ahead

So far, we have worked with expectation. One thing that we mentioned is that the expectation serves as a deterministic quantity that can *well* represent a random variable with uncertainty. However, it might not be quite representative especially when the interested random variable takes many different values with high chances. One such extreme example might be the number X chosen

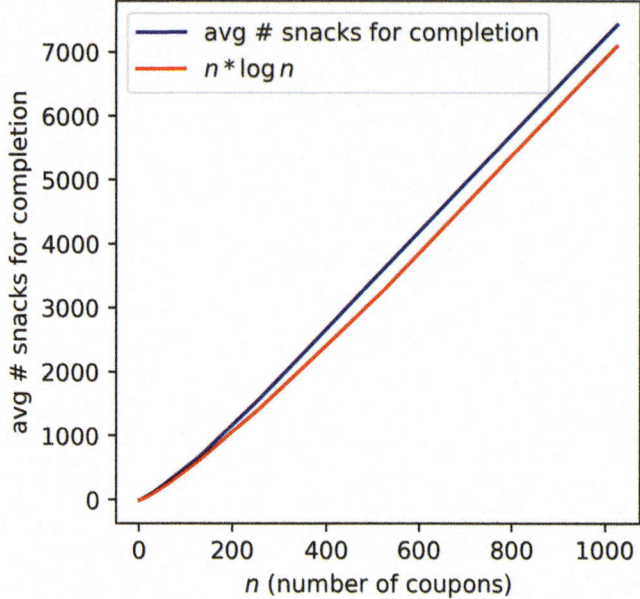

Fig. 1.36 An empirical average of the number of snacks purchased for coupon completion

from $\{1, 2, \ldots, 100\}$ uniformly at random. In this case, $\mathbb{E}[X] = \frac{1+2\cdots+100}{100} = 50.5$ might be far away from one particular realization, say 13, which has the same chance among 100 possibilities. Then, can we say $\mathbb{E}[X]$ *well represent* a random number X? It turns out there is a measure that quantifies the degree of well representing. In the next section, we will investigate the measure: *variance*.

Problem Set 3

Problem 3.1 (*Coupon collector problem*) Consider the coupon collection problem with n distinct coupons and k snacks purchased. Let X be the number of coupons acquired. Compute the distribution $\mathbb{P}_X(x)$ and the mean $\mathbb{E}[X]$.

Problem 3.2 (*Geometric distribution*) Consider an experiment where a coin is flipped repeatedly until a "Head" is obtained. It is assumed that the coin flips are independent across different trials, and the probability of showing a "Head" is p.

(*a*) Construct a sample space.
(*b*) Let X be a random variable that indicates the number of flips until we see "Head". What is its range $\mathcal{X}$? Compute its probability distribution $\mathbb{P}_X(x)$, $\forall x \in \mathcal{X}$.
(*c*) Compute $\mathbb{E}[X]$.
   Hint: Think about $\frac{df(q)}{dq}$ where $f(q) := \sum_{m=0}^{\infty} q^m$.

Problem 3.3 (*RAID* Patterson et al. 1988) Data storage in the cloud, such as Dropbox and OneDrive, carries the risk of server failure due to malfunctioning hard disks. To prevent such failures, data storage systems often employ a technique, called RAID (Redundant Array of Inexpensive Disks), which is based on error-correcting codes. In this approach, data is divided into k chunks, and $n - k$ redundant chunks (called "parities" in the coding theory literature) are generated according to a specific rule. Figure 1.37 illustrates a simple example with $(n, k) = (3, 2)$. A crucial aspect of the RAID technology is its ability to recover the original k chunks of data by accessing any k out of the n chunks stored across n servers. In the example depicted in Figure 1.37, we can retrieve the original data (x_1, x_2) by using any two chunks, such as x_2 and $x_1 \oplus x_2$, to recover x_1.

   Consider a scenario where the coded data is stored according to RAID technology, where n servers are used to store the data in k chunks. Let the probability of failure for each server be p, and assume that the servers fail independently of each other. In this case, we can define the random variable X as the number of servers that have failed.

(*a*) What is the range $\mathcal{X}$ of X? Compute its probability distribution $\mathbb{P}_X(x)$, $\forall x \in \mathcal{X}$.

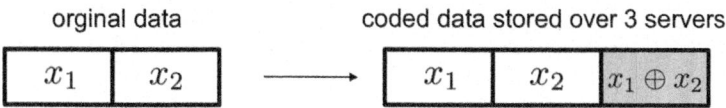

Fig. 1.37 An example of RAID: $(n, k) = (3, 2)$. The original data is divided into $k = 2$ binary chunks, denoted as $x_1 \in \{0, 1\}$ and $x_2 \in \{0, 1\}$. To generate a single parity, the modulo-2 summation $x_1 \oplus x_2$ is applied to these chunks. This single parity together with the original data form the coded data consisting of $n = 3$ chunks, which are stored over three servers. An important feature of this system is that the original data (x_1, x_2) can be recovered by accessing *any two* out of the three chunks. For example, if we have x_2 and $x_1 \oplus x_2$, we can obtain x_1 and therefore recover (x_1, x_2)

(b) What is the probability that the original data can be successfully recovered?
     Hint: You can express it simply in terms of summation.
(c) Let $p = 0.1$ and n be the number of servers. Let $k^*(n)$ be the largest k such that
     the probability of successful recovery (derived in part (b)) is at least above 98%.
     Using a **Python** code, plot $k^*(n)$ for a proper range of n, say $50 \leq n \leq 150$.
     Hint: You may want to use math.comb and/or math.factorial.

Problem 3.4 (*Family planning*) Mr and Mrs Brown decide to continue having
children until they either have their first boy or until they have five children. Assume
that each child is equally likely to be a boy or a girl, independent of all other children,
and that there are no multiple births. Let B and G denote the numbers of boys and
girls respectively that the Browns have.

(a) Write down the sample space together with the probability of each outcome.
(b) Compute and plot the distributions of the random variables B and G.
(c) Compute the mean of B and G using a direct calculation.

Problem 3.5 (*Some pmf computations*) Suppose X_1 and X_2 represent two independent rolls of a 6 sided die.

(a) Calculate the probability mass function and the expectation of $Y = \max\{X_1, X_2\}$.
(b) Calculate the probability mass function and the expectation of $Z = Y - X_1$.

Problem 3.6 (*Flipping coins*) Suppose we flip n coins to obtain $X_1, \ldots, X_n \in \{H, T\}$. Let Y be the number of pairs of consecutive Heads (i.e., number of i's such
that $X_i = X_{i+1} = H$). Compute $\mathbb{E}[Y]$.

Problem 3.7 (*Linearity of expectation*) Suppose you put m balls randomly in n
boxes. Each box can hold an arbitrarily large number of balls. What is the expected
number of empty boxes?

Problem 3.8 (*Conditional expectation*) The conditional expectation of a random
variable X given an event A is defined as:

$$\mathbb{E}[X|A] := \sum_{x \in \mathcal{X}} x \cdot \mathbb{P}(X = x|A) \tag{1.124}$$

where $\mathcal{X}$ is the range of the random variable X. Suppose we have two random
variables X and Y defined on the same sample space. Show that

$$\mathbb{E}[X] = \sum_{y \in \mathcal{Y}} \mathbb{E}[X|Y = y]\mathbb{P}(Y = y). \tag{1.125}$$

Problem 3.9 (*Conditional expectation*) James Bond is imprisoned in a cell from which there are three possible ways to escape: (i) an air-conditioning duct; (ii) a sewer pipe; and (iii) the door (which is unlocked). The air-conditioning duct leads him on a two-hour trip whereupon he falls through a trap door onto his head, much to the amusement of his captors. The sewer pipe is similar but takes five hours to traverse. Each fall produces temporary amnesia and he is returned to the cell immediately after each fall. Assume that he always immediately chooses one of the three exits from the cell with probability $\frac{1}{3}$.

(a) What is the distribution of the number X of times James Bond returns to the cell before he exits? What is $\mathbb{E}[X]$?
(b) What is the expected duration (in hours) until James Bond exits the cell?

Problem 3.10 (*Conditional expectation*) Assume $X_1, \ldots, X_n$ are mutually independent and identically distributed (i.i.d.).

(a) For $1 \leq m \leq n$, show that

$$\mathbb{E}[X_1 + \cdots + X_m | X_1 + \cdots + X_n] = \frac{m}{n}(X_1 + \cdots + X_n). \qquad (1.126)$$

(b) For $1 \leq n \leq m$, show that

$$\mathbb{E}[X_1 + \cdots + X_m | X_1 + \cdots + X_n] = X_1 + \cdots + X_n + (m - n)\mathbb{E}[X_1]. \qquad (1.127)$$

Problem 3.11 (*True or False?*)

(a) Let $X \in \mathcal{X}$ be a random variable and $\mathbb{P}_X(x)$ be its probability distribution where $x \in \mathcal{X}$. The sum-up-to-one property $\sum_{x \in \mathcal{X}} \mathbb{P}_X(x) = 1$ holds because: (i) events $\{X = x\}$'s are disjoint for different x's; and (ii) the random variable is a function that takes the sample space Ω as the domain.
(b) If $X \sim \text{Bin}(n, p)$ and $X \sim \text{Bin}(m, p)$ are independent binomial random variables with the same probability p, then $Z = X + Y$ is again a binomial random variable; its distribution is $Z \sim \text{Bin}(n + m, p)$.

1.10   Variance

Recap
Over the past sections, we have explored the concept of expectation, which can act as a representative for a random variable. The definition of expectation is as follows:

$$\mathbb{E}[X] := \sum_{x \in \mathcal{X}} x \cdot \mathbb{P}_X(x). \qquad (1.128)$$

We also investigated two important properties. The first is the *function invariance* property which enables the computation of $\mathbb{E}[Y]$ directly without the need to calculate $\mathbb{P}_Y(y)$ for $Y = g(X)$. The property is stated as:

$$\mathbb{E}[Y] = \sum_{x \in \mathcal{X}} g(x)\mathbb{P}_X(x). \qquad (1.129)$$

The second is the *linearity* property consisting of two sub-properties:

(Additivity): $\mathbb{E}[X + Y] = \mathbb{E}[X] + \mathbb{E}[Y]$;
(Homogeneity): $\mathbb{E}[cX] = c\mathbb{E}[X]$ for any constant c. $\qquad (1.130)$

Finally, we posed a question: Can the expectation well represent the uncertain random variable? We subsequently asserted that there exists a measure that can quantify the degree of representation. The measure is *variance*.

Outline
This section will explore details on variance and be divided into four parts. Firstly, we will introduce the definition of variance and explain the rationale behind it. In the second part, we will uncover a valuable fact that allows for efficient computation of variance. The third part will concentrate on two essential properties that are critical for handling challenging scenarios where computing variance using the definition alone is difficult. Lastly, we will examine the prominent inequality known as *Chebyshev's inequality*, which can act as a more precise measure of quantifying the degree of representation of a random variable, relative to variance.

Definition of variance For a random variable X, the variance is denoted by $\mathsf{var}(X)$ and defined as:

$$\mathsf{var}(X) := \mathbb{E}[(X - \mu)^2] \qquad (1.131)$$

where $\mu := \mathbb{E}[X]$. As is typical with mathematical definitions, there is a rationale behind the variance being defined as above. A more natural definition might be $\mathbb{E}[X - \mu]$, as it represents the averaged deviation from the center μ. However, this definition always evaluates to zero, regardless of $\mathbb{P}_X(x)$. The sign of $X - \mu$ is equally likely to be positive or negative. To address this issue, one can think of another measure that always takes a positive value, namely $\mathbb{E}[|X - \mu|]$. It serves as a proper measure, but its computation turns out to be often challenging. Therefore, the third candidate was introduced, $\text{var}(X) := \mathbb{E}[(X - \mu)^2]$, and its computation was found to be more tractable in many cases. This is the main reason why variance is defined as in (1.131). Had the computation of $\mathbb{E}[|X - \mu|]$ been more feasible, it would have been the preferred choice.

However, there was a problem with the definition given in (1.131). The use of exponent 2 in the definition causes the *unit* of $\text{var}(X)$ to differ from that of X. For instance, if the unit of X is meter, then the unit of $\text{var}(X)$ would be meter^2. This discrepancy results in a failure to accurately capture the *deviation* from the center. To address this issue, another measure is commonly used, which is known as the *standard deviation*. It is defined as the square root of the variance:

$$\sigma(X) := \sqrt{\text{var}(X)}. \tag{1.132}$$

In fact, this can be a frustrating situation. At times, mathematicians introduce multiple seemingly repetitive concepts, such as variance and standard deviation, simply for the sake of the "beauty of math". They strive for elegance and ease of computation, and variance is one of the notions of this mindset.

A useful fact There exists a useful fact that enables us to compute the variance in a more efficient manner. The fact is:

$$\text{var}(X) = \mathbb{E}[X^2] - \mu^2. \tag{1.133}$$

The proof of this is straightforward. Starting with the definition of variance, we get:

$$\begin{aligned}
\text{var}(X) &:= \mathbb{E}[(X - \mu)^2] \\
&= \mathbb{E}[X^2 - 2\mu X + \mu^2] \\
&= \mathbb{E}[X^2] + \mathbb{E}[-2\mu X] + \mathbb{E}[\mu^2] \\
&= \mathbb{E}[X^2] - 2\mu\mathbb{E}[X] + \mathbb{E}[\mu^2] \\
&= \mathbb{E}[X^2] - \mu^2
\end{aligned}$$

where the third and fourth steps are due to the additivity and homogeneity properties of expectation (1.130), respectively. It has been observed that computing $\mathbb{E}[X^2]$ is often simpler than computing the translated version $\mathbb{E}[(X - \mu)^2]$ in many cases. Hence, people frequently use (1.133) instead of the original definition (1.131).

Here is an example where computing $\mathbb{E}[X^2]$ is indeed easier. Consider a uniformly distributed random variable X that takes one value from the set $\mathcal{X} := \{1, 2, \ldots, n\}$. Let $\mathbb{P}_X(x) = \frac{1}{n}$ be the corresponding probability distribution where $x \in \mathcal{X}$. The expectation can be calculated as $\mathbb{E}[X] = \frac{1}{n}(1 + 2 + \cdots + n) = \frac{n+1}{2}$. The expectation of X^2 (also known as the second moment) can be expressed as:

$$
\begin{aligned}
\mathbb{E}[X^2] &= \frac{1}{n}(1^2 + 2^2 + \cdots + n^2) \\
&= \frac{1}{n} \cdot \frac{n(n+1)(2n+1)}{6} \\
&= \frac{(n+1)(2n+1)}{6}
\end{aligned}
$$

where the second equality comes from a well-known calculus fact: $\sum_{i=1}^{n} i^2 = \frac{n(n+1)(2n+1)}{6}$. If you have forgotten the proof of this fact, one way to approach it is to consider the function $f(n) = n^3$ and compute the difference $f(n+1) - f(n)$. Using the second moment, we can then compute:

$$
\begin{aligned}
\text{var}(X) &= \mathbb{E}[X^2] - (\mathbb{E}[X])^2 \\
&= \frac{(n+1)(2n+1)}{6} - \frac{(n+1)^2}{4} \\
&= (n+1)\left(\frac{n-1}{12}\right) = \frac{n^2 - 1}{12}.
\end{aligned}
$$

Two properties As previously mentioned, two important properties aid in tackling situations where computing variance directly is not feasible. These properties pertain to *independent* random variables. For the sake of illustration, let's consider a simple scenario where there exist two independent random variables, say X and Y. The first property, known as *uncorrelatedness*, states that the expectation of the product is equal to the product of the individual expectations:

$$
\mathbb{E}[XY] = \mathbb{E}[X]\mathbb{E}[Y]. \tag{1.134}
$$

The second property is known as *additivity*, which states that the variance of the sum of two independent random variables is equal to the sum of their individual variances:

$$
\text{var}(X + Y) = \text{var}(X) + \text{var}(Y). \tag{1.135}
$$

The proof of these properties is not too difficult. Let's first prove (1.134). Starting with the definition of expectation, we have:

$$\mathbb{E}[XY] = \sum_{x \in \mathcal{X}} \sum_{y \in \mathcal{Y}} xy \mathbb{P}_{X,Y}(x, y)$$

$$= \sum_{x \in \mathcal{X}} \sum_{y \in \mathcal{Y}} xy \mathbb{P}_X(x) \mathbb{P}_Y(y)$$

$$= \sum_{x \in \mathcal{X}} x \mathbb{P}_X(x) \sum_{y \in \mathcal{Y}} y \mathbb{P}_Y(y)$$

$$= \mathbb{E}[X] \mathbb{E}[Y]$$

where the second equality is due to the independence of (X, Y). Proving (1.135) is also straightforward, especially with the help of (1.134). Let $\mu_X = \mathbb{E}[X]$ and $\mu_Y = \mathbb{E}[Y]$. Starting with the useful fact (1.133), we have:

$$\text{var}(X + Y) = \mathbb{E}[(X + Y)^2] - (\mu_X + \mu_Y)^2$$

$$= \mathbb{E}[X^2] + \mathbb{E}[Y^2] + 2\mathbb{E}[XY] - (\mu_X + \mu_Y)^2$$

$$= \mathbb{E}[X^2] + \mathbb{E}[Y^2] + 2\mathbb{E}[X]\mathbb{E}[Y] - (\mu_X + \mu_Y)^2$$

$$= \mathbb{E}[X^2] - \mu_X^2 + \mathbb{E}[Y^2] - \mu_Y^2$$

$$= \text{var}(X) + \text{var}(Y)$$

where the second equality is due to the linearity of expectation (1.130); and the third equality follows from the uncorrelateness property (1.134).

Example: Tossing a p-biased coin n times Here is an example that demonstrates the usefulness of the two properties, especially the second property. This example is related to tossing a p-biased coin n times and computing the total number S of "Heads". In this example, we aim to calculate $\text{var}(S)$. By applying the useful fact (1.133), we get:

$$\text{var}(S) = \mathbb{E}[S^2] - (\mathbb{E}[S])^2. \tag{1.136}$$

As we have seen in Sect. 1.7, S follows the well-known binomial distribution. One approach to computing $\text{var}(S)$ is to first compute the first and second moments, i.e., $\mathbb{E}[S]$ and $\mathbb{E}[S^2]$, and then use the formula (1.136). However, this approach turns out to be much more complicated than the alternative approach we will take in the sequel.

The alternative approach is to make use of the additivity property (1.135). To do so, as we did before, we will represent S as the sum of the following elementary random variables:

$$S = X_1 + X_2 + \cdots + X_n$$

where $X_i = \mathbf{1}\{i\text{th flip shows "Head"}\}$. Here the key observation is that X_i's are *independent*. Hence, by applying (1.135) several times ($n - 1$ times precisely), we get:

$$\begin{aligned} \text{var}(S) &= \text{var}(X_1) + \text{var}(X_2 + \cdots + X_n) \\ &= \text{var}(X_1) + \text{var}(X_2) + \text{var}(X_3 + \cdots + X_n) \\ &\quad\vdots \\ &= \text{var}(X_1) + \text{var}(X_2) + \cdots + \text{var}(X_n). \end{aligned}$$
(1.137)

Here, the variance of one particular random variable, say X_1, is:

$$\text{var}(X_1) = \mathbb{E}[X_1^2] - (\mathbb{E}[X_1])^2 = p - p^2 = p(1-p). \tag{1.138}$$

By symmetry, $\text{var}(X_1) = \cdots = \text{var}(X_n)$. This together with (1.137) yields:

$$\text{var}(S) = \text{var}(X_1) + \cdots + \text{var}(X_n) = np(1-p).$$

A more precise measure for capturing how well $\mathbb{E}[X]$ represents X We introduced the variance (more accurately, the standard deviation to match units) as a measure that indicates how well the uncertain random variable X is represented by its expectation $\mathbb{E}[X]$:

$$\sigma(X) := \sqrt{\text{var}(X)}, \quad \text{var}(X) := \mathbb{E}[(X - \mu)^2].$$

However, the variance has a limitation in that it only measures the degree of spread around the center and does not provide a concrete probabilistic number. For instance, someone may be interested in the explicit probabilities of how often X deviates from μ by a certain distance d, known as *tail probability*. This probability can be written formally as:

$$\mathbb{P}(|X - \mu| \geq d). \tag{1.139}$$

The measure of tail probability can serve as a more precise measure since it quantifies how often a random variable deviates from its mean by a certain distance. However, computing this measure can be difficult. In such cases, it is helpful to relax the goal by obtaining a bound on the tail probability. An upper bound is preferable since a small upper bound reduces the tail probability. *Chebyshev's inequality* is a prominent inequality that provides an upper bound on the tail probability.

$$\mathbb{P}(|X - \mu| \geq d) \leq \frac{\text{var}(X)}{d^2}. \tag{1.140}$$

The intuitive idea behind Chebyshev's inequality is that a small variance will result in a small upper bound, which in turn decreases the tail probability. If X deviates significantly from its mean μ, there should be a small chance that $|X - \mu|$ is greater than some deviation d. This is clearly reflected in the inequality.

Proof of Chebyshev's inequality (1.140) The proof becomes straightforward if we make use of another popular inequality called *Markov's inequality*. This inequality applies to a non-negative random variable Y. It is stated as:

$$\mathbb{P}(Y \geq d) \leq \frac{\mathbb{E}[Y]}{d} \tag{1.141}$$

where $d > 0$. Using Markov's inequality, we get:

$$\mathbb{P}(|X - \mu| \geq d) = \mathbb{P}((X - \mu)^2 \geq d^2)$$
$$\leq \frac{\mathbb{E}[(X - \mu)^2]}{d^2} = \frac{\text{var}(X)}{d^2}$$

where the first equality follows from the fact that $\{(X - \mu)^2 \geq d^2\}$ is equivalent to $\{|X - \mu| \geq d\}$; and the last step is due to the definition of variance.

Proof of Markov's inequality (1.141) We start with a key observation:

$$Y \geq d \cdot \mathbf{1}\{Y \geq d\}. \tag{1.142}$$

This is immediate because the right-hand side equals 0 when $Y < d$ (while the left-hand side is $Y \geq 0$), and when $Y \geq d$, the right-hand side equals d while the left-hand side is $Y \geq d$. Since (1.142) is valid for any value of Y, the inequality remains true even after taking the expectation on both sides.

$$\mathbb{E}[Y] \geq d \cdot \mathbb{E}[\mathbf{1}\{Y \geq d\}]$$
$$= d \cdot \mathbb{P}(Y \geq d).$$

Hence, we complete the proof.

Look ahead
Up to this point, we have focused exclusively on *discrete* random variables. However, in real-world situations, there are numerous scenarios that involve *continuous-valued* random quantities. One such scenario is *communication*, which we will delve into in detail in Part III. As mentioned in Sect. 1.1, there is an adversary in communication systems, namely *noise*, which can be represented as a continuous-valued random signal. A pertinent concept related to such continuous random signals is the notion of a *continuous* random variable. Hence, in the next section, we will explore continuous random variables.

1.11   Continuous Random Variables

Recap
In the preceding section, we introduced another important measure in probability: *variance*. This quantity is essential in measuring the degree to which a random variable X differs from its expected value $\mathbb{E}[X]$. The definition of variance is as follows:

$$\text{var}(X) := \mathbb{E}[(X - \mu)^2] \tag{1.143}$$

where $\mu := \mathbb{E}[X]$. Subsequently, we learned a useful fact that frequently simplifies the process of calculating variance:

$$\text{var}(X) = \mathbb{E}[X^2] - \mu^2. \tag{1.144}$$

We also investigated two properties w.r.t. independent random variables X and Y:

$$\begin{aligned} \text{(Uncorrelatedness): } \mathbb{E}[XY] = \mathbb{E}[X]\mathbb{E}[Y]; \\ \text{(Additivity): } \text{var}(X + Y) = \text{var}(X) + \text{var}(Y). \end{aligned} \tag{1.145}$$

Thus far, we have only discussed a specific type of random variable that involves taking *discrete* real numbers. In reality, however, there are numerous scenarios that involve *continuous*-valued random quantities. One situation in which continuous random variables play a significant role is in the field of *communication*. In communication systems, *noise* is introduced, and this can be represented as a continuous-valued random quantity. Finally, we mentioned that the concept of a *continuous* random variable is pertinent to understanding such continuous random signals.

Outline
This section is dedicated to exploring the concept of a continuous random variable. It consists of four parts. First, we will introduce the concept of a continuous random variable within the context of a new type of sample space that we have yet to explore. Next, we will delve into a new concept called the *probability density function*, which arises when dealing with continuous random variables. Following that, we will examine the *cumulative density function*, a relevant concept that is closely related to continuous random variables. Finally, we will investigate the definitions of expectation and variance in the context of a continuous random variable.

A new type of sample space Up to this point, we have focused on a specific type of sample space that consists of a *countable* set, which is either a *finite* set or a *countably infinite* set. In reality, however, there are numerous situations in which the sample space is not countable. To illustrate this, let us consider an experiment in which we select a point uniformly at random from the interval [0, 1], as depicted in Figure 1.38. Note that the sample space is not countable, as an element ω in Ω is a *continuous* value. As a result, a *continuous* random variable can naturally arise in this context. One straightforward example of a continuous random variable is a mapping that simply outputs the input element: $X(\omega) = \omega$. This example helps to illustrate the concept of a continuous random variable, which has the same definition as its discrete counterpart but takes a *continuous* value in the range.

How to define probability distribution? When dealing with continuous random variables, defining probability distribution is not straightforward. One might consider using a discrete probability distribution, $\mathbb{P}_X(x) = \mathbb{P}(X = x)$, but this approach raises an issue. The problem is that for all $x \in \mathcal{X}$, we would need to set $\mathbb{P}(X = x) = 0$, regardless of how X behaves. The reason is that if we assign a positive value, say ϵ, to $\mathbb{P}(X = x)$ for all $x \in \mathcal{X}$, the sum of all probabilities $\mathbb{P}(X = x)$ would be infinite, which violates the constraint that probabilities should sum up to one. Therefore, $\mathbb{P}(X = x)$ must be zero for all $x \in \mathcal{X}$. Obviously this is not a proper assignment.

To avoid encountering such uninteresting scenarios, we may instead consider the probability of X falling within an *interval*:

$$\mathbb{P}(X \in [a, b]) \tag{1.146}$$

where $0 \le a \le b \le 1$. In this case, the most natural probability assignment would be to calculate the length of the corresponding interval $[a, b]$ and normalize it by the entire interval [0, 1]:

$$\mathbb{P}(X \in [a, b]) = \frac{\text{length of } [a, b]}{\text{length of } [0, 1]} = b - a. \tag{1.147}$$

This is a valid choice because the probability of an interval increases with its length, which intuitively makes sense. Moreover, it satisfies the sum-up-to-one constraint that applies to the probability distribution of a discrete random variable. To see this, consider *disjoint* intervals I_i's such that $\bigcup_i I_i = [0, 1]$. Then,

$$\Omega = [0, 1]$$

$$0 \qquad \omega \qquad 1 \qquad X(\omega) = \omega$$

Fig. 1.38 A new type of sample space in which an element in Ω is a *continuous* value between 0 and 1. In this context, $X(\omega) = \omega$ is the simplest example of a *continuous* random variable

$$\sum_i \mathbb{P}(X \in I_i) = \mathbb{P}\left(\bigcup_i I_i\right) = 1 \qquad (1.148)$$

where the first equality follows from the fact that I_i's are disjoint.

Probability density function What about the general case when the distribution is not necessarily uniform? In order to handle this, we need to specify the probability $\mathbb{P}(X \in [a, b])$ for all intervals $[a, b]$, not just limited to the $[0, 1]$ interval. Moreover, $\mathbb{P}(X \in [a, b])$ may not solely be a function of the interval length $b - a$, but may also depend on how often X falls within the interval. This is where the concept of a probability density function (pdf) comes in. The pdf formally specifies the probability distribution, and is denoted by $f(x)$. It is defined as a function that satisfies:

$$\mathbb{P}(a \leq X \leq b) = \int_a^b f(x)dx \qquad \forall a, b \in \mathbb{R} \text{ such that } a \leq b \qquad (1.149)$$

where $f(x)$ is assumed to be continuous everywhere and can be integrable. With this definition, we can express the probabilities of interest, such as $\mathbb{P}(a \leq X \leq b)$, in terms of the pdf $f(x)$. Some may prefer to use a different notation, such as $f_X(x)$, to explicitly display the associated random variable X. However, many people use the simpler notation $f(x)$. The definition involves integrating $f(x)$, which has a geometric interpretation as the area under the function $f(x)$ over the corresponding interval. Thus, the meaning of $\mathbb{P}(a \leq X \leq b)$ in (1.149) can be interpreted as the *area* below the pdf within the interval $[a, b]$, as shown in Figure 1.39. Thus, we can observe that the pdf serves a similar purpose as the histogram, which was employed to visualize the probability distribution of a *discrete* random variable.

Two properties of the pdf Similar to the histogram in the discrete counterpart, there are two properties w.r.t. the pdf. The first is the non-negativity property:

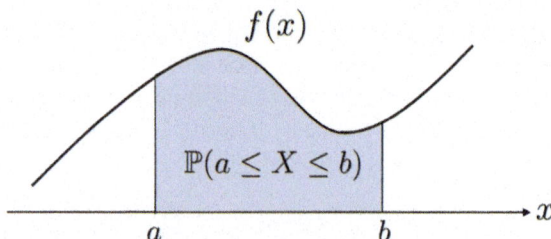

Fig. 1.39 A visual representation of the relationship between a probability density function $f(x)$ and $\mathbb{P}(a \leq X \leq b)$ is shown in the figure. The area under the curve $f(x)$ between a and b corresponds to the probability of X taking a value in that interval

$$f(x) \geq 0. \tag{1.150}$$

The reason for the non-negativity property is straightforward. If $f(x) < 0$ for some x, say t, we can find an interval containing t where the integral over the interval in (1.149) is negative. This implies that we have a negative probability for this event, which contradicts the non-negativity property of the probability distribution. The second is the sum-up-to-one property:

$$\int_{-\infty}^{\infty} f(x)dx = \mathbb{P}(-\infty \leq X \leq \infty) = 1 \tag{1.151}$$

where the first equality is due to the definition of the pdf (1.149); and the second equality is because X must take on some value in the real line $\mathbb{R}$.

Two caveats However, there are significant differences compared to the histogram, which come with the following two caveats. The first is that the integration in (1.149) may not always be well-defined. Nevertheless, this is a rare and practically irrelevant case, so we will not worry about it throughout this book. In fact, it is okay to disregard this unless you are interested in exploring hardcore probability theory. We assume that many of the readers are interested in application-driven probability rather than advanced topics. However, if you do wish to study hardcore probability, you may want to consider a graduate-level course on measure theory.

The second caveat is that the pdf is not a probability value, which can be easily demonstrated with the following example. Let us take a uniformly distributed random variable X that assumes continuous values in the interval $[0, 0.5]$. In this case, we have $f(x) = 2$ as per (1.151). As the pdf $f(x) = 2$ exceeds 1, it is evident that it does not represent a probability quantity. This raises the following questions: What does the pdf signify? How is it related to the probability quantity? To answer these, consider a very small interval $[x, x + \delta]$ and approximate the probability of X being in the interval as follows:

$$\mathbb{P}(x \leq X \leq x + \delta) = \int_{x}^{x+\delta} f(x)dx \approx \delta f(x).$$

As the interval $[x, x + \delta]$ becomes smaller, this approximation becomes increasingly accurate, since the integration more closely approximates the area of the rectangle with width δ and height $f(x)$. More formally, in the limit as $\delta \to 0$, we obtain:

$$f(x) = \lim_{\delta \to 0} \frac{\mathbb{P}(x \leq X \leq x + \delta)}{\delta}. \tag{1.152}$$

Hence, the pdf can be interpreted as the *probability per unit length*. This is why the term "density" is used in its name.

Example: Throwing darts Let's practice computing $f(x)$ through an example. Consider the experiment of throwing a dart into a circle-shaped target with a unit radius, as shown in Figure 1.40. Let X be a continuous random variable representing the distance from the center to the point where the dart lands. The range of a continuous random variable is always the real line $\mathcal{X} = [-\infty, \infty]$, but in this case, X cannot be negative, so we can assume $\mathbb{P}(X < 0) = 0$. We can also assume that the dart always lands within the circle, so $\mathbb{P}(X > 1) = 0$.

How can we calculate the probability density function $f(x)$ within the range of interest $x \in [0, 1]$? To begin with, we must compute $\mathbb{P}(x \le X \le x + \delta)$ according to (1.152). This probability should be proportional to the area of the ring formed between the x-radius circle and the $(x + \delta)$-radius circle, as shown in the green-colored ring in Figure 1.41. Hence, it should read:

$$
\begin{aligned}
\mathbb{P}(x \le X \le x + \delta) &= \frac{\text{green ring area}}{\pi \cdot 1^2} \\
&= \frac{\pi(x + \delta)^2 - \pi x^2}{\pi} \\
&= 2\delta x + \delta^2
\end{aligned}
$$

where the first equality is due to the normalization by the unit-circle area $\pi \cdot 1^2 = \pi$. Why normalized? This is because of the sum-up-to-one constraint. This together with (1.152) and the assumption $\mathbb{P}(X > 1) = \mathbb{P}(X < 0) = 0$ gives:

$$
f(x) = \begin{cases} 0, & \text{if } x < 0; \\ 2x, & \text{if } 0 \le x \le 1; \\ 0, & \text{if } x > 1. \end{cases} \tag{1.153}
$$

Fig. 1.40 Experiment of throwing darts. Let X be a continuous random variable that indicates the distance from the center to the point where the dart lands

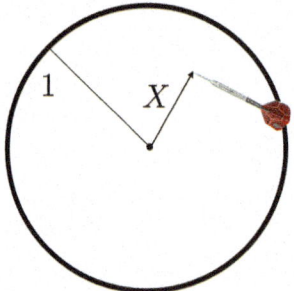

Fig. 1.41 The probability
$\mathbb{P}(x \leq X \leq x + \delta)$ can be
visualized as the area of the
ring, colored in green, that
lies between the circles with
radii x and $x + \delta$

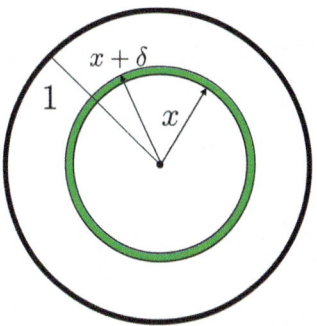

It intuitively makes sense that $f(x) = 2x$ in the interested range. Why? Well, a ring farther away from the center has a greater area than a ring closer to the center with the same width δ.

Cumulative density function In the case of discrete random variables, the histogram provides a complete description of the statistical behavior. This may lead us to assume that in the case of continuous random variables, only the pdf (which plays a similar role as the histogram) is necessary. However, it turns out that the situation is somewhat different in the continuous case, and there is another relevant concept to consider. To understand this, let's first review the relationship between the pdf and the probability measure:

$$f(x) = \lim_{\delta \to 0} \frac{\mathbb{P}(x \leq X \leq x + \delta)}{\delta}. \tag{1.154}$$

Here a key observation that one can make is:

$$\begin{aligned} \mathbb{P}(x \leq X \leq x + \delta) &= \mathbb{P}(X \leq x + \delta) - \mathbb{P}(X < x) \\ &= \mathbb{P}(X \leq x + \delta) - \mathbb{P}(X \leq x) \end{aligned} \tag{1.155}$$

where the first equality is because the event $\{X \leq x + \delta\}$ can be decomposed into the two disjoint events $\{X < x\}$ and $\{x \leq X \leq x + \delta\}$; and the second equality follows from $\mathbb{P}(X = x) = 0$ for a continuous random variable X. Defining $F(x) := \mathbb{P}(X \leq x)$, we can then rewrite (1.154) as:

$$f(x) = \lim_{\delta \to 0} \frac{F(x + \delta) - F(x)}{\delta}.$$

What does this remind you of? Yes, it is the *derivative* $\frac{d}{dx} F(x)$. Hence, we get:

$$f(x) = \frac{d}{dx} F(x). \tag{1.156}$$

Fig. 1.42 Illustration of the
cumulative density function,
usually denoted by $F(x)$.
The pictorial meaning of
$F(x)$ is the area below $f(x)$
from $-\infty$ up to x

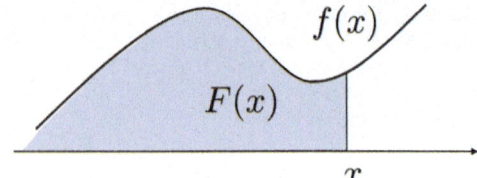

Taking integration on both sides, we also obtain another equivalent representation:

$$F(x) = \int_{-\infty}^{x} f(t)dt. \tag{1.157}$$

The importance of the function $F(x)$ in the context of continuous random variables needs to be addressed. However, before delving into its significance, let us interpret what $F(x)$ represents from (1.157). Visually, $F(x)$ indicates the area under $f(x)$ covered by the interval $[-\infty, x]$. This is illustrated in Figure 1.42.   Therefore, the cdf $F(x)$ can be understood as the accumulated area under $f(x)$ up to the point x, which is why it is called the cumulative density function.

Importance of the cdf   Pondering

$$f(x) = \frac{d}{dx}F(x), \qquad F(x) = \int_{-\infty}^{x} f(t)dt, \tag{1.158}$$

we observe that the pdf and cdf contain identical information: one can be deduced from the other, and vice versa. Why? The equation $f(x) = \frac{d}{dx}F(x)$ implies that $F(x) = \int_{-\infty}^{x} f(t)dt + C$ where C is a constant. However, we know that C must be zero, since $F(\infty) := \mathbb{P}(X \leq \infty) = 1 = \int_{-\infty}^{\infty} f(t)dt$. The converse direction is straightforward.

You may be wondering why the concept of cdf is important when we already have the pdf. The reason is that for many problems, it is much easier to compute the cdf than the pdf. Therefore, we often first compute the cdf and then derive the pdf from it. Additionally, the cdf shares similar properties as the pdf:
1. (Non-negativity): $0 \leq F(x) \leq 1$;
2. (Terminal point): $F(\infty) = \int_{-\infty}^{\infty} f(t)dt = 1$;
3. (Initial point): $F(-\infty) = 0$.

Expectation and variance   For a continuous random variable X, the expectation is defined as:

$$\mathbb{E}[X] := \int_{-\infty}^{\infty} xf(x)dx. \tag{1.159}$$

The reason behind this definition is analogous to that of the discrete case. In the continuous case, the term $f(x)dx$ serves as a suitable weight that reflects the frequency of the occurrence of x, similar to how $\mathbb{P}_X(x)$ works in the discrete case. The variance is defined similarly:

$$\text{var}(X) := \mathbb{E}[(X - \mu)^2] \tag{1.160}$$

where $\mu := \mathbb{E}[X]$. We also have a useful fact as in the discrete case:

$$\begin{aligned} \text{var}(X) &= \mathbb{E}[(X - \mu)^2] \\ &= \mathbb{E}[X^2] - \mu^2 \\ &= \int_{-\infty}^{\infty} x^2 f(x)dx - \mu^2 \end{aligned}$$

where the second equality is due to the linearity of expectation.

Look ahead
As previously stated, the relevance of continuous random variables lies in their application to the noise in communication systems, which we will explore in Part III. In fact, one of the most common continuous random variables used to model the noise signal is the Gaussian random variable. Therefore, in the following section, we will examine the Gaussian random variable.

1.12   Gaussian Random Variables

Recap

In the previous section, we introduced another type of random variable: *continuous* random variables. Like *discrete* random variables, they also map an element in Ω to a real-valued number. However, the key difference is that they take on a *continuous* range of values, while their discrete counterparts take on specific and countable values. Since the probability of a continuous random variable taking on a particular value is zero, the relevant probability distribution involves the probability of the variable belonging to an *interval* of interest. Thus, in this context, we must specify $\mathbb{P}(a \leq X \leq b)$ for all allowable intervals $[a, b]$. To formally specify such probabilities, a new concept analogous to the histogram in the discrete case has been introduced: the probability density function (pdf) $f(x)$. It is defined as a function that satisfies:

$$\mathbb{P}(a \leq X \leq b) = \int_a^b f(x)dx \qquad \forall a, b \in \mathbb{R} \text{ such that } a \leq b. \qquad (1.161)$$

The term "density" in its name is derived from the fact that it can be understood as the "probability per unit length", as evidenced by another expression: $f(x) = \lim_{\delta \to 0} \frac{\mathbb{P}(x \leq X \leq x+\delta)}{\delta}$. We have also introduced another concept that may seem redundant: the cumulative density function (cdf) $F(x) := \mathbb{P}(X \leq x)$. It has the following relationship with $f(x)$:

$$f(x) = \frac{d}{dx}F(x), \quad F(x) = \int_{-\infty}^x f(t)dt. \qquad (1.162)$$

The motivation for introducing the cdf is that, in many cases, it is much simpler to compute than the pdf. Therefore, it is common practice to compute the cdf first and then derive the pdf from it.

As mentioned earlier, the Gaussian random variable is a widely used continuous random variable in many applications. One example is the modeling of a noise signal in communication systems, which will be a focal topic in Part III.

Outline

In this section, we will delve into the Gaussian random variable. The section consists of four parts. We will first introduce the definition of the Gaussian random variable. The definition is given in terms of a specific form of the probability density function (pdf). In the second part, we will demonstrate that the pdf satisfies the constraint of summing up to one and calculate its mean and variance. The third part focuses on an important property of the Gaussian random variable, known as *normality preservation*. This property states that any linear transformation of a Gaussian random variable remains Gaussian. It turns out this property enables efficient computation of the cdf for any Gaussian random variable. In the final part, we will study how to compute the cdf of a Gaussian random variable with this property.

Definition of a Gaussian random variable (Gauß 1809) We say that a random variable X is *Gaussian* if its pdf reads:

$$f(x) = \frac{1}{\sqrt{2\pi}\sigma} e^{-\frac{(x-\mu)^2}{2\sigma^2}} \qquad \forall x \in \mathbb{R} \tag{1.163}$$

where constants μ and $\sigma^2 > 0$ will be demonstrated to be the mean and variance of the Gaussian random variable, respectively (this will be confirmed shortly). We exclude the uninteresting case of $\sigma^2 = 0$, which results in a deterministic X. As previously mentioned, this random variable is commonly encountered in various scenarios and is sometimes referred to as the *normal* distribution. The function $f(x)$ is named the Gaussian or normal distribution.

Why Gaussian random variables become popular? The popularity of the Gaussian distribution can be explained by two sequential facts. First, in many scenarios such as communication, a signal of interest, like a noise signal, can be expressed as the sum of many independent random variables. Second, there is a well-known theorem, the *Central Limit Theorem (CLT)*, which states that the sum, properly scaled, can be approximated as a Gaussian random variable as the number of variables involved in the summation increases. This theorem is a key reason why the Gaussian distribution is widely used. More details about this will be discussed later in Part III, where we will focus on the communication application.

Sum-up-to-one constraint We will now confirm that the Gaussian distribution given in (1.163) satisfies the sum-up-to-one constraint:

$$\int_{-\infty}^{\infty} \frac{1}{\sqrt{2\pi}\sigma} e^{-\frac{(x-\mu)^2}{2\sigma^2}} dx = 1. \tag{1.164}$$

To simplify the pdf form, which involves parameters such as μ and σ^2, we can use the well-known technique called the *change of variable*. By setting $t = \frac{x-\mu}{\sigma}$, we can find that $dt = \frac{dx}{\sigma}$. Substituting these into the left-hand side of the equation above, we obtain:

$$\int_{-\infty}^{\infty} \frac{1}{\sqrt{2\pi}\sigma} e^{-\frac{(x-\mu)^2}{2\sigma^2}} dx = \int_{-\infty}^{\infty} \frac{1}{\sqrt{2\pi}} e^{-\frac{t^2}{2}} dt. \qquad (1.165)$$

After simplification, we are left with an integral that involves $e^{-t^2/2}$. However, there is no known formula for this integral, which seems to make further progress difficult. Fortunately, we can use a technique from calculus called *converting into polar coordinates*. To understand this technique, note that the integral in (1.165) is positive since the pdf is positive. Therefore, it is enough to prove that the square of the integral is 1:

$$\left(\int_{-\infty}^{\infty} \frac{1}{\sqrt{2\pi}} e^{-\frac{t^2}{2}} dt \right)^2 = 1.$$

The left-hand side in the above equation can be rewritten by using two different dummy variables, x and y, in the double integral:

$$\left(\int_{-\infty}^{\infty} \frac{1}{\sqrt{2\pi}} e^{-\frac{t^2}{2}} dt \right)^2 = \frac{1}{2\pi} \int_{-\infty}^{\infty} e^{-\frac{x^2}{2}} dx \int_{-\infty}^{\infty} e^{-\frac{y^2}{2}} dy$$

$$= \frac{1}{2\pi} \int_{-\infty}^{\infty} \int_{-\infty}^{\infty} e^{-\frac{x^2+y^2}{2}} dxdy.$$

The polar coordinate (r, θ) comes into play here:

$$x = r\cos\theta, \ y = r\sin\theta.$$

By utilizing the identity $(\sin\theta)^2 + (\cos\theta)^2 = 1$, we can obtain $x^2 + y^2 = r^2$. The task at hand is to find an expression for $dxdy$ in terms of $(dr, d\theta)$. This can be accomplished by observing the change in area when transitioning from (r, θ) to $(r + dr, \theta + d\theta)$.

$$\text{area change} \approx dr \times (\text{width change due to } d\theta)$$
$$= dr \times (rd\theta) = rdrd\theta. \qquad (1.166)$$

As a result, the change in area $dxdy$ in Cartesian coordinates can be converted to that in polar coordinates: $rdrd\theta$. Combining this with the equation $x^2 + y^2 = r^2$, we obtain:

$$\left(\int_{-\infty}^{\infty} \frac{1}{\sqrt{2\pi}} e^{-\frac{t^2}{2}} dt \right)^2 = \frac{1}{2\pi} \int_{-\infty}^{\infty} \int_{-\infty}^{\infty} e^{-\frac{x^2+y^2}{2}} dxdy$$

$$= \frac{1}{2\pi} \int_{0}^{2\pi} \int_{0}^{\infty} e^{-\frac{r^2}{2}} rdrd\theta$$

$$= \int_0^\infty e^{-\frac{t^2}{2}} r \, dr$$

$$= \int_0^\infty e^{-u} \, du$$

$$= \left[-e^{-u}\right]_0^\infty = 1$$

where the third equality is due to the integration over θ from 0 to 2π; and the second last equality follows from the change of variable: $u = \frac{r^2}{2}$ and $du = r \, dr$. This completes the proof of (1.164).

Computation of $\mathbb{E}[X]$ and $\mathrm{var}(X)$ The expectation calculation is straightforward. By linearity, we first get:

$$\mathbb{E}[X - \mu] = \mathbb{E}[X] - \mu. \tag{1.167}$$

We then manipulate $\mathbb{E}[X - \mu]$ as:

$$\mathbb{E}[X - \mu] = \int_{-\infty}^\infty (x - \mu) \cdot \frac{1}{\sqrt{2\pi}\sigma} e^{-\frac{(x-\mu)^2}{2\sigma^2}} \, dx$$

$$= \int_{-\infty}^\infty t \cdot \frac{1}{\sqrt{2\pi}\sigma} e^{-\frac{t^2}{2\sigma^2}} \, dt$$

$$= 0$$

where the second equality is due to the change of variable $t = x - \mu$; and the last equality is because the pdf is symmetric around $t = 0$ and hence the interested function multiplied by t is an *odd* function. Applying this into (1.167), we obtain the expectation as:

$$\mathbb{E}[X] = \mu.$$

Now, we will proceed to compute the variance, which is a bit challenging. Utilizing the definition of $\mathrm{var}(X)$ and employing the change of variable $t = \frac{x-\mu}{\sigma}$, we obtain:

$$\mathrm{var}(X) := \int_{-\infty}^\infty (x - \mu)^2 \cdot \frac{1}{\sqrt{2\pi}\sigma} e^{-\frac{(x-\mu)^2}{2\sigma^2}} \, dx$$

$$= \frac{\sigma^2}{\sqrt{2\pi}} \int_{-\infty}^\infty t^2 e^{-\frac{t^2}{2}} \, dt.$$

The question is how to integrate $t^2 e^{-t^2/2}$. You may find this challenging as there is no specific integral formula for this function. However, there is another familiar method from Calculus that can be utilized in this situation. That is, *integration by parts*:

$$\int f \cdot g' = fg - \int f' \cdot g. \tag{1.168}$$

Defining $f = -t$ and $g' = -te^{-t^2/2}$, and applying the "integration by parts" (1.168), we get:

$$\text{var}(X) := \int_{-\infty}^{\infty} (x - \mu)^2 \cdot \frac{1}{\sqrt{2\pi}\sigma} e^{-\frac{(x-\mu)^2}{2\sigma^2}} dx$$

$$= \frac{\sigma^2}{\sqrt{2\pi}} \int_{-\infty}^{\infty} t^2 e^{-\frac{t^2}{2}} dt$$

$$= \frac{\sigma^2}{\sqrt{2\pi}} \int_{-\infty}^{\infty} \underbrace{-t}_{f} \cdot \underbrace{\left(-te^{-\frac{t^2}{2}}\right)}_{g'} dt$$

$$= \frac{\sigma^2}{\sqrt{2\pi}} \left[-t \cdot e^{-\frac{t^2}{2}}\right]_{-\infty}^{\infty} - \frac{\sigma^2}{\sqrt{2\pi}} \int_{-\infty}^{\infty} (-1) \cdot e^{-\frac{t^2}{2}} dt$$

$$= \sigma^2 \cdot \frac{1}{\sqrt{2\pi}} \int_{-\infty}^{\infty} e^{-\frac{t^2}{2}} dt$$

$$= \sigma^2$$

where the last equality is due to the sum-up-to-one constraint of the Gaussian distribution. As $\mathbb{E}[X] = \mu$ and $\text{var}(X) = \sigma^2$, the Gaussian distribution appears as shown in Figure 1.43. The Gaussian distribution has a somewhat "bell-shaped" curve, with symmetry around and centered at $x = \mu$, and a width that is determined by the standard deviation σ.

Normality preservation Up until now, we have covered some tedious material that required complex integration. Let's shift gears and explore some exciting and highly useful concepts, such as the important property of "normality preservation". Suppose we have a Gaussian random variable X with mean μ and variance σ^2, denoted as

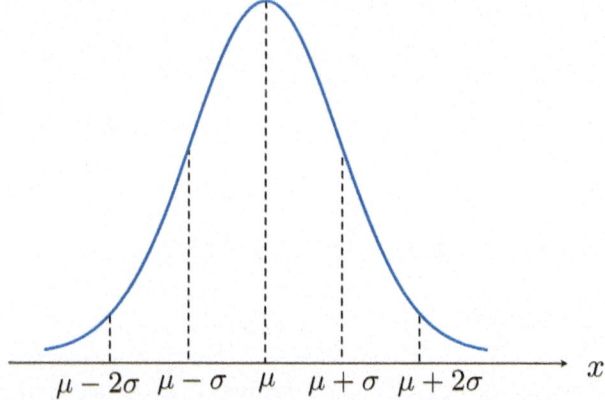

Fig. 1.43 The Gaussian distribution with mean μ and variance σ^2

$X \sim \mathcal{N}(\mu, \sigma^2)$, where $\mathcal{N}$ represents "normal". The normality preservation property states that any *linear transformation* of X will also retain the normal distribution. In other words, for any constants (c_1, c_2),

$$Y = c_1 X + c_2 \sim \mathcal{N}(c_1\mu + c_2, c_1^2\sigma^2). \tag{1.169}$$

Here is the proof for this property. The case where $c_1 = 0$ is straightforward as $Y = c_2$ in this situation. Let's first examine the case where $c_1 > 0$. By utilizing the definition of the cumulative distribution function (cdf), we have:

$$
\begin{aligned}
F_Y(y) &:= \mathbb{P}(Y \le y) \\
&= \mathbb{P}(c_1 X + c_2 \le y) \\
&= \mathbb{P}\left(X \le \frac{y - c_2}{c_1}\right) \\
&= F_X\left(\frac{y - c_2}{c_1}\right)
\end{aligned}
$$

where the third equality is due to $c_1 > 0$. Taking derivatives w.r.t. y on both sides, we obtain:

$$
\begin{aligned}
\text{Case } c_1 > 0 : \quad f_Y(y) &= \frac{d}{dy} F_X\left(\frac{y - c_2}{c_1}\right) \\
&= \frac{d}{dx} F_X\left(\frac{y - c_2}{c_1}\right) \cdot \frac{d}{dy}\left(\frac{y - c_2}{c_1}\right) \\
&= \frac{1}{c_1} f_X\left(\frac{y - c_2}{c_1}\right) \\
&= \frac{1}{c_1} \cdot \frac{1}{\sqrt{2\pi}\sigma} e^{-\frac{\left(\frac{y-c_2}{c_1} - \mu\right)^2}{2\sigma^2}} \\
&= \frac{1}{\sqrt{2\pi}c_1\sigma} e^{-\frac{(y-(c_1\mu+c_2))^2}{2c_1^2\sigma^2}}
\end{aligned}
\tag{1.170}
$$

where the second equality is due to the chain rule ($\frac{d}{dy}g(x) = \frac{d}{dx}g(x) \cdot \frac{dx}{dy}$); and the second-to-last equality follows from the formula of the Gaussian distribution (1.163). The case $c_1 < 0$ is similar. In this case, one can readily verify that:

$$\text{Case } c_1 < 0 : \quad f_Y(y) = -1 \cdot \frac{1}{\sqrt{2\pi}c_1\sigma} e^{-\frac{(y-(c_1\mu+c_2))^2}{2c_1^2\sigma^2}}. \tag{1.171}$$

If you are not convinced, please check this. This together with (1.170) gives:

$$Y \sim \mathcal{N}(c_1\mu + c_2, c_1^2\sigma^2). \tag{1.172}$$

We can also double-check in part via the following calculations:

$$
\begin{aligned}
\mathbb{E}[Y] &= c_1 \mathbb{E}[X] + c_2 = c_1 \mu + c_2; \\
\mathsf{var}(Y) &:= \mathbb{E}[(Y - (c_1 \mu + c_2))^2] \\
&= \mathbb{E}[c_1^2 (X - \mu)^2] \\
&= c_1^2 \mathbb{E}[(X - \mu)^2] = c_1^2 \sigma^2.
\end{aligned}
\tag{1.173}
$$

Computation of the cdf of a Gaussian random variable The normality preservation property (1.169) has an important implication which is that a Gaussian random variable, say $X \sim \mathcal{N}(\mu, \sigma^2)$, can be expressed as a linear transformation of the *standard Gaussian* random variable with a mean of 0 and a variance of 1, say $Z \sim \mathcal{N}(0, 1)$:

$$
X = \sigma Z + \mu.
\tag{1.174}
$$

This property (1.174) allows for efficient computation of the cumulative distribution function (cdf) of any Gaussian random variable. To understand this, we can start by examining the cdf of $X \sim \mathcal{N}(\mu, \sigma^2)$:

$$
\begin{aligned}
F_X(x) &= \mathbb{P}(X \leq x) \\
&= \int_{-\infty}^{x} \frac{1}{\sqrt{2\pi}\sigma} e^{-\frac{(t-\mu)^2}{2\sigma^2}} \, dt.
\end{aligned}
\tag{1.175}
$$

We encounter a challenge when computing the cdf as it involves integration, which often has no closed form solution. However, the property (1.174) comes in handy. By applying this property, we can transform the integration into a function of tractable integration related to the standard Gaussian $\mathcal{N}(0, 1)$. The advantage of working with $\mathcal{N}(0, 1)$ is that we can numerically evaluate the integration even when there is no closed formula. The details are explained below:

$$
\begin{aligned}
F_X(x) &= \mathbb{P}(X \leq x) \\
&= \mathbb{P}\left(\frac{X - \mu}{\sigma} \leq \frac{x - \mu}{\sigma}\right) \\
&= \int_{-\infty}^{\frac{x-\mu}{\sigma}} \frac{1}{\sqrt{2\pi}} e^{-\frac{z^2}{2}} \, dz =: \Phi\left(\frac{x - \mu}{\sigma}\right)
\end{aligned}
\tag{1.176}
$$

where $Z = \frac{X-\mu}{\sigma} \sim \mathcal{N}(0, 1)$. The cdf of the standard Gaussian distribution is denoted by $\Phi(z) := \mathbb{P}(Z \leq z)$. It has already been computed numerically and can be found in various references, including books and Wikipedia, in tabulated form. The numerical values of $\Phi(z)$ can be easily accessed in Python using the function "scipy.stats.norm.cdf".

$$\text{norm.cdf}(z) = \Phi(z). \tag{1.177}$$

Applying this into (1.176), we get:

$$F_X(x) = \Phi\left(\frac{x-\mu}{\sigma}\right) = \text{norm.cdf}\left(\frac{x-\mu}{\sigma}\right). \tag{1.178}$$

Look ahead

We have thus far covered a wide range of concepts, laws and techniques: sample space, probability model, events, conditional probability, total probability law, Bayes' law, independence, union bound, discrete random variables, pmf, expectation, variance, Chebyshev's inequality, Markov's inequality, continuous random variables, pdf, cdf, and Gaussian random variables. These form the contents of Part I.

However, to fully understand the two key principles (MAP and ML estimation) that we introduced in Sect. 1.1, it is necessary to delve into the concept of *random processes*. Thus, the upcoming section will be dedicated to exploring random processes, which marks the beginning of Part II. The rest of Part II will center around the two key principles and significant theorems, including the Law of Large Numbers and the Central Limit Theorem.

Problem Set 4

Problem 4.1 (*Revisit: Coupon collector problem*) Consider the coupon collector problem discussed in Sect. 1.6 and Problem 3.1. Suppose that we have n different coupons and buy k snacks. Let X be the number of coupons acquired. Compute $\text{var}(X)$.

Problem 4.2 (*Revisit: Homework matching problem*) Consider the homework matching problem discussed in Sect. 1.7. In this problem, the number n of students submit their homework, and the homework is then shuffled and distributed across the students. Each student gets one single homework, but there is no guarantee that the homework they receive is their own. We define S as the number of students who receive their own homework. Let $X_i = \mathbf{1}\{i\text{th student receives her own homework}\}$. Then we can express S as the sum of all X_i's:

$$S = X_1 + X_2 + \cdots + X_n. \tag{1.179}$$

(a) Compute $\mathbb{E}[S]$.
(b) Compute $\mathbb{E}[X_1 X_2]$.
(c) Compute $\text{var}(S)$.

Problem 4.3 (*The sum of independent random variables*) Suppose that continuous random variables X and Y are independent. Let $Z = X + Y$.

(a) Show that $\mathbb{E}[XY] = \mathbb{E}[X]\mathbb{E}[Y]$ and $\text{var}(Z) = \text{var}(X) + \text{var}(Y)$.
(b) Show that

$$\text{var}(aX) = a^2 \text{var}(X), \quad \forall a \in \mathbb{R}.$$

(c) Show that the probability density function (pdf) $f_Z(\cdot)$ is given by

$$f_Z(a) = f_X(a) * f_Y(a) := \int_{-\infty}^{\infty} f_X(t) f_Y(a - t) dt$$

for $a \in \mathbb{R}$.
(d) Consider the (double-sided) Laplace transform of $f_Z(\cdot)$:

$$F_Z(s) = \int_{-\infty}^{\infty} e^{-sa} f_Z(a) da.$$

Show that

$$F_Z(s) = F_X(s) F_Y(s)$$

where $F_X(s)$ and $F_Y(s)$ indicate the Laplace transforms of $f_X(\cdot)$ and $f_Y(\cdot)$, respectively.

Problem 4.4 (*Revisit: Geometric distribution*) Recall the experiment in Problem 3.2 where a coin is flipped repeatedly until a "Head" is obtained. The assumption was that the coin flips are independent across different trials, and the probability of showing a "Head" is p. Let X be a random variable that indicates the number of flips until we see "Head". Compute $\mathsf{var}(X)$.

Hint: First express $\mathbb{E}[X^2] = \mathbb{E}[X(X-1)] + \mathbb{E}[X]$ and then compute $\mathbb{E}[X(X-1)]$ by a similar trick that you applied while computing $\mathbb{E}[X]$ in Problem 3.2(*c*).

Problem 4.5 (*Revisit: Flipping coins*) Revisit the experiment in Problem 3.6 where we flip n coins to obtain $X_1, \ldots, X_n \in \{H, T\}$. Let Y be the number of pairs of consecutive "Heads" (i.e., the number of i's such that $X_i = X_{i+1} = H$). Compute $\mathsf{var}(Y)$.

Problem 4.6 (*Functions of a random variable*)

(*a*) Consider a discrete random variable X that is uniformly distributed over the set $\{1, 2, \ldots, n\}$. What is the pmf of the random variable $Y = X^2$? Does the pmf of Y remain uniform over its range?

(*b*) Consider a continuous random variable X that is uniformly distributed over the interval $[1, n]$. What is the pdf of the random variable $Y = X^2$? Does the pdf of Y remain uniform over its range? Also, provide a sketch of the pdf and offer an intuitive explanation for the shape of the curve.

(*c*) Consider a continuous random variable X that is uniformly distributed over the interval $[-n, n]$. What is the pdf of the random variable $Y = X^2$? Also, provide a sketch of the pdf.

Problem 4.7 (*Some pdf calculations*) Suppose X is uniformly distributed over $[0, 1]$.

(*a*) Let $Y = \sqrt{X}$. Find the cdf and pdf of Y.

(*b*) Let $Z = \log X$. Find the cdf and pdf of Z.

Problem 4.8 (*Inequality techniques*)

(*a*) (*Markov's inequality*) Show that for any non-negative random variable X and $d > 0$,

$$\mathbb{P}(X \geq d) \leq \frac{\mathbb{E}[X]}{d}. \tag{1.180}$$

(*b*) (*Chebyshev's inequality*) Let X be a discrete random variable with mean μ. Using part (*a*), show that for any $d > 0$,

$$\mathbb{P}(|X - \mu| \geq d) \leq \frac{\mathsf{var}(X)}{d}. \tag{1.181}$$

Problem 4.9 (*Exponential distribution*) For $\lambda > 0$, consider a continuous random variable X with the following pdf:

$$f(x) = \begin{cases} ce^{-\lambda x} & \text{if } x \geq 0; \\ 0 & \text{if } x < 0. \end{cases} \qquad (1.182)$$

(*a*)   Find the constant c.

(*b*)   Compute $\mathbb{E}[X]$.

　　　Hint: Think about "integration by parts":

$$\int g(x)h'(x)dx = g(x)h(x) - \int g'(x)h(x)dx.$$

(*c*)   Compute $\text{var}(X)$.

Problem 4.10 (*Exponential vs. geometric distributions*) This problem will guide you to understand that the exponential distribution is a continuous variant of the geometric distribution through an application called the "call arrivals problem".

　　　Consider a 1-min time window that starts at $t = 0$. We will divide this time window into smaller intervals of length $\frac{1}{n}$, as shown in Figure 1.44. Assume that the duration of the interval is extremely small such that only one call can be initiated if there is a call. Consequently, there are only two possible events that can occur in the small interval: either no initiation is made or precisely one. Let λ represent the average number of call arrivals per minute. We define X_i to be an indicator variable that shows whether there is a call arrival in interval $i \in \{1, 2, \ldots, n\}$:

$$X_i = \begin{cases} 1 \text{ with probability } \frac{\lambda}{n}; \\ 0 \text{ with probability } 1 - \frac{\lambda}{n}. \end{cases} \qquad (1.183)$$

It is assumed that the arrival of calls is independent across intervals, meaning that the values of X_i are independent of each other. We define S as the number of intervals

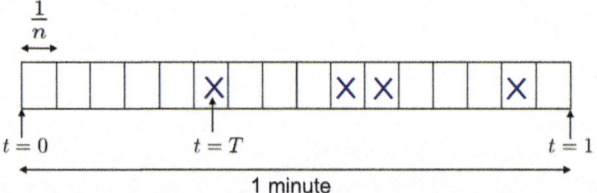

Fig. 1.44 Call arrivals problem: Assume that the one-minute time window is divided into n tiny intervals of duration $\frac{1}{n}$. It is further assumed that each interval either has none or at most one call. The probability that each interval has a call is $\frac{\lambda}{n}$, where λ is the rate of call arrivals per minute, and this probability is independent of the other intervals. Let T denote the time of arrival of the first call

we need to wait for the first call arrival, while T represents the continuous-time equivalent of S, specifically referring to the time of the first call arrival:

$$\frac{S}{n} \leq T < \frac{S+1}{n}.$$ (1.184)

(a) What is the probability distribution $\mathbb{P}_S(s)$ of S?

(b) Assume that S is roughly equal to nT when n is sufficiently large. Show that the cdf of T can then be approximated as:

$$F(t) \approx 1 - \left(1 - \frac{\lambda}{n}\right)^{nt}.$$ (1.185)

(c) In the limit of $n \to \infty$, show that the pdf of T is:

$$f(t) = \lambda e^{-\lambda t} \text{ for } t > 0.$$ (1.186)

Hint: You may want to use a well-known fact: $\lim_{n\to\infty}(1 + \frac{x}{n})^n = e^x$.

Problem 4.11 (*Laplace transform of a Gaussian random variable*) Let $X \sim \mathcal{N}(\mu, \sigma^2)$ and $f_X(x)$ be the pdf of X. Compute the Laplace transform of the pdf $f_X(x)$:

$$F_X(s) := \int_{-\infty}^{\infty} e^{-sx} f_X(x)dx.$$ (1.187)

Problem 4.12 (*Gaussian distribution*) Consider a set of grades on a probability exam in a class that has an approximately Gaussian distribution with a mean of 64 and a standard deviation of 7.1.

(a) Determine the minimum passing score for the class if the lowest 5% of students are considered to have failed.

(b) Determine the highest grade of B if the top 10% of the students are awarded an A grade.

Note: Assuming that X has a Gaussian distribution with mean 0 and variance 1, it can be assumed that $\mathbb{P}(X \leq 1.3) \approx 0.9$ and $\mathbb{P}(X \leq 1.65) \approx 0.95$.

Problem 4.13 (*Total expectation*) Consider a sequence of independent and identically distributed (i.i.d.) random variables U_i, each having a mean of μ_U and variance of σ_U^2. Let N be a random variable that takes only positive integer values and is independent of U_i's, with a mean of μ_N and variance of σ_N^2. Let

$$S = \sum_{i=1}^{N} U_i.$$ (1.188)

(a) Assume that $\mu_U = 0$. Calculate the mean and variance of S using σ_U^2, μ_N, and σ_N^2.
(b) Repeat part (a) for general μ_U.

Problem 4.14 (*True or False?*)

(a) If $\text{var}(X + Y) = \text{var}(X) + \text{var}(Y)$, then X and Y are independent.
(b) Suppose that X and Y are independent random variables and both follow a uniform distribution over the interval $[0, 1]$. Then, $Z = X + Y$ is also uniformly distributed.
(c) Let X be a continuous random variable with the pdf f_X, and let $Y = X - 3$. The pdf of Y is $f_X(y - 3)$.
(d) Let $X \sim \mathcal{N}(\mu, \sigma^2)$ and $f_X(x)$ be its pdf. Let $F_X(s)$ be the Laplace transform of $f_X(x)$:

$$F_X(s) := \int_{-\infty}^{\infty} f_X(x)e^{-sx}dx. \qquad (1.189)$$

Then, $F_X(s)$ is solely determined by σ^2 and is independent of μ.

Chapter 2
Introductory Random Processes and Key Principles

*Three notable random processes are the Bernoulli,
Gaussian and Markov processes.*

2.1 Introduction to Random Processes

Recap

Part I has covered a range of fundamental concepts in probability, including
sample space, events, conditional probability, independence, random variables,
probability mass function (pmf), probability density function (pdf), and cumulative distribution function (cdf). We have also studied important laws such as
total probability law, Bayes' law, and De Morgan's law, along with techniques
like the complement counting trick, the union bound, Markov's inequality, and
Chebyshev's inequality.

Part II will focus on the two key principles introduced in Sect. 1.1, namely
MAP and ML estimation. However, before we delve into these topics, we must
first learn about another important notion: *random processes*.

Outline

This section marks the beginning of Part II, where we will explore the first
topic: random processes. It consists of three parts. Firstly, we will introduce
the definition of random processes, along with the rationale behind it. It will
become apparent that many interesting signals that arise in the real world can be

© The Author(s), under exclusive license to Springer Nature Singapore Pte Ltd. 2025   111
C. Suh, *Probability for Information Technology*,
https://doi.org/10.1007/978-981-97-4032-1_2

modeled using a particular type of random process called the stationary process. In the second part, we will examine the stationary process and its special, well-known version known as the i.i.d. process. Finally, we will explore three notable examples of such stationary processes that are practically relevant, and which we will study in greater detail in Part III. These examples are: (i) *Bernoulli* process; (ii) *Gaussian* process; and (iii) *Markov* process.

Definition of a random process (Gallager 2013) The definition is simple to state. A random process is simply a sequence of random variables. Typically, this sequence represents a time-series sequence, i.e., the associated random variables often evolve over time. The sequence can be finite or infinite, denoted by $(X_1, X_2, \ldots, X_n)$ or $(X_1, X_2, \ldots, X_n, \ldots)$. Alternatively, the simpler notation $\{X_i\}_{i=1}^n$ or $\{X_i\}_{i=1}^\infty$ can be used. If X_i belongs to the space $\mathcal{X}_i$ for all i, then the range is:

$$\mathcal{X} := \mathcal{X}_1 \times \mathcal{X}_2 \times \cdots \times \mathcal{X}_n \times \cdots$$

Generally, there are two types of notations for the probability distribution depending on whether X_i is discrete or continuous. In the case of discrete random variables, we use the bold $\mathbb{P}$ notation. For any finite subset of $\{x_i\}_{i=1}^\infty$, e.g., for $\{x_i\}_{i=1}^n$:

$$\mathbb{P}(x_1, x_2, \ldots, x_n). \tag{2.1}$$

One may prefer the following more formal notation to distinguish it from the probability distribution with respect to a sample space Ω:

$$\mathbb{P}_{X_1, \ldots, X_n}(x_1, x_2, \ldots, x_n).$$

However, this notation may appear complicated as it puts many stuffs in the subscript. Therefore, a simpler notation (2.1) is preferred. It is called the *joint distribution*, as it involves multiple random variables. On the other hand, for continuous cases, we use the notation f. For all finite subsets of $\{x_i\}_{i=1}^\infty$, e.g., for $\{x_i\}_{i=1}^n$:

$$f(x_1, x_2, \ldots, x_n). \tag{2.2}$$

Applications of random processes What is the reason for our interest in random processes? It is primarily due to their numerous practical applications, highlighted in Sect. 1.1. These include: (i) communication, (ii) community detection in social networks, (iii) machine learning, and (iv) speech recognition. In communication systems, the noise signals that change over time can be modeled as a random process, specifically the Gaussian process, which we will soon explore. In machine learning, the data samples such as input samples denoted as $\{x^{(i)}\}_{i=1}^m$ can be regarded as a random process, specifically the i.i.d. process, which we will also study. Similarly, in speech recognition, the voice signals that are input to the system can be modeled as a random process, specifically the Markov process, which we will investigate further. Of course, this is just the tip of the iceberg. There are countless real-world signals that can be modeled as random processes, such as a sequence of daily stock prices, a sequence of daily temperature measurements, or a sequence of call inter-arrival times.

Two important types The stationary process is a type of random process that is extremely valuable for modeling a variety of real-world signals. This process can be defined as follows: if a sequence of random variables $(X_1, X_2, \ldots, X_n, \ldots)$ remains statistically unchanged after being shifted by a constant integer ℓ, i.e., $(X_{1+\ell}, X_{2+\ell}, \ldots, X_{n+\ell}, \ldots)$, then it is said to be stationary. This means that each shift ℓ produces a statistical copy of the original sequence, with the same joint distribution as the original.

The *i.i.d.* process is a simpler type of process that is a well-known and special version of the stationary process. We say that $(X_1, X_2, \ldots, X_n, \ldots)$ is *i.i.d.* if the associated random variables are mutually independent, and each has the same distribution (identically distributed). The acronym *i.i.d.* stands for "independent and identically distributed", which means that for every possible finite subset of variables, all random variables are independent of each other.

Example #1: Bernoulli process (Bernoulli 1713) We discuss three notable examples that belong to either the i.i.d. process or the stationary process. The first example is a well-known and simple one called the *Bernoulli process*, which is named after Jacob Bernoulli, a famous mathematician in history. Please refer to Figure 2.1 for his portrait. The Bernoulli process is characterized by a sequence of i.i.d. binary random variables $(X_1, X_2, \ldots, X_n, \ldots)$, where $X_i \in \{0, 1\}$ with $\mathbb{P}(X_i = 1) = p$. Each random variable is represented by $X_i \sim \mathsf{Bern}(p)$. Jacob Bernoulli, who made a significant contribution to the field of mathematics by discovering the Law of Large Numbers (LLN), employed this simple process in his research. As a tribute to this

Fig. 2.1 The image depicts Jacob Bernoulli (1655–1705), a Swiss mathematician who lived in the 17th century. He is renowned for his discovery of the number e and one of the fundamental laws known as the Law of Large Numbers (LLN)

discovery, the random process he used was named the Bernoulli process. We will delve deeper into the LLN in Sect. 2.5. Please bear with us until we reach that point.

As you may recall, we have already encountered the Bernoulli process in Part I. Specifically, we considered the experiment of flipping a coin with probability p of landing on "head" n times. In this experiment, the component random variables were defined as follows:

$$X_i = \mathbf{1}\{i\text{th flip shows "Head"}\} \quad \forall i \in \{1, 2, \ldots, n\}.$$

This is indeed the Bernoulli process.

The Bernoulli process is used as the foundation for several other interesting random variables. One of these variables is the total number S of "Head"s that appear in n coin flips. We have already discussed this variable earlier, and its probability distribution is binomial:

$$\mathbb{P}_S(s) = \binom{n}{s} p^s (1 - p)^{n-s} \quad \forall s \in \{0, 1, \ldots, n\}.$$

Another random variable that is relevant to the Bernoulli process is the total number X of flips needed to get the first "Head". A similar random variable was discussed in the context of *BitTorrent* system. Remember the number of servers requested in order for a user to obtain a new chunk, given that the user has already collected $i - 1$ chunks. Also we encountered exactly the same variable in Problem 3.2, following the *geometric distribution*:

$$\mathbb{P}_X(x) = (1 - p)^{x-1} p \quad \forall x \in \{1, 2, 3, \ldots\}.$$

An application of the Bernoulli process The Bernoulli process has various practical uses beyond coin flipping. One such application is the *call arrivals problem* demonstrated in Figure 2.2. Assume that the entire time window is divided into n small intervals, and each interval can receive at most one call. In other words, there are only two possibilities: either no call is initiated in an interval or exactly one call is initiated. Additionally, we assume that each interval has a call arrival probability of p, and that the call arrivals are independent from interval to interval. Let

$$X_i = \mathbf{1}\{i\text{th interval has a call arrival}\}. \tag{2.3}$$

Fig. 2.2 Call arrivals problem

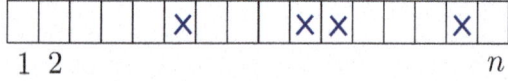

Then, it is evident that $\{X_i\}_{i=1}^{n}$ follows the Bernoulli process, where each random variable X_i is distributed as $X_i \sim \text{Bern}(p)$.

Example #2: Gaussian process (Gauß 1809) The second notable example is the Gaussian process, which is particularly tailored for the i.i.d. scenario. It is referred to as the i.i.d. Gaussian process. We define $(X_1, X_2, \ldots, X_n, \ldots)$ to be i.i.d. Gaussian if the random variables X_i are independent and identically distributed according to the Gaussian distribution, i.e., $X_i \sim \mathcal{N}(\mu, \sigma^2)$:

$$f_{X_i}(x) = \frac{1}{\sqrt{2\pi}\sigma} e^{-\frac{(x-\mu)^2}{2\sigma^2}} \qquad \forall x \in \mathbb{R} \text{ and } \forall i = \{1, 2, \ldots\}.$$

As previously mentioned, the i.i.d. Gaussian process has an important practical application in communication. Specifically, it can be used to model noise signals in communication, a topic that we will delve into in Part III.

Example #3: Markov process (Markov 1906; Gagniuc 2017) The final example we will discuss is the Markov process, which is widely regarded as the most renowned and valuable stationary process. It is named in honor of Andrey Markov, a Russian mathematician from the 19th and early 20th centuries. See Figure 2.3 for his portrait. The assumption of i.i.d. does not often hold in real-world scenarios. A concrete example of this is English text. If we assume that the first and second letters of a word are "t" and "h" respectively, then the next letter is highly likely to be "e" because "the" is a commonly used word in English.

In contrast, the *stationarity* assumption is often more relevant in practice. Consider the example of English text again. We can observe that the statistical properties of a text from 10 years ago are likely to be very similar to those of a current text. For instance, the frequency of the word "the" in an old text would be roughly the same as that in a modern text. The Markov process is a type of stationary process that is particularly well-suited for modeling this scenario.

Fig. 2.3 A Russian mathematician, Andrey Markov (1856–1922), is credited with important contributions to the development of a random process that would later become known as the Markov process

We say that $(X_1, X_2, \ldots, X_n, \ldots)$ is a Markov process if it satisfies the following condition:

$$\mathbb{P}(x_{m+1}|x_m, x_{m-1}, \ldots, x_1) = \mathbb{P}(x_{m+1}|x_m). \tag{2.4}$$

The condition referred to here implies that, given the current value x_m, the future x_{m+1} and the past values $(x_{m-1}, \ldots, x_1)$ are independent of each other. This is known as the *Markov* property. However, the dependency of random variables in real-world scenarios can be much more complex than what is captured by the simple Markov property (2.4). To handle such complex dependencies that are more representative of reality, we use a *generalized Markov process* characterized by:

$$\mathbb{P}(x_{m+1}|x_m, \ldots, x_{m-\ell+1}, x_{m-\ell}, \ldots, x_1) = \mathbb{P}(x_{m+1}|x_m, \ldots, x_{m-\ell+1}). \tag{2.5}$$

The dependence of x_{m+1} on the past is now through a possibly larger number of past values, denoted by $x_m, \ldots, x_{m-\ell+1}$, where ℓ is a positive integer. This process is known as a generalized Markov process, as it includes the Markov process as a special case when $\ell = 1$.

As previously mentioned, voice signals in speech recognition (Vaseghi 2008) can be modeled as a generalized Markov process, and this will be discussed further in Part III.

Visualization of the Markov process The Markov property (2.4) is a concept commonly used in statistics and machine learning to describe the relationship among random variables. This relationship can be visualized using a graphical model (Koller & Friedman 2009), which has been introduced to better understand a generic random process, not limited to the Markov process. A graphical model consists of two components: (1) nodes, which correspond to random variables, and (2) edges, which capture the dependence between pairs of random variables. A graph can be interpreted as follows: if a node, say X_i, is removed and the remaining nodes can be separated into two subgraphs $\mathcal{G}_1$ and $\mathcal{G}_2$, then the random variables in $\mathcal{G}_1$ are independent of those in $\mathcal{G}_2$, conditioned on X_i. For example, consider (X_1, X_2, X_3) with probability $\mathbb{P}(x_1, x_2, x_3) = \mathbb{P}(x_1)\mathbb{P}(x_2|x_1)\mathbb{P}(x_3|x_2)$. It is easy to see that $\mathbb{P}(x_3|x_2, x_1) = \mathbb{P}(x_3|x_2)$, which in turn implies that

$$\begin{aligned} \mathbb{P}(x_1, x_3|x_2) &= \mathbb{P}(x_1|x_2)\mathbb{P}(x_3|x_2, x_1) \\ &= \mathbb{P}(x_1|x_2)\mathbb{P}(x_3|x_2) \end{aligned} \tag{2.6}$$

where the first equality is due the definition of condition probability; and the second equality follows from $\mathbb{P}(x_3|x_2, x_1) = \mathbb{P}(x_3|x_2)$. This then implies that X_1 are X_3 are independent conditioned on X_2. So the graph is illustrated as:

$$X_1 - X_2 - X_3. \tag{2.7}$$

Notice that when X_2 is removed, X_1 and X_3 become disconnected. By applying this reasoning to the Markov process $(X_1, X_2, \ldots, X_n, \ldots)$, we can represent the graphical model as follows:

$$X_1 - X_2 - X_3 - \cdots - X_n - \cdots \tag{2.8}$$

If we remove any x_m, then x_{m+1} and $(x_1, \ldots, x_{m-1})$ become disconnected, indicating that x_{m+1} and $(x_1, \ldots, x_{m-1})$ are independent when conditioned on x_m. This is referred to as the *Markov chain*, which resembles a chain due to its sequential nature. Some curious readers may question whether the lack of directionality in the chain (2.8) implies that given the current value x_m, the one-step past value is independent of all future values:

$$\mathbb{P}(x_{m-1}|x_m, x_{m+1}, x_{m+2}, \ldots) = \mathbb{P}(x_{m-1}|x_m). \tag{2.9}$$

It turns out this is the case. You can verify this in Problem 5.4.

Look ahead
While there are many other interesting examples and concepts related to random processes, we will conclude our discussion here as delving further into the topic may cause you to lose interest. It is essential to note that there is much to learn about random processes, and having a solid understanding of them is crucial. You will have an opportunity to explore more by taking a graduate-level course on random processes. We recommend to taking such a course if you plan on pursuing a related field.

Moving on, let us now focus on the two primary objectives: the MAP and ML estimation principles. In the next section, we will delve into the MAP estimation.

2.2   The MAP Principle

Recap

In the preceding section, we delved into the concept of random processes. While the concept is challenging to grasp, its definition is very simple to state. It is nothing but a sequence of random variables. We focused on two types of random processes: (i) the stationary process; and (ii) the i.i.d. process, the latter being a special case of the former where the random variables are independent and identically distributed. We then highlighted three examples - the Bernoulli process, i.i.d. Gaussian process, and Markov process - that have relevance to the forthcoming applications.

With this groundwork laid, we can now move on to the two essential principles that we have mentioned several times earlier: MAP (Maximum A Posteriori) estimation and ML (Maximum Likelihood) estimation.

Outline

This section will focus on the first: the MAP principle. The section comprises four parts. In fact, the cancer testing problem explored in Sect. 1.4 is a good example which helps us to figure out the essence of the MAP principle. In the first part, we will revisit the problem setup, and introduce appropriate notations. We will then reinterpret several concepts that we have already encountered in Sect. 1.4 using different terminology that is typically used in the context of the MAP principle. The next crucial concept we will explore is the "a posteriori probability", which plays a central role in MAP. Finally, we will investigate the MAP estimation.

Revisit: The cancer testing problem The cancer testing problem is illustrated in Figure 2.4. In Sect. 1.4, we focused on a question regarding a person undergoing a cancer test. The test result indicates either a positive or negative reading, and we sought to determine the probability that the person truly has cancer when the test result is positive:

$$\mathbb{P}(\text{has cancer}|\text{positive test}). \tag{2.10}$$

We will now introduce some notation to connect the probability of interest, as given in (2.10), to the MAP principle. To this end, we define two binary random variables: X denotes whether the person has indeed cancer ($X = 1$ for cancer, $X = 0$ otherwise), while Y indicates the outcome of the test ($Y = 1$ if positive, $Y = 0$ if negative). Figure 2.5 provides a visual representation of these variables. Before delving into the details of (2.10), we will begin by explaining various quantities that arise during the calculation of the probability (2.10).

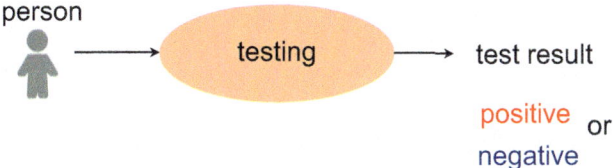

Fig. 2.4 The cancer testing problem

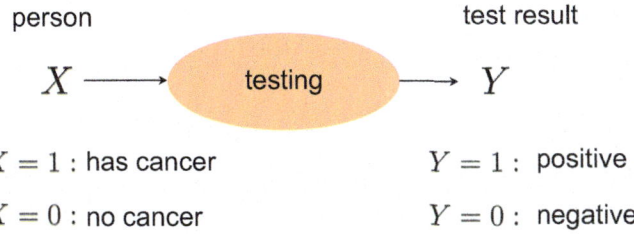

Fig. 2.5 The cancer testing problem can be reformulated using our new notation: X represents whether the person has cancer, while Y indicates the test result

A priori probability The first quantity is known as the *prevalence* of cancer. Expressed in terms of X, it can be defined as follows:

$$\mathbb{P}(X = 1) =: p. \tag{2.11}$$

The prevalence of cancer is a crucial concept, also known as the *a priori probability*. The term "a priori" comes from Latin and means "before". It refers to the probability we have prior knowledge of. In Sect. 1.4, we made a reasonable assumption that the a priori probability is a small value, such as $p = 0.1$.

True Positive Rate (TPR) and False Negative Rate (FNR) As we recall, we examined two quantities that can be deduced from numerous clinical trials. The first quantity is the True Positive Rate (TPR), which is the probability that a person with cancer will test positive. Using the (X, Y) notation, we can express it as follows:

$$\mathbb{P}(Y = 1|X = 1). \tag{2.12}$$

The other is the False Negative Rate (FNR), or called the misdetection rate:

$$\mathbb{P}(Y = 0|X = 1). \tag{2.13}$$

Naturally, we aim for a very high TPR, which corresponds to a very low FNR. In our previous example, we assumed a misdetection rate of 5%, which then yields TPR = 0.95.

False Positive Rate (FPR) and True Negative Rate (TNR) There were two other statistical quantities related to the *normal* population. The first one is the False Positive Rate (FPR), also called the false alarm rate. It is the probability that a normal person will test positive:

$$\mathbb{P}(Y = 1|X = 0). \tag{2.14}$$

The other is the True Negative Rate (TNR): $\mathbb{P}(Y = 0|X = 0)$.

Tradeoff between FPR and TPR Having a small FPR (or a large TNR) is certainly desirable. However, in reality, aiming for a very small FPR can be problematic. Why? To understand this, let us consider the extreme case where we aim for *exactly zero FPR*. In order to achieve this, the test result must always be negative, no matter what and whatsoever. As a consequence, the TPR would be zero as well, which is obviously not desirable. In fact, there is a *tradeoff* relationship between FPR and TPR:

$$\text{Tradeoff: } \mathbb{P}(Y = 1|X = 0) \downarrow \Longrightarrow \mathbb{P}(Y = 1|X = 1) \downarrow .$$

To reduce FPR, one would need to design the test such that the probability of a positive result ($Y = 1$) is lower. However, this would have a negative impact on TPR, resulting in a lower TPR. Conversely, to increase TPR, the test should be designed to yield a higher probability of a positive result, which would negatively affect FPR, causing an increase in FPR.

Then, what can we do? One advantage of FPR is that it does not have to be extremely small because false alarms for cancer are acceptable in reality. While false alarms can be annoying, they are endurable because they are not related to life-or-death matters. Conversely, TPR must be high enough because misdetection is disastrous for a person with cancer. Therefore, it is crucial to find a good balance between the two. A typical rule-of-thumb is to increase FPR (sacrifice accuracy for a non-fatal measure) up to a point where TPR degradation is minimized. In the previous example, the desired values for FPR and TPR were:

FPR: $\mathbb{P}(Y = 1|X = 0) = 0.2$ not very small, but somewhat small;

TPR: $\mathbb{P}(Y = 1|X = 0) = 0.95$ very close to 1.

In general, it is important to design the test in a way that respects the following configuration:

$$\text{TPR} : \mathbb{P}(Y = 1 | X = 1) = 1 - \epsilon_1$$
$$\text{FNR} : \mathbb{P}(Y = 0 | X = 1) = \epsilon_1 \text{ very small}$$
$$\text{FPR} : \mathbb{P}(Y = 1 | X = 0) = \epsilon_2 \text{ somewhat small} \qquad (2.15)$$
$$\text{TNR} : \mathbb{P}(Y = 0 | X = 0) = 1 - \epsilon_2.$$

A posteriori probability We are now prepared to tackle the objective probability (2.10), which can be expressed in terms of the (X, Y) notations as:

$$\mathbb{P}(X = 1 | Y = 1). \qquad (2.16)$$

Note that the expression in (2.16) resembles the "a priori probability" $\mathbb{P}(X = 1)$. The distinction is that it is conditioned on the event $Y = 1$. In fact, one can interpret the test result $Y = 1$ as an observation. Therefore, the probability (2.16) can be viewed as the probability *after* making an observation, and is referred to as the "a posteriori probability", since "a posteriori" comes from Latin meaning "after". This is the last key concept that plays a crucial role in the MAP principle.

An inference problem The MAP principle arises in the context of an *inference problem*, where the objective is to infer a specific entity that is probabilistically related to the given observation. In the case of cancer testing, a natural inference problem is depicted in Figure 2.6. In this setup, the goal of the problem is to make an inference about X, which indicates whether a person has cancer, based on the observation Y, which has a probabilistic relationship with X. The conditional probabilities $\mathbb{P}(Y = y | X = x)$ in (2.15) (for $x, y \in \{0, 1\}$) capture this statistical relationship, and the conditional probabilities, which are determined by ϵ_1 and ϵ_2, are fixed once the test mechanism is designed. The block labeled as "inference" in Figure 2.6 takes Y as an input and produces an estimate $\hat{X}$. Note that $\hat{X}$ is a function of the random variable Y and therefore, is also a random variable. However, once the test result is revealed as $Y = y \in \{0, 1\}$, it becomes a fixed value, provided that no new randomness is introduced in the inference block. We refer to such y as a realization of the random variable Y.

The MAP estimation Suppose that the test result is revealed as $Y = y \in \{0, 1\}$. Given the realization $Y = y$, the optimal inference (estimation) for X is defined as the one that maximizes the probability of making a correct decision:

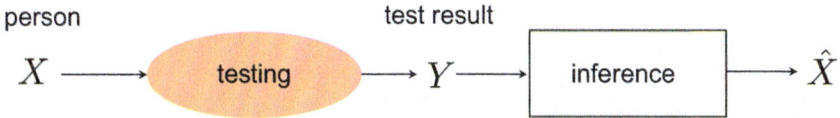

Fig. 2.6 An inference problem

$$\mathbb{P}(X = \hat{X}|Y = y).$$

Notice that $X = \hat{X}$ is the event that we made a correct decision. Since the observation $Y = y$ is provided in the problem, we need to take this into account. Here $\hat{X}$ refers to the choice we make using our inference mechanism. In the case of cancer testing, there are only two possible choices for X, either 0 or 1. Hence, the optimal estimate for X can be expressed as:

$$\hat{X}_{opt} := \arg \max_{\hat{x} \in \{0,1\}} \underbrace{\mathbb{P}(X = \hat{x}|Y = y)}_{\text{A Posteriori probability}} \qquad (2.17)$$

where the notation "arg max" means "is the one that maximizes", and $\hat{x}$ refers to a dummy variable that serves as a candidate that $\hat{X}$ can take on.

Note that the objective probability is the "a posteriori probability". Hence, one can interpret the optimal estimator as the one that *Maximizes A Posteriori probability* (MAP). So, the name of the optimal estimator is the MAP estimator:

$$\hat{X}_{opt} = \hat{X}_{MAP} = \arg \max_{\hat{x} \in \{0,1\}} \mathbb{P}(X = \hat{x}|Y = y). \qquad (2.18)$$

Derivation of the MAP estimator How to compute $\hat{X}_{MAP}$? First observe that the "a posteriori probability" can be written as:

$$\mathbb{P}(X = \hat{x}|Y = y) = \frac{\mathbb{P}(X = \hat{x}, Y = y)}{\mathbb{P}(Y = y)}.$$

This is due to the definition of conditional probability. One crucial observation is that the denominator $\mathbb{P}(Y = y)$ is independent of the variation of $X = \hat{x}$ and is not a function of $X = \hat{x}$. Therefore, we can simplify the MAP estimator (2.18) as:

$$\begin{aligned}\hat{X}_{MAP} &= \arg \max_{\hat{x} \in \{0,1\}} \mathbb{P}(X = \hat{x}, Y = y) \\ &= \arg \max_{\hat{x} \in \{0,1\}} \mathbb{P}(X = \hat{x})\mathbb{P}(Y = y|X = \hat{x})\end{aligned} \qquad (2.19)$$

where the second equality is due to the definition of conditional probability. Using (2.15), we can express the objective probability as:

$$\mathbb{P}(X = 1)\mathbb{P}(Y = y|X = 1) = \begin{cases} p(1 - \epsilon_1), & \text{if } y = 1; \\ p\epsilon_1, & \text{if } y = 0, \end{cases}$$

$$\mathbb{P}(X = 0)\mathbb{P}(Y = y|X = 0) = \begin{cases} p\epsilon_2, & \text{if } y = 1; \\ p(1 - \epsilon_2), & \text{if } y = 0. \end{cases}$$

A more succinct way to express the above is:

$$\mathbb{P}(X = 1)\mathbb{P}(Y = y|X = 1) = p(1 - \epsilon_1)^y \epsilon_1^{1-y};$$
$$\mathbb{P}(X = 0)\mathbb{P}(Y = y|X = 0) = (1 - p)(1 - \epsilon_2)^{1-y} \epsilon_2^y. \tag{2.20}$$

Applying this to (2.19), we can obtain the MAP decision rule:

$$p(1 - \epsilon_1)^y \epsilon_1^{1-y} \underset{\hat{X}_{\text{MAP}}=0}{\overset{\hat{X}_{\text{MAP}}=1}{\gtrless}} (1 - p)(1 - \epsilon_2)^{1-y} \epsilon_2^y. \tag{2.21}$$

You may wonder what happens if the left-hand side and right-hand side of the MAP decision rule are identical. In such a rare scenario, we can either flip a fair coin to decide or arbitrarily choose one of the two options, such as $\hat{X}_{\text{MAP}} = 1$. Even though this may seem like a foolish decision, it still satisfies the MAP decision rule of selecting the option that maximizes the a posteriori probability, as both probabilities are equal.

A different expression of the MAP decision rule (2.21) One might be interested in finding an alternative expression for (2.21) that could offer valuable insights. To do so, we can apply an increasing function $\log(\cdot)$ to both sides of (2.21) to obtain:

$$\log p + y \log(1 - \epsilon_1) + (1 - y) \log \epsilon_1$$
$$\underset{\hat{X}_{\text{MAP}}=0}{\overset{\hat{X}_{\text{MAP}}=1}{\gtrless}} \log(1 - p) + (1 - y) \log(1 - \epsilon_2) + y \log \epsilon_2. \tag{2.22}$$

Massaging the above a bit, we obtain:

$$y \log \frac{(1 - \epsilon_1)(1 - \epsilon_2)}{\epsilon_1 \epsilon_2} \underset{\hat{X}_{\text{MAP}}=0}{\overset{\hat{X}_{\text{MAP}}=1}{\gtrless}} \log \frac{1 - p}{p} + \log \frac{1 - \epsilon_2}{\epsilon_1}. \tag{2.23}$$

In a typical test design scenario where $\epsilon_1 < \frac{1}{2}$ and $\epsilon_2 < \frac{1}{2}$, the term multiplied by $\log \frac{(1-\epsilon_1)(1-\epsilon_2)}{\epsilon_1 \epsilon_2}$ on the left-hand side is positive. Hence, we obtain:

$$y \underset{\hat{X}_{\text{MAP}}=0}{\overset{\hat{X}_{\text{MAP}}=1}{\gtrless}} \frac{\log \frac{1-p}{p}}{\log \frac{(1-\epsilon_1)(1-\epsilon_2)}{\epsilon_1 \epsilon_2}} + \frac{\log \frac{1-\epsilon_2}{\epsilon_1}}{\log \frac{(1-\epsilon_1)(1-\epsilon_2)}{\epsilon_1 \epsilon_2}}. \tag{2.24}$$

Example: $p = 0.1$, $\epsilon_1 = 0.05$, $\epsilon_2 = 0.2$ Let us see how the MAP estimator (2.24) works in our earlier example where $p = 0.1$, $\epsilon_1 = 0.05$, and $\epsilon_2 = 0.2$. In this case,

$$\frac{\log \frac{1-p}{p}}{\log \frac{(1-\epsilon_1)(1-\epsilon_2)}{\epsilon_1 \epsilon_2}} \approx 0.5074; \quad \frac{\log \frac{(1-\epsilon_2)}{\epsilon_1}}{\log \frac{(1-\epsilon_1)(1-\epsilon_2)}{\epsilon_1 \epsilon_2}} \approx 0.6402.$$

Putting this into (2.24) yields:

$$y \underset{0}{\overset{1}{\gtrless}} 0.5074 + 0.6402 = 1.1476. \tag{2.25}$$

In this case, we declare $\hat{X}_{\mathsf{MAP}} = 0$ regardless of the test result y. Even though it may seem like a foolish decision, this decision is actually the optimal decision rule, and there are no mistakes in the derivation. Then, why does this happen? This is because the "a priori probability" $p = 0.1$ (the prevalence of cancer) has a significant impact on the decision. In (2.25), the value of 0.6402 is a quantity that depends only on the test accuracy, as it is a function of ϵ_1 and ϵ_2. On the other hand, the value of 0.5074 is a quantity that is affected by the a priori probability p. If there were no bias in cancer population, i.e., $p = 0.5$, the first quantity would be 0. In this case, we would make a reasonable decision: declare 1 if the result is positive, and declare 0 otherwise. However, in the biased yet realistic situation where $p = 0.1$, the value of 0.5074 makes the decision threshold above 1, resulting in the seemingly foolish decision of $\hat{X}_{\mathsf{MAP}} = 0$.

While this decision provides some relief to the person who received a positive result, it may not completely alleviate the concern because the decision may seem foolish—declaring 0 regardless of the result.

Look ahead
A natural question that arises in this context is whether there is a more reliable and trustworthy approach. This question will be addressed in the next section.

2.3 MAP: Multiple Observations

Recap

In the previous section, we examined one of the two fundamental principles: the MAP estimation. We uncovered the core idea of this principle in the context of the inference problem for cancer prediction, as shown in Figure 2.7.

The optimal estimator for inferring whether an interested person has cancer $X = 1$ given a test result $Y = y \in \{0, 1\}$ is proven to be the MAP estimator that maximizes the a posteriori probability:

$$\hat{X}_{\text{opt}} = \hat{X}_{\text{MAP}} = \arg \max_{\hat{x} \in \{0,1\}} \underbrace{\mathbb{P}(X = \hat{x}|Y = y)}_{\text{A Posteriori probability}}$$

$$= \arg \max_{\hat{x} \in \{0,1\}} \mathbb{P}(X = \hat{x})\mathbb{P}(Y = y|X = \hat{x}) \qquad (2.26)$$

where the second equality is because $\mathbb{P}(X = \hat{x}|Y = y) = \frac{\mathbb{P}(X=\hat{x},Y=y)}{\mathbb{P}(Y=y)}$ and the denominator $\mathbb{P}(Y = y)$ is not a function of $X = \hat{x}$. Under the following reasonable setting:

$$\begin{aligned}
&\text{TPR} : \mathbb{P}(Y = 1|X = 1) = 1 - \epsilon_1 \\
&\text{FNR} : \mathbb{P}(Y = 0|X = 1) = \epsilon_1 \text{ very small} \\
&\text{FPR} : \mathbb{P}(Y = 1|X = 0) = \epsilon_2 \text{ somewhat small} \\
&\text{TNR} : \mathbb{P}(Y = 0|X = 0) = 1 - \epsilon_2,
\end{aligned} \qquad (2.27)$$

we then obtained the MAP decision rule:

$$y \underset{\hat{X}_{\text{MAP}}=0}{\overset{\hat{X}_{\text{MAP}}=1}{\gtrless}} \frac{\log \frac{1-p}{p}}{\log \frac{(1-\epsilon_1)(1-\epsilon_2)}{\epsilon_1 \epsilon_2}} + \frac{\log \frac{1-\epsilon_2}{\epsilon_1}}{\log \frac{(1-\epsilon_1)(1-\epsilon_2)}{\epsilon_1 \epsilon_2}}. \qquad (2.28)$$

We noted that this decision rule is unreliable as it consistently yields the same result of $\hat{X}_{\text{MAP}} = 0$ regardless of the situation. This holds true even in practical situations such as when $p = 0.1$, $\epsilon_1 = 0.05$, and $\epsilon_2 = 0.2$:

$$y \overset{1}{\underset{0}{\gtrless}} 0.5074 + 0.6402 = 1.1476 \longrightarrow \hat{X}_{\text{MAP}} = 0. \qquad (2.29)$$

If the decision indicates "no cancer" despite a positive test result, individuals are unlikely to have confidence in such a simplistic conclusion. This is because the estimator produces identical outcomes for *every* person tested, regardless of her/his unique circumstances or test results. So a natural question is: Is there a reliable alternative approach to take?

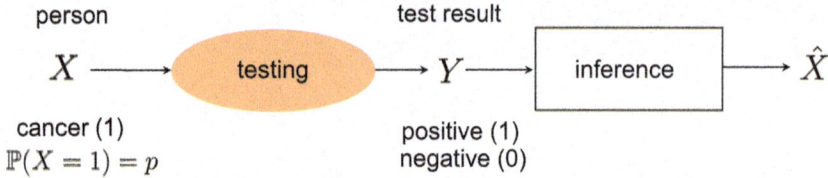

Fig. 2.7 The inference problem for cancer prediction involves a priori knowledge about the prevalence of cancer $\mathbb{P}(X = 1) = p$. The objective is to determine whether a person has cancer $X = 1$ based on a given test result Y

Outline

In this section, we aim to address this question by presenting a simple solution for ensuring the test's reliability. The solution involves conducting *multiple tests*, and we will demonstrate how this enhances the trustworthiness of the results. The section consists of three parts. Firstly, we will introduce a reasonable assumption concerning multiple testing. Under the assumption, we will then derive the MAP decision rule. Finally, we will discuss the relationship between the reliability of the MAP decision rule and the number of tests conducted (Figure 2.8).

Multiple tests The scenario we consider involves conducting multiple tests, which is depicted in Figure 2.8. In the context of multiple testing, we have several observations of the test results, denoted as $(Y_1, \ldots, Y_n) = (y_1, \ldots, y_n)$, where each $y_i \in \{0, 1\}$. Our objective is to infer the true value of X. As previously demonstrated, the optimal estimator is always the MAP estimator, which is defined as the one that maximizes the

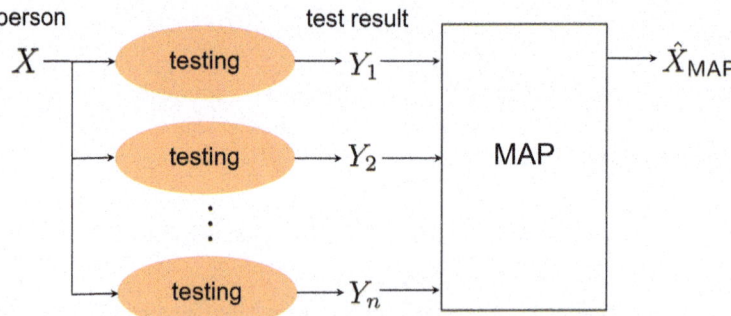

Fig. 2.8 The inference problem for cancer prediction under multiple test results $(Y_1, \ldots, Y_n)$

probability of making a correct decision, or equivalently, the a posteriori probability. Therefore, we will focus on the MAP estimator, as illustrated in Figure 2.8.

As previously stated, we will rely on a crucial assumption. The assumption is grounded on the observation that, for a given individual, the test results are determined solely by the *testing environment*, such as device variation during each trial. It is reasonable to assume that such environments are *independent* across trials and have the *same statistical behavior*. Therefore, we will make this assumption and formalize it as follows: given $X = x$, Y_i's are *conditionally independent* and *identically distributed*.

$$\text{(Assumption): Given } X = x, \{Y_i\}_{i=1}^n \text{ are } i.i.d. \tag{2.30}$$

Referring back to (2.27), which states that $\mathbb{P}(Y_i = 1 | X = 1) = 1 - \epsilon_1$ and $\mathbb{P}(Y_i = 1 | X = 0) = \epsilon_2$, the assumption implies:

$$\begin{aligned} Y_i &\sim \text{Bern}(1 - \epsilon_1) & \text{when } X = 1; \\ Y_i &\sim \text{Bern}(\epsilon_2) & \text{when } X = 0. \end{aligned} \tag{2.31}$$

Derivation of the MAP estimator Given $(Y_1, \ldots, Y_n) = (y_1, \ldots, y_n)$, the MAP estimator reads:

$$\hat{X}_{\text{MAP}} = \arg\max_{\hat{x} \in \{0,1\}} \mathbb{P}(X = \hat{x} | Y_1 = y_1, \ldots, Y_n = y_n). \tag{2.32}$$

The primary difference in the multiple-test scenario compared to the single-test case is that the conditioned event now encompasses multiple test results. Again, due to the definition of conditional probability,

$$\mathbb{P}(X = \hat{x} | Y_1 = y_1, \ldots, Y_n = y_n) = \frac{\mathbb{P}(X = \hat{x}, Y_1 = y_1, \ldots, Y_n = y_n)}{\mathbb{P}(Y_1 = y_1, \ldots, Y_n = y_n)}.$$

Since $\mathbb{P}(Y_1 = y_1, \ldots, Y_n = y_n)$ is independent of $X = \hat{x}$, the estimator (2.32) can be written as:

$$\hat{X}_{\text{MAP}} = \arg\max_{\hat{x} \in \{0,1\}} \mathbb{P}(X = \hat{x}) \mathbb{P}(Y_1 = y_1, \ldots, Y_n = y_n | X = \hat{x}). \tag{2.33}$$

Due to the i.i.d. assumption (2.30) of $\{Y_i\}_{i=1}^n$ given $X = \hat{x}$, $\mathbb{P}(Y_1 = y_1, \ldots, Y_n = y_n | X = \hat{x}) = \prod_{i=1}^n \mathbb{P}(Y_i = y_i | X = \hat{x})$. This then yields:

$$\hat{X}_{\text{MAP}} = \arg\max_{\hat{x} \in \{0,1\}} \mathbb{P}(X = \hat{x}) \prod_{i=1}^n \mathbb{P}(Y_i = y_i | X = \hat{x}). \tag{2.34}$$

Using (2.27), we can then compute the objective probability as:

$$\mathbb{P}(X=1)\prod_{i=1}^{n}\mathbb{P}(Y_i=y_i|X=1) = p(1-\epsilon_1)^{\sum_{i=1}^{n}y_i}\epsilon_1^{n-\sum_{i=1}^{n}y_i};$$

$$\mathbb{P}(X=0)\prod_{i=1}^{n}\mathbb{P}(Y_i=y_i|X=0) = (1-p)(1-\epsilon_2)^{n-\sum_{i=1}^{n}y_i}\epsilon_2^{\sum_{i=1}^{n}y_i}.$$

(2.35)

The expression for the MAP estimator in the multiple-test scenario is quite similar to its single-test counterpart (2.21). The crucial difference lies in the exponents on the right-hand side of the equation. Specifically, instead of y, we now read $\sum_{i=1}^{n}y_i$ in the exponent. To simplify the notation, we introduce $s := \sum_{i=1}^{n}y_i$. With this notation, the MAP estimator can be expressed as:

$$p(1-\epsilon_1)^s\epsilon_1^{n-s}\mathop{\gtrless}\limits_{\hat{X}_{\mathrm{MAP}}=0}^{\hat{X}_{\mathrm{MAP}}=1}(1-p)(1-\epsilon_2)^{n-s}\epsilon_2^s.$$

(2.36)

Similar to the single-test scenario, we can manipulate this equation to obtain a more insightful expression. Specifically, by taking the logarithmic function on both sides, we arrive at:

$$\log p + s\log(1-\epsilon_1) + (n-s)\log\epsilon_1$$
$$\mathop{\gtrless}\limits_{0}^{1}\log(1-p) + (n-s)\log(1-\epsilon_2) + s\log\epsilon_2.$$

(2.37)

By summing the terms in the left-hand side with respect to s, we can obtain:

$$s\log\frac{(1-\epsilon_1)(1-\epsilon_2)}{\epsilon_1\epsilon_2}\mathop{\gtrless}\limits_{0}^{1}n\log\frac{1-\epsilon_2}{\epsilon_1} + \log\frac{1-p}{p}.$$

(2.38)

As long as the values of ϵ_1 and ϵ_2 are less than $\frac{1}{2}$, the term $\log\frac{(1-\epsilon_1)(1-\epsilon_2)}{\epsilon_1\epsilon_2}$ is positive in the setting. Hence, we get:

$$\underbrace{\frac{s}{n}}_{\text{fraction of positive tests}}\mathop{\gtrless}\limits_{0}^{1}\underbrace{\frac{\log\frac{1-\epsilon_2}{\epsilon_1}}{\log\frac{(1-\epsilon_1)(1-\epsilon_2)}{\epsilon_1\epsilon_2}} + \frac{1}{n}\cdot\frac{\log\frac{1-p}{p}}{\log\frac{(1-\epsilon_1)(1-\epsilon_2)}{\epsilon_1\epsilon_2}}}_{\text{threshold}}.$$

(2.39)

We can interpret the left-hand side as the *empirical average* $\frac{s}{n} = \frac{\sum_{i=1}^{n}y_i}{n}$, which represents the fraction of positive test results. With this interpretation, the decision rule becomes more intuitive: if the fraction of positive results is greater than the threshold on the right-hand side, we declare $\hat{X}_{MAP} = 1$; otherwise, $\hat{X}_{MAP} = 0$.

Reliability of the MAP estimator (2.39) To gain a deeper understanding of the threshold (the right-hand side in (2.39)) in relation to the reliability of the MAP estimator, let us consider the previous example where $p = 0.1$, $\epsilon_1 = 0.05$, and $\epsilon_2 = 0.2$. In this case, the two quantities on the right-hand side are approximately:

$$\frac{\log \frac{1-\epsilon_2}{\epsilon_1}}{\log \frac{(1-\epsilon_1)(1-\epsilon_2)}{\epsilon_1 \epsilon_2}} \approx 0.6402. \quad \frac{\log \frac{1-p}{p}}{\log \frac{(1-\epsilon_1)(1-\epsilon_2)}{\epsilon_1 \epsilon_2}} \approx 0.5074.$$

Putting this into (2.39), the rule can be simply stated as:

$$\frac{s}{n} \underset{0}{\overset{1}{\gtrless}} 0.6402 + \frac{0.5074}{n}. \tag{2.40}$$

The value of 0.6402 is determined only by the *test accuracy*, as it is a function of ϵ_1 and ϵ_2. However, the second value 0.5074 is also influenced by the *a priori probability* p.

When $n = 1$ (the single-test case), the right-hand-side in (2.40) is $0.6402 + \frac{0.5074}{1} = 1.1476$, which is greater than 1, causing the MAP rule to always declare $\hat{X}_{MAP} = 0$. In the multiple-test case, on the other hand, the term $\frac{0.5074}{n}$, which depends on the a priori probability p, becomes smaller as n increases, and the MAP rule becomes dominated solely by the first term 0.6402 that reflects the test accuracy (determined by ϵ_1 and ϵ_2). If the test accuracies for the cancer and normal populations were identical, that is, $\epsilon_1 = \epsilon_2$, then the value of 0.6402 would become 0.5:

$$\frac{\log \frac{1-\epsilon_2}{\epsilon_1}}{\log \frac{(1-\epsilon_1)(1-\epsilon_2)}{\epsilon_1 \epsilon_2}} = \frac{\log \frac{1-\epsilon_1}{\epsilon_1}}{2 \log \frac{1-\epsilon_1}{\epsilon_1}} = \frac{1}{2}. \tag{2.41}$$

This corresponds to exactly "majority voting".

Because the test is designed to allow for a higher ϵ_2 (compared to ϵ_1) due to the tradeoff relationship between true positive rate (TPR) and false positive rate (FPR), positive outcomes are more likely. Therefore, the threshold of 0.6402 is set to be higher than 0.5. As ϵ_2 increases, the threshold becomes higher, as shown in Figure 2.9 for $\epsilon_1 = 0.05$. Notably, when $\epsilon_1 = \epsilon_2 = 0.05$, the threshold is precisely 0.5, and it grows as ϵ_2 increases. In the given example where $\epsilon_2 = 0.2$, the MAP estimator provides a kind of "slightly biased majority voting" as n approaches infinity: declaring $\hat{X}_{MAP} = 1$ for more than about 64% positive test results.

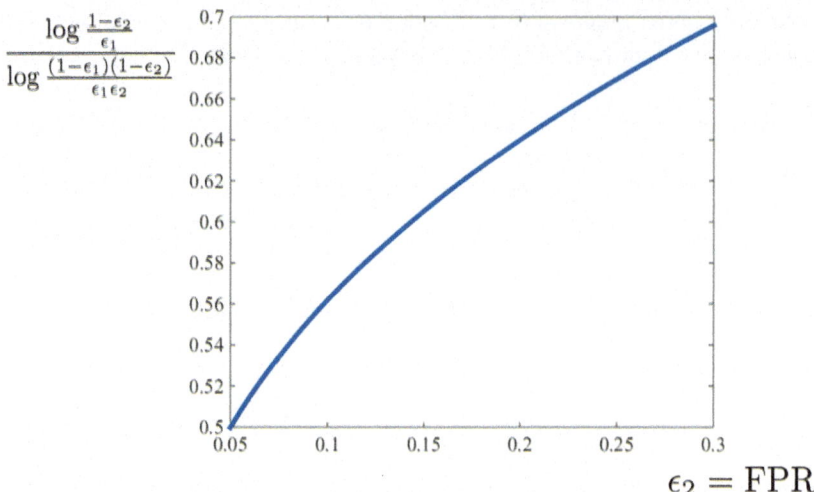

$\epsilon_2 = \text{FPR}$

Fig. 2.9 Decision threshold as a function of ϵ_2 (False Positive Rate) when $1 - \text{TPR} = \epsilon_1 = 0.05$ and $n \to \infty$

The above example demonstrates that as n increases, the MAP rule depends exclusively on test accuracy, and therefore one can anticipate enhanced reliability in the test, and the decision rule will undoubtedly outperform random guessing. However, the extent of this improvement in reliability as n increases is not precisely quantified.

Look ahead

In the forthcoming section, we will quantitatively evaluate the reliability of the MAP decision rule and demonstrate how accuracy improves with an increase in n.

2.4 MAP: Performance and **Python** Simulation

Recap

In the preceding section, we derived the optimal MAP estimator in the multiple-test setting for cancer prediction, illustrated in Figure 2.10. Given test results $(Y_1, \ldots, Y_n) = (y_1, \ldots, y_n)$, the MAP decision rule is shown to be:

$$\underbrace{\frac{s}{n}}_{\text{fraction of positive tests}} \underset{0}{\overset{1}{\gtrless}} \underbrace{\frac{\log \frac{1-\epsilon_2}{\epsilon_1}}{\log \frac{(1-\epsilon_1)(1-\epsilon_2)}{\epsilon_1 \epsilon_2}} + \frac{1}{n} \cdot \frac{\log \frac{1-p}{p}}{\log \frac{(1-\epsilon_1)(1-\epsilon_2)}{\epsilon_1 \epsilon_2}}}_{\text{threshold}} \qquad (2.42)$$

where $s = \sum_{i=1}^{n} y_i$ and (ϵ_1, ϵ_2) are FNR and FPR, respectively:

$$\begin{aligned} \epsilon_1 &= \text{FNR} = \mathbb{P}(Y = 0 | X = 1); \\ \epsilon_2 &= \text{FPR} = \mathbb{P}(Y = 1 | X = 0). \end{aligned} \qquad (2.43)$$

In the single-test case, the decision rule yields the same answer no matter what and hence it is believed to be unreliable. In the multiple-test scenario, on the other hand, the MAP rule depends on test results y_i's, precisely the faction of test results $\frac{s}{n}$. Therefore, one can expect that the accuracy of the decision rule would improve as n increases.

Outline

In this section, we will substantiate the observed improvement and quantify it as a function of the number of observations. The section is divided into four parts. Initially, we will introduce a significant measure, based on probability concepts, that quantifies the accuracy of the decision rule. Subsequently, we will calculate this measure under a proper assumption. Following that, we will demonstrate the enhancement of performance measures with an increasing value of n. Finally, we will validate this result by conducting a Python simulation.

A performance measure Recall the MAP decision rule:

$$\frac{s}{n} \underset{0}{\overset{1}{\gtrless}} \frac{\log \frac{1-\epsilon_2}{\epsilon_1}}{\log \frac{(1-\epsilon_1)(1-\epsilon_2)}{\epsilon_1 \epsilon_2}} + \frac{1}{n} \cdot \frac{\log \frac{1-p}{p}}{\log \frac{(1-\epsilon_1)(1-\epsilon_2)}{\epsilon_1 \epsilon_2}} =: g(n). \qquad (2.44)$$

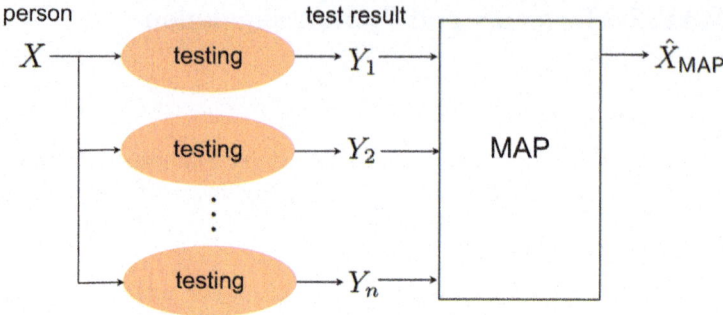

Fig. 2.10 Cancer prediction under multiple test results $(Y_1, \ldots, Y_n)$

The information provided comprises test results: $(Y_1, \ldots, Y_n) = (y_1, \ldots, y_n)$. Given $(Y_1, \ldots, Y_n) = (y_1, \ldots, y_n)$, a performance measure that we will utilize is the probability of a correct decision event, denoted by $\mathcal{C}$ (the cancer presence of an individual that we denote by X is equal to its estimate $\hat{X}$):

$$\mathbb{P}(\mathcal{C}|Y_1 = y_1, \ldots, Y_n = y_n) = \mathbb{P}(X = \hat{X}|Y_1 = y_1, \ldots, Y_n = y_n). \qquad (2.45)$$

However, an issue arises in adopting this quantity. The issue is that the quantity is a function of $(y_1, \ldots, y_n)$ and hence it varies depending on a particular realization of the test results $(Y_1, \ldots, Y_n)$. So it cannot serve as a representative quantity that is obviously preferred to be fixed. In an effort to find a representative quantity that does not rely on a specific realization, we consider its *expectation*: a weighted sum of the measures over all of the realizations $(y_1, \ldots, y_n)$'s with the weight determined by the joint distribution $\mathbb{P}_{Y_1, \ldots, Y_n}(y_1, \ldots, y_n)$:

$$\sum_{y_1, \ldots, y_n} \mathbb{P}_{Y_1, \ldots, Y_n}(y_1, \ldots, y_n)\mathbb{P}(\mathcal{C}|Y_1 = y_1, \ldots, Y_n = y_n)$$

$$= \sum_{y_1, \ldots, y_n} \mathbb{P}(\mathcal{C}, Y_1 = y_1, \ldots, Y_n = y_n) \qquad (2.46)$$

$$= \mathbb{P}(\mathcal{C})$$

where the first equality is due to the definition of conditional probability; and the second equality is because of the total probability law.

Analysis of error probability Alternatively we may consider the average *error* probability:

$$\mathbb{P}(\mathcal{E}) = 1 - \mathbb{P}(\mathcal{C}) \qquad (2.47)$$

where $\mathcal{E}$ denotes an error event. Here we will focus on evaluating $\mathbb{P}(\mathcal{E})$ instead. An error occurs if $\hat{X} = 1$ when $X = 0$, or vice versa. The average error is a weighted sum of the probabilities of these two types of error events, with the weights being equal to the a priori probabilities of X:

$$
\begin{aligned}
\mathbb{P}(\mathcal{E}) &= \mathbb{P}(\mathcal{E}, X = 0) + \mathbb{P}(\mathcal{E}, X = 1) \\
&= \mathbb{P}(X = 0)\mathbb{P}(\mathcal{E}|X = 0) + \mathbb{P}(X = 1)\mathbb{P}(\mathcal{E}|X = 1) \\
&= (1 - p)\mathbb{P}(\mathcal{E}|X = 0) + p\mathbb{P}(\mathcal{E}|X = 1)
\end{aligned}
\tag{2.48}
$$

where the first equality is due to the total probability law and the second comes from the definition of conditional probability.

First consider $\mathbb{P}(\mathcal{E}|X = 0)$. Under the MAP decision rule (2.44), we get:

$$
\begin{aligned}
\mathbb{P}(\mathcal{E}|X = 0) &= \mathbb{P}(\hat{X} = 1|X = 0) \\
&= \mathbb{P}\left(S > ng(n)|X = 0\right).
\end{aligned}
\tag{2.49}
$$

Given $X = 0$, $Y_i \sim \mathsf{Bern}(\epsilon_2)$ and hence $S := Y_1 + \cdots + Y_n$ follows a binomial distribution with ϵ_2:

$$
\mathbb{P}_S(s) = \binom{n}{s} \epsilon_2^{s} (1 - \epsilon_2)^{n-s} \quad \forall s \in \{0, 1, \ldots, n\} =: \mathcal{S}.
\tag{2.50}
$$

Applying this to (2.49) yields:

$$
\mathbb{P}(\mathcal{E}|X = 0) = \sum_{s \in \mathcal{S}: s > ng(n)} \binom{n}{s} \epsilon_2^{s} (1 - \epsilon_2)^{n-s}.
\tag{2.51}
$$

Next, consider the other error probability:

$$
\mathbb{P}(\mathcal{E}|X = 1) = \mathbb{P}(\hat{X} = 0|X = 1) = \mathbb{P}\left(S < ng(n)|X = 1\right).
\tag{2.52}
$$

Given $X = 1$, $Y_i \sim \mathsf{Bern}(1 - \epsilon_1)$ and hence S follows a binomial distribution with $1 - \epsilon_1$, yielding:

$$
\mathbb{P}(\mathcal{E}|X = 1) = \sum_{s \in \mathcal{S}: s < ng(n)} \binom{n}{s} (1 - \epsilon_1)^{s} \epsilon_1^{n-s}.
\tag{2.53}
$$

Plugging (2.51) and (2.53) into (2.48), we obtain:

$$
\begin{aligned}
\mathbb{P}(\mathcal{E}) =&(1 - p) \sum_{s \in \mathcal{S}: s > ng(n)} \binom{n}{s} \epsilon_2^{s} (1 - \epsilon_2)^{n-s} \\
&+ p \sum_{s \in \mathcal{S}: s < ng(n)} \binom{n}{s} (1 - \epsilon_1)^{s} \epsilon_1^{n-s}.
\end{aligned}
\tag{2.54}
$$

Our interest is in understanding the behavior of $\mathbb{P}(\mathcal{E})$ as a function of n. As the formula in (2.54) is not in a closed form, discerning its behavior is challenging. Therefore, we will conduct a numerical simulation using Python to observe the behavior numerically.

Python simulation We consider a simple yet practical scenario with parameters set as before: $p = 0.1$, $\epsilon_1 = 0.05$, and $\epsilon_2 = 0.2$. Let us first construct a function that calculates the error probability (2.54) for a specified value of n. The following presents the code implementation of this function.

```python
import numpy as np
from scipy.stats import binom

p=0.1 # prevalence of cancer
e1=0.05 # FNR
e2=0.2  # FPR
n=10    # num of observations

def Pe(n,p,e1,e2):
    # compute threshold
    a = np.log((1-e2)/e1)/np.log((1-e1)*(1-e2)/e1/e2)
    b = np.log((1-p)/p)/np.log((1-e1)*(1-e2)/e1/e2)
    g_n = a + b/n
    # indices for P(E|X=0)
    i_values = list(range(int(np.ceil(n*g_n)),n+1))
    # indices for P(E|X=1)
    j_values = list(range(int(np.floor(n*g_n))+1))
    # Binom(n,e2) when X=0
    dist1=[binom.pmf(i,n,e2) for i in i_values]
    # Binom(n,1-e1) when X=1
    dist2=[binom.pmf(j,n,1-e1) for j in j_values]
    # probability of error
    Pe = (1-p)*sum(dist1)+p*sum(dist2)
    return Pe

print(Pe(n,p,e1,e2))
```

0.0008807723537890629

In the given instance with $n = 10$, the error probability is approximately 8.808×10^{-4}.

Expanding on the Pe function, we now plot the error probability as a function of the number of observations n. The following is the code implementation for this plotting.

```python
import matplotlib.pyplot as plt

p=0.1 # prevalence of cancer
e1=0.05 # FNR
e2=0.2  # FPR
n=np.arange(1,32,2) # [1,3,5,...,31]
```

```
Pe_tot = np.zeros(len(n))
for i in range(len(n)): Pe_tot[i] = Pe(n[i],p,e1,e2)

plt.figure(figsize=(4,4),dpi=500)
plt.plot(n,Pe_tot)
plt.xlabel('$n$ (number of observations)')
plt.title('probability of error')
plt.yscale('log')
plt.show()
```

It is apparent from Figure 2.11 that the probability of error experiences a substantial reduction, exhibiting an *exponential* decrease with an increase in n.

Look ahead

Over the previous sections, we have examined an application of the MAP principle in the context of cancer prediction, which served as a prominent and illustrative example. It is worth noting that the MAP principle has numerous practical applications beyond this example. In Part III, we will explore some of these applications. For now, let us turn our attention to the second key principle: the Maximum Likelihood (ML) estimation.

Fig. 2.11 The probability of error as a function of n

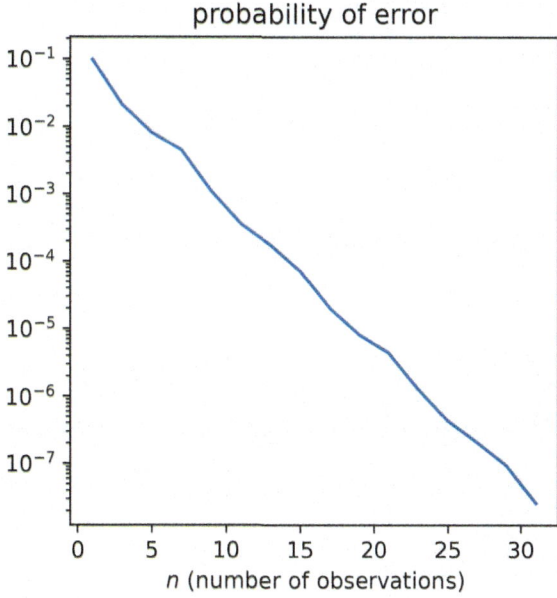

Problem Set 5

Problem 5.1 (*The Bernoulli process*) Suppose that two binary random variables X_1 and X_2 satisfy: $\mathbb{P}(X_1 = i_1, X_2 = i_2) = \frac{1}{4}$ for all possible sequence patterns $(i_1, i_2) \in \{(0,0), (0,1), (1,0), (1,1)\}$. Show that X_1 and X_2 are the Bernoulli process with $X_i \sim \text{Bern}(\frac{1}{2})$.

Problem 5.2 (*Distribution of an infinite-length random process*) Let $\{X_i\}_{i=1}^{\infty}$ be an i.i.d. process where X_i is uniformly distributed in the $(0, 1)$ open interval:

$$f_{X_i}(x) = 1, \qquad \forall x \in (0, 1). \tag{2.55}$$

(a) Compute the joint cdf of X_1 and X_2: $F_{X_1, X_2}(x_1, x_2)$.
(b) Define:

$$F_{X_1, \ldots, X_n, \ldots}(x_1, \ldots, x_n, \ldots) := \lim_{n \to \infty} \mathbb{P}(X_1 \leq x_1, \ldots, X_n \leq x_n), \forall x_i \in (0, 1). \tag{2.56}$$

Compute $F_{X_1, \ldots, X_n, \ldots}(x_1, \ldots, x_n, \ldots)$.
Remark: Think about why an infinite-length random process takes a set of joint distributions for all finite subsets of $\{X_i\}_{i=1}^{\infty}$.

Problem 5.3 (*Markov's inequality and Chebyshev's inequality*) Let $\{X_i\}_{i=1}^n$ be the Bernoulli process with $\text{Bern}(p)$ where $0 \leq p \leq 1$. Let

$$S_n = \frac{X_1 + X_2 + \cdots + X_n}{n}. \tag{2.57}$$

(a) (*Markov's inequality*) Consider a non-negative random variable X and a positive value d. Prove that

$$\mathbb{P}(X \geq d) \leq \frac{\mathbb{E}[X]}{d}. \tag{2.58}$$

(b) (*Chebyshev's inequality*) Let Y be a random variable with mean μ and variance σ^2. Show that for any $t > 0$,

$$\mathbb{P}(|Y - \mu| \geq t) \leq \frac{\sigma^2}{t^2}. \tag{2.59}$$

(c) Show that for any $\epsilon > 0$,

$$\lim_{n \to \infty} \mathbb{P}(|S_n - p| \geq \epsilon) = 0. \tag{2.60}$$

Problem 5.4 (*Markov processes*)

(a) Consider a generalized Markov process $\{X_i\}_{i=1}^{\infty}$ with $\ell = 2$ memories:

$$\mathbb{P}(x_{m+1}|x_m, x_{m-1}, x_{m-2}, \ldots, x_1) = \mathbb{P}(x_{m+1}|x_m, x_{m-1}). \qquad (2.61)$$

Let $S_m := (X_m, X_{m-1})$ where $m \geq 2$. Show that

$$\mathbb{P}(s_{m+1}|s_m, s_{m-1}, \ldots, s_2) = \mathbb{P}(s_{m+1}|s_m) \qquad (2.62)$$

where $m \geq 2$.

(b) Consider a Markov process (Y_1, Y_2, Y_3):

$$\mathbb{P}(y_3|y_2, y_1) = \mathbb{P}(y_3|y_2). \qquad (2.63)$$

Show that

$$\mathbb{P}(y_1|y_2, y_3) = \mathbb{P}(y_1|y_2). \qquad (2.64)$$

Remark: Think about why a Markov chain has no directionality in edges.

Problem 5.5 (*The MAP principle: Exercise #1*)

(a) Let $X \sim \text{Bern}(p)$ where $0 \leq p \leq 1$. Suppose that another binary random variable Y satisfies $\mathbb{P}(Y = 0|X = 1) = \mathbb{P}(Y = 1|X = 0) = \epsilon$ where $0 \leq \epsilon \leq 1$. Compute $\mathbb{P}(X = 1|Y = 1)$. Given $Y = 1$, find the MAP estimator $\hat{X}_{\text{MAP}}$ for X.

(b) Let $X \sim \text{Bern}(0.6)$. Given $X = x$, another random variable Y follows the Gaussian distribution with mean $0.1 + x$ and variance $0.1 + x$: $Y \sim \mathcal{N}(0.1 + x, 0.1 + x)$. Given $Y = y$, find the MAP estimator $\hat{X}_{\text{MAP}}$ for X.

Remark: This model is inspired by optical communication links.

Problem 5.6 (*The MAP principle: Exercise #2*) Let $X \sim \text{Bern}(\frac{3}{4})$. Consider:

$$Y = 2X - 1 + Z \qquad (2.65)$$

where $Z \sim \mathcal{N}(0, \sigma^2)$.

(a) Given $Y = y$, derive the MAP estimator $\hat{X}_{\text{MAP}}$.

(b) Derive the error probability when using the MAP estimator: $P_e = \mathbb{P}(\hat{X}_{\text{MAP}} \neq X)$.

Hint: You may want to use the Q-function:

$$Q(z) := \int_z^{\infty} \frac{1}{\sqrt{2\pi}} e^{-\frac{t^2}{2}} dt. \qquad (2.66)$$

Problem 5.7 (*The MAP principle: Exercise #3*) Let $X \in \{+1, -1\}$ be a discrete random variable with $\mathbb{P}(X = +1) = p$. Let (Y_1, Y_2) be random variables with:

$$\begin{aligned} Y_1 &= X + W_1; \\ Y_2 &= X + W_2 \end{aligned} \tag{2.67}$$

where (W_1, W_2) are independent and $W_i \sim \mathcal{N}(0, \sigma_i^2)$ for $i \in \{1, 2\}$. Assume that W_i's are independent of X. Here we define:

$$\mathbb{P}(X = x | Y_1 = y_1) := \lim_{\delta \to 0} \mathbb{P}(X = x | Y_1 \in [y_1, y_1 + \delta]). \tag{2.68}$$

(*a*) Show that

$$\mathbb{P}(X = x | Y_1 = y_1) = \frac{\mathbb{P}(X = x) f(y_1 | X = x)}{f(y_1)} \tag{2.69}$$

where $f(y_1)$ represents the Gaussian pdf of Y, and $f(y_1 | X = x)$ denotes the conditioned version given $X = x$.

(*b*) Consider the case $n = 1$. Given $Y_1 = y_1$, derive the MAP estimator $\hat{X}_{\mathsf{MAP}}$ for X. Show all the detailed derivations.

(*c*) Given $(Y_1, Y_2) = (y_1, y_2)$, derive the MAP estimator $\hat{X}_{\mathsf{MAP}}$ for X. Show all the detailed derivations.

Problem 5.8 (*True or False?*)

(*a*) A generalized Markov process with $\ell \geq 2$ memories can always be transformed into an equivalent Markov process with a single memory.

(*b*) Suppose $X \sim \mathsf{Binom}(n, p)$ and $Y \sim \mathsf{Binom}(m, q)$ where $q \neq p$. Then, $Z = X + Y$ follows a binomial distribution.

2.5 The Maximum Likelihood (ML) Principle

Recap

Over the past sections, we have explored the MAP principle in the context of the inference problem illustrated in Figure 2.12. The key insight we have uncovered is that the optimal inference, defined as the one that maximizes the probability of making a correct decision, is equivalent to the Maximum A Posteriori (MAP) estimation, which seeks to maximize the posterior probability:

$$\hat{X}_{\text{opt}} = \hat{X}_{\text{MAP}} = \underset{\hat{x} \in \mathcal{X}}{\arg\max} \underbrace{\mathbb{P}(X = \hat{x} | Y = y)}_{\text{A Posteriori probability}} \tag{2.70}$$

$$= \underset{\hat{x} \in \mathcal{X}}{\arg\max} \, \mathbb{P}(X = \hat{x})\mathbb{P}(Y = y | X = \hat{x}).$$

In the above, the second equality is due to $\mathbb{P}(X = \hat{x} | Y = y) = \frac{\mathbb{P}(X=\hat{x}, Y=y)}{\mathbb{P}(Y=y)}$ and the denominator $\mathbb{P}(Y = y)$ is not a function of $X = \hat{x}$.

Here, the critical assumption that we have made is as follows:

(Assumption): The "a priori probability" $\mathbb{P}(X = \hat{x})$ is known.   (2.71)

In practice, however, numerous inference scenarios may lack this prior knowledge about the entity of interest. This is where the Maximum Likelihood (ML) principle becomes essential.

Outline

In this section, we will investigate the ML principle in depth. The section consists of four parts. We will first introduce a common scenario in which we lack prior knowledge of an entity's statistics. Interestingly, it will become evident that optimal inference in such a setting leads to the Maximum Likelihood estimation. Subsequently, we will derive the ML estimator. Finally, we will illustrate that the ML estimator offers a reasonably good performance as the number of observations increases.

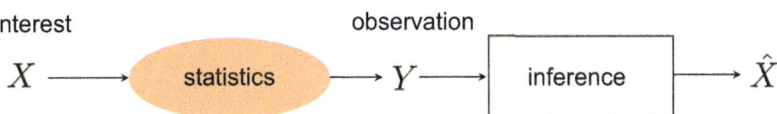

Fig. 2.12 The objective of the inference problem is to infer the state of an entity of interest, typically represented as X, based on observation, say Y, having a statistical association with X

A parameter estimation setting The prominent scenario that will be our primary focus is *parameter estimation*. Let us begin by explaining what parameter estimation is, using a simple example that involves the Bernoulli process with the parameter p. See Figure 2.13. Regarding the Bernoulli parameter p, a set of i.i.d. samples, denoted as $\{Y_i\}_{i=1}^n$, are generated. The objective of this problem is to estimate the parameter p based on these observations. In this problem scenario, a notable distinction compared to the inference problem for cancer prediction, where we possess prior knowledge of the cancer population ratio, is that we generally lack any information about the statistics of the parameter p. Since p is a continuous value, this means that the pdf $f_p(\cdot)$ of p is unknown.

What is the optimal inference to estimate p in this context? To answer this question, let us begin our exploration by considering the "correct decision probability", which plays a key role in deriving the optimal estimate.

Correct decision probability Consider the correct decision probability: $\mathbb{P}(p = \hat{p})$. An issue becomes apparent here, since with p being a continuous value, this probability always equals zero:

$$\mathbb{P}(p = \hat{p}) = 0. \tag{2.72}$$

To obtain a meaningful non-zero quantity, we should focus on its *density*, just as we do for continuous random variables:

$$\lim_{\delta \to 0} \frac{\mathbb{P}(p \in [\hat{p}, \hat{p} + \delta])}{\delta} =: f_p(\hat{p}). \tag{2.73}$$

Given that we have observations from the problem, we should consider its conditional counterpart:

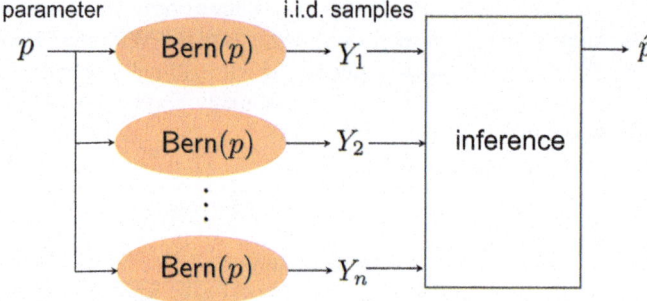

Fig. 2.13 Estimating the Bernoulli parameter p from a collection of i.i.d. samples generated according to the Bernoulli distribution with parameter p

$$\lim_{\delta \to 0} \frac{\mathbb{P}(p \in [\hat{p}, \hat{p} + \delta]) | Y_1 = y_1, \ldots, Y_n = y_n)}{\delta} =: f_p(\hat{p} | Y_1 = y_1, \ldots, Y_n = y_n).$$

$$(2.74)$$

The optimal estimator With the conditional density (2.74), we can define the optimal estimator as:

$$\hat{p}_{\text{opt}} := \arg \max_{t \in [0,1]} f_p(t | Y_1 = y_1, \ldots, Y_n = y_n) \qquad (2.75)$$

where t represents a dummy variable for the choice of $\hat{p}$ and $t \in [0, 1]$ as it is a candidate for the Bernoulli parameter. By utilizing Bayes' law (i.e., applying the definition of conditional probability twice), we get:

$$\begin{aligned} f_p(t | Y_1 = y_1, \ldots, Y_n = y_n) &= \frac{f_{p,Y_1,\ldots,Y_n}(t, y_1, \ldots, y_n)}{\mathbb{P}(Y_1 = y_1, \ldots, Y_n = y_n)} \\ &= \frac{f_p(t) \mathbb{P}(Y_1 = y_1, \ldots, Y_n = y_n | p = t)}{\mathbb{P}(Y_1 = y_1, \ldots, Y_n = y_n)} \end{aligned} \qquad (2.76)$$

where $f_{p,Y_1,\ldots,Y_n}(t, y_1, \ldots, y_n)$ denotes the joint distribution w.r.t. p and $\{Y_i\}_{i=1}^n$. The joint distribution is defined as:

$$f_{p,Y_1,\ldots,Y_n}(t, y_1, \ldots, y_n) := \lim_{\delta \to 0} \frac{\mathbb{P}(p \in [t, t + \delta], Y_1 = y_1, \ldots, Y_n = y_n)}{\delta}. \qquad (2.77)$$

In (2.76), the denominator $\mathbb{P}(Y_1 = y_1, \ldots, Y_n = y_n)$ is not a function of t. Therefore, the optimal estimator can be simplified to:

$$\hat{p}_{\text{opt}} = \arg \max_{t \in [0,1]} f_p(t) \mathbb{P}(Y_1 = y_1, \ldots, Y_n = y_n | p = t). \qquad (2.78)$$

Considering that we do not have knowledge of $f_p(t)$, a reasonable approach is to remain equally open to all possible choices. In an attempt to avoid any bias toward a specific value, it is common for people to assume:

$$f_p(t) \text{ is } uniformly \text{ distributed.} \qquad (2.79)$$

Maximum Likelihood Estimation (MLE) Since $f_p(t)$ is irrelevant to t under the assumption (2.79), the optimal estimator (2.78) reads:

$$\hat{p}_{\text{opt}} = \arg \max_{t \in [0,1]} \underbrace{\mathbb{P}(Y_1 = y_1, \ldots, Y_n = y_n | p = t)}_{\text{Likelihood}}. \qquad (2.80)$$

The objective conditional probability underbraced is a well-known concept, referred to as the *likelihood*. Therefore, one can perceive the estimator as the one that maximizes likelihood, and it is consequently termed the *ML estimator*.

$$\hat{p}_{\mathrm{opt}} = \hat{p}_{\mathrm{ML}} = \arg \max_{t \in [0,1]} \mathbb{P}(Y_1 = y_1, \ldots, Y_n = y_n | p = t). \qquad (2.81)$$

MLE derivation Now, let's explore how to compute the MLE (2.81). Initially, consider that when $p = t$ is given, the set $\{Y_i\}_{i=1}^{n}$ comprises i.i.d. variables, each following a Bernoulli distribution with parameter t. Hence, we can observe that:

$$
\begin{aligned}
\mathbb{P}(Y_1 = y_1, \ldots, Y_n = y_n | p = t) &= \prod_{i=1}^{n} \mathbb{P}(Y_i = y_i | p = t) \\
&= t^{(\# \text{ of 1's})} (1 - t)^{(\# \text{ of 0's})} \\
&= t^{\sum_{i=1}^{n} y_i} (1 - t)^{n - \sum_{i=1}^{n} y_i}.
\end{aligned}
\qquad (2.82)
$$

Let $s := \sum_{i=1}^{n} y_i$. Using this and applying (2.81) to (2.82), the ML estimator can be simplified as:

$$\hat{p}_{\mathrm{ML}} = \arg \max_{t \in [0,1]} t^s (1 - t)^{n-s}. \qquad (2.83)$$

How do we determine the maximizer, say t^*, in the above? It's important to note that the objective function is non-negative and attains a value of 0 at both the extremes, namely when $t = 0$ and when $t = 1$. One can easily visualize that this function increases and decreases as t varies from 0 to 1. This is indeed the case, as depicted in Figure 2.14. The curve in Figure 2.14 is an illustrative example when $(n, s) = (4, 2)$.

Some curious readers might wonder how we can ensure that there is only one "up-and-down" movement, rather than multiple such fluctuations. To address this, we need to check if there is just one stationary point within the open interval $t \in (0, 1)$. The term stationary point is a well-known concept in the field of optimization, denoting a point where the slope, or derivative, is equal to 0. Thus, let's compute its derivative:

Fig. 2.14 A shape of the objective $t^s (1 - t)^{n-s}$ as a function of $t \in [0, 1]$ when $(n, s) = (4, 2)$

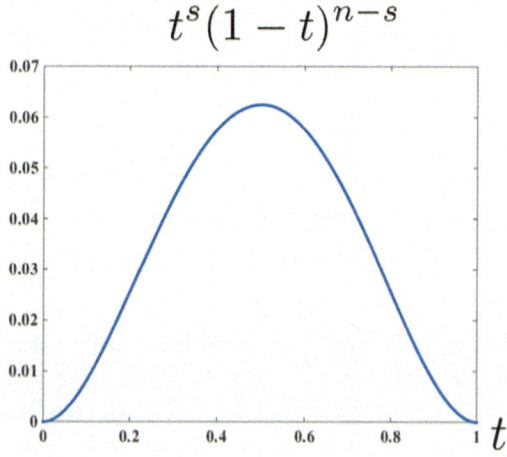

$$\frac{d}{dt} t^s (1-t)^{n-s} = st^{s-1}(1-t)^{n-s} + t^s (n-s)(1-t)^{n-s-1} \cdot (-1)$$

$$= t^{s-1}(1-t)^{n-s-1} \{s(1-t) - t(n-s)\} \tag{2.84}$$

where the first equality is due to the product rule for derivative: $(f \cdot g)' = f' \cdot g + f \cdot g'$. Observe that except for $t = 0$ and $t = 1$, there is only one stationary point t^* that satisfies:

$$s(1 - t^*) - t^*(n - s) = 0. \tag{2.85}$$

Consequently, the shape indeed resembles the one depicted in Figure 2.14, and therefore, the maximum point of this curve occurs at the stationary point:

$$t^* = \frac{s}{n} = \frac{\sum_{i=1}^{n} y_i}{n}. \tag{2.86}$$

The ML estimator is precisely as described above. Notice that this solution applies to a specific realization $\{y_i\}_{i=1}^{n}$. For the random process $\{Y_i\}_{i=1}^{n}$, the ML estimator should be written as:

$$\hat{p}_{\text{ML}} = \frac{s}{n} = \frac{\sum_{i=1}^{n} Y_i}{n}. \tag{2.87}$$

This solution makes intuitive sense. It coincides with what you might initially guess as the optimal estimate, namely the *sample mean*.

Law of Large Numbers (LLN) (Bernoulli 1713; Bertsekas & Tsitsiklis 2008) Now, a natural question that may arise is whether $\hat{p}_{\text{ML}}$ converges to the true value of p as the sample size n tends to infinity. It turns out that this is indeed the case, and the foundation for this assertion lies in a fundamental law in the history of probability, known as the Law of Large Numbers (LLN). Here's what the LLN states. Suppose we have a sequence of i.i.d. random variables $\{X_i\}_{i=1}^{n}$ with an expected value of μ and finite variance (often denoted by $\text{var}(X_i) < \infty$). Then, the sample mean of $\{X_i\}_{i=1}^{n}$ converges to its true mean μ in probability as n tends to infinity:

$$S_n := \frac{\sum_{i=1}^{n} X_i}{n} \overset{\text{in probability}}{\longrightarrow} \mathbb{E}[X_i] = \mu. \tag{2.88}$$

It's important to be cautious when stating "convergence in probability". Note that S_n is another random variable, as it is a function of the random variables $\{X_i\}_{i=1}^{n}$. Therefore, the convergent quantity is not guaranteed to be deterministic. To compare the potentially non-deterministic converged quantity with the deterministic value μ, we should make a *probabilistic* statement. In this context, "convergence in probability" means:

$$\mathbb{P}(|S_n - \mu| \geq \epsilon) \longrightarrow 0 \qquad \text{for any } \epsilon > 0. \tag{2.89}$$

Applying the LLN (2.88) to the Bernoulli process case of our interest, we can conclude:

$$\hat{p}_{\text{ML}} = \frac{\sum_{i=1}^{n} Y_i}{n} \xrightarrow{\text{in prob}} p. \tag{2.90}$$

Proof of the LLN (2.88) The proof of the Law of Large Numbers (LLN) is straightforward, especially when we leverage an inequality technique we have encountered, namely Chebyshev's inequality. By applying Chebyshev's inequality, we obtain:

$$\mathbb{P}(|S_n - \mu| \geq \epsilon) = \mathbb{P}(|S_n - \mathbb{E}[S_n]| \geq \epsilon)$$
$$\leq \frac{\text{var}(S_n)}{\epsilon^2} \tag{2.91}$$

where the first equality is due to $\mathbb{E}[S_n] = \frac{\mathbb{E}[X_1 + \cdots + X_n]}{n} = \frac{n\mu}{\mu} = \mu$. Notice that

$$\text{var}(S_n) = \text{var}\left(\frac{1}{n}\sum_{i=1}^{n} X_i\right)$$

$$= \mathbb{E}\left[\frac{1}{n^2}\left(\sum_{i=1}^{n} X_i\right)^2\right] - \left(\mathbb{E}\left[\frac{1}{n}\sum_{i=1}^{n} X_i\right]\right)^2$$

$$= \frac{1}{n^2}\left(\mathbb{E}\left[\left(\sum_{i=1}^{n} X_i\right)^2\right] - \left(\mathbb{E}\left[\sum_{i=1}^{n} X_i\right]\right)^2\right)$$

$$= \frac{1}{n^2}\text{var}\left(\sum_{i=1}^{n} X_i\right)$$

$$= \frac{1}{n^2}\sum_{i=1}^{n} \text{var}(X_i)$$

$$= \frac{\text{var}(X_1)}{n}$$

where the second and fourth equalities are due to the useful fact for the variance; the second-to-last equality comes from the independence of X_i's; and the last is because X_i's are identically distributed. Putting this to (2.91),

$$\mathbb{P}(|S_n - \mu| \geq \epsilon) \leq \frac{\text{var}(S_n)}{\epsilon^2}$$

$$= \frac{\text{var}(X_1)}{\epsilon^2 n} \longrightarrow 0 \qquad \text{as } n \to \infty.$$

Look ahead

Parameter estimation is a crucial problem that arises in various scenarios. One particularly significant problem that frequently surfaces across diverse domains is the estimation of Gaussian distribution. In our next discussion, we will delve into the MLE for Gaussian distribution.

2.6 ML: Gaussian Parameter Estimation

Recap

In the previous section, we have studied the ML principle in the context of a parameter estimation problem for the Bernoulli process. According to the Bernoulli parameter $p \in [0, 1]$, the i.i.d. samples $\{Y_i\}_{i=1}^n$ are generated. Given the samples, we wish to estimate the parameter p, assuming no prior knowledge on the statistics of p. We showed that the optimal estimate under the setting reduces to the ML estimate:

$$\hat{p}_{\text{opt}} = \hat{p}_{\text{ML}} = \arg \max_{t \in [0,1]} \underbrace{\mathbb{P}(Y_1 = y_1, \ldots, Y_n = y_n | p = t)}_{\text{likelihood}}$$

$$= \arg \max_{t \in [0,1]} t^{\sum_{i=1}^n y_i} (1 - t)^{n - \sum_{i=1}^n y_i} \tag{2.92}$$

where the last equality follows from the fact that $\{Y_i\}_{i=1}^n$ are i.i.d. $\sim \text{Bern}(t)$ given $p = t$. Finding the stationary point in the open interval $(0, 1)$ by taking the derivative w.r.t. t, we showed that the maximizer aligns with our intuition: the *sample mean*:

$$\hat{p}_{\text{ML}} = \frac{\sum_{i=1}^n Y_i}{n}.$$

With the aid of the Law of Large Numbers (LLN), we have also demonstrated that the ML estimate converges to the true parameter p (in probability) as the number n of observations approaches infinity:

$$\frac{\sum_{i=1}^n Y_i}{n} \xrightarrow{\text{in prob}} \mathbb{E}[Y_i] = p \quad \text{as } n \to \infty,$$

$$\text{i.e., } \mathbb{P}\left(\left| \frac{\sum_{i=1}^n Y_i}{n} - p \right| \geq \epsilon \right) \xrightarrow{\text{as } n \to \infty} 0 \quad \text{for any } \epsilon > 0.$$

We emphasized the significance of parameter estimation. It is a fundamental problem that frequently arises in statistics and machine learning. Notably, parameter estimation for Gaussian distribution is a common practical challenge, making it a topic of considerable interest across a diverse range of contexts. This topic is the focus of this section.

Outline

In this section, we will explore the optimal estimation for Gaussian distribution parameters: mean and variance. The section consists of four parts. We will commence by introducing the problem setting and establishing reasonable assumptions that facilitate the derivation of the optimal estimator. We will then show that the optimal estimator simplifies to the ML estimator. Next we will derive the ML estimator for the mean and the variance. Lastly we will demonstrate that the ML estimate converges to the true value as the number of observations grows.

Gaussian parameter estimation Consider the problem setting for Gaussian parameter estimation; see Figure 2.15. Let (μ, σ^2) represent the mean and the variance of Gaussian distribution. We have a set of i.i.d. samples $\{Y_i\}_{i=1}^n$ generated according to the parameters, and the samples serve as input for the estimation process. Just as in the Bernoulli case, we assume no prior knowledge about the statistics of (μ, σ^2).

As before, let us begin by computing the correct decision probability and determine our course of action in the absence of such prior knowledge.

Probability density function Since the interested entities to estimate are continuous values, the correct decision probability is always zero:

$$\mathbb{P}(\mu = \hat{\mu}, \sigma = \hat{\sigma}) = 0. \tag{2.93}$$

Hence, we should consider its *density* counterpart instead:

$$\lim_{\delta_1, \delta_2 \to 0} \frac{\mathbb{P}(\mu \in [\hat{\mu}, \hat{\mu} + \delta_1], \sigma \in [\hat{\sigma}, \hat{\sigma} + \delta_2])}{\delta_1 \delta_2} =: f_{\mu,\sigma}(\hat{\mu}, \hat{\sigma}). \tag{2.94}$$

Why do we divide by $\delta_1 \delta_2$ on the left-hand side? Keep in mind that the pdf is defined as:

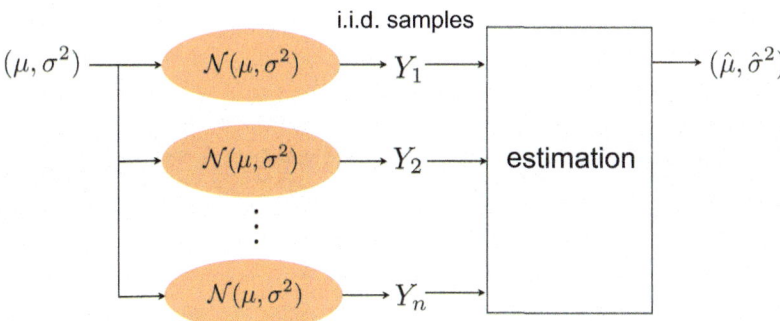

Fig. 2.15 Estimating the mean μ and the variance σ^2 of Gaussian distribution, given i.i.d. samples generated as per the Gaussian parameters

$$\mathbb{P}(\mu \in [\hat{\mu}, \hat{\mu} + \delta_1], \sigma \in [\hat{\sigma}, \hat{\sigma} + \delta_2]) = \int_{\hat{\mu}}^{\hat{\mu}+\delta_1} \int_{\hat{\sigma}}^{\hat{\sigma}+\delta_2} f_{\mu,\sigma}(a, b) da db. \qquad (2.95)$$

For infinitesimally small values of δ_1 and δ_2, the probability of μ being in the interval $[\hat{\mu}, \hat{\mu} + \delta_1]$ and σ in $[\hat{\sigma}, \hat{\sigma} + \delta_2]$ is approximately equal to $f_{\mu,\sigma}(\hat{\mu}, \hat{\sigma})\delta_1\delta_2$. This is why the pdf is defined as in (2.94). The conditioned version of the pdf is as follows:

$$f_{\mu,\sigma}(\hat{\mu}, \hat{\sigma}|Y_1 \in [y_1, y_1 + \delta_1], \ldots, Y_n \in [y_n, y_n + \delta_n]). \qquad (2.96)$$

In this context, we consider an interval for the event with respect to Y_i's because $\{Y_i\}_{i=1}^n$ constitute a continuous random process. For the sake of illustration, we assume that the intervals are of the same size: $\delta = \delta_1 = \cdots = \delta_n$. This assumption is acceptable, as we will later approach the limit as $\delta \to 0$.

The optimal estimator Based on (2.96), we define the optimal estimator as:

$$(\hat{\mu}_{\text{opt}}, \hat{\sigma}_{\text{opt}}) := \arg \max_{a,b \in \mathbb{R}} f_{\mu,\sigma}(a, b|Y_1 \in [y_1, y_1 + \delta], \ldots, Y_n \in [y_n, y_n + \delta])$$

$$(2.97)$$

where (a, b) are dummy variables for $(\hat{\mu}, \hat{\sigma})$. Let δ be an infinitesimally small value. Then, using the definition of conditional probability,

$$\begin{aligned} &f_{\mu,\sigma}(a, b|Y_1 \in [y_1, y_1 + \delta], \ldots, Y_n \in [y_n, y_n + \delta]) \\ &= \frac{f_{\mu,\sigma,Y_1,\ldots,Y_n}(a, b, [y_1, y_1 + \delta], \ldots, [y_n, y_n + \delta])}{\mathbb{P}(Y_1 \in [y_1, y_1 + \delta], \ldots, Y_n \in [y_n, y_n + \delta])} \qquad (2.98) \end{aligned}$$

where $f_{\mu,\sigma,Y_1,\ldots,Y_n}$ is the joint distribution w.r.t. (μ, σ^2) and $(Y_1, \ldots, Y_n)$ defined as:

$$\begin{aligned} &f_{\mu,\sigma,Y_1,\ldots,Y_n}(a, b, [y_1, y_1 + \delta], \ldots, [y_n, y_n + \delta]) \\ &:= \lim_{\epsilon \to 0} \frac{\mathbb{P}(\mu \in [a, a + \epsilon], \sigma \in [b, b + \epsilon], Y_1 \in [y_1, y_1 + \delta], \ldots, Y_n \in [y_n, y_n + \delta])}{\epsilon^2}. \end{aligned}$$

Once more, for the sake of illustration, we consider the same interval size: $\epsilon = \epsilon_1 = \epsilon_2$. By applying the definition of conditional probability to (2.98), we obtain:

$$\begin{aligned} &f_{\mu,\sigma}(a, b|Y_1 \in [y_1, y_1 + \delta], \ldots, Y_n \in [y_n, y_n + \delta]) \\ &= \frac{f_{\mu,\sigma,Y_1,\ldots,Y_n}(a, b, [y_1, y_1 + \delta], \ldots, [y_n, y_n + \delta])}{\mathbb{P}(Y_1 \in [y_1, y_1 + \delta], \ldots, Y_n \in [y_n, y_n + \delta])} \\ &= \frac{f_{\mu,\sigma}(a, b)\mathbb{P}(Y_1 \in [y_1, y_1 + \delta], \ldots, Y_n \in [y_n, y_n + \delta]|\mu = a, \sigma = b)}{\mathbb{P}(Y_1 \in [y_1, y_1 + \delta], \ldots, Y_n \in [y_n, y_n + \delta])} \\ &\approx \frac{f_{\mu,\sigma}(a, b)f(y_1, \ldots, y_n|\mu = a, \sigma = b)\delta^n}{f(y_1, \ldots, y_n)\delta^n} \end{aligned}$$

where the approximation is made assuming a very small value for δ. It is important to observe that $f(y_1, \ldots, y_n)$ is independent of the values of $(\mu, \sigma) = (a, b)$. Consequently, the optimal estimator can be expressed as:

$$(\hat{\mu}_{opt}, \hat{\sigma}^2_{opt}) = \arg \max_{a,b \in \mathbb{R}} f_{\mu,\sigma}(a, b) f(y_1, \ldots, y_n | \mu = a, \sigma = b). \qquad (2.99)$$

Since we are assuming no prior knowledge about $f_{\mu,\sigma}(a, b)$, a reasonable assumption we can make is:

$$f_{\mu,\sigma}(a, b) \text{ is } \textit{uniformly} \text{ distributed.} \qquad (2.100)$$

Maximum Likelihood Estimator (MLE) Since $f_{\mu,\sigma}(a, b)$ remains constant with respect to (a, b) under the uniform distribution assumption (2.100), we can simplify the optimal estimator as:

$$(\hat{\mu}_{opt}, \hat{\sigma}_{opt}) = (\hat{\mu}_{ML}, \hat{\sigma}_{ML}) = \arg \max_{a,b \in \mathbb{R}} \underbrace{f(y_1, \ldots, y_n | \mu = a, \sigma = b)}_{\text{likelihood}}. \qquad (2.101)$$

Again, as in the Bernoulli case, the optimal estimator is the MLE.

MLE derivation Now, let's derive the MLE (2.101). Utilizing the fact that $\{Y_i\}_{i=1}^n$ are i.i.d. with a normal distribution $\mathcal{N}(a, b^2)$ given $(\mu, \sigma) = (a, b)$, we can derive:

$$\begin{aligned} f(y_1, \ldots, y_n | \mu = a, \sigma = b) &= \prod_{i=1}^n f(y_i | \mu = a, \sigma = b) \\ &= \frac{1}{(\sqrt{2\pi}b)^n} e^{-\frac{1}{2b^2} \sum_{i=1}^n (y_i - a)^2}. \end{aligned}$$

This expression appears somewhat intricate, involving a complex term in the exponent. To simplify this expression, we can employ a common technique by using the natural logarithm function $\log(\cdot)$. By applying the logarithm to the right-hand side of the equation, we can simplify the intricate term while retaining the maximizer. Taking the logarithm yields:

$$\log \left(\frac{1}{(\sqrt{2\pi}b)^n} e^{-\frac{1}{2b^2} \sum_{i=1}^n (y_i - a)^2} \right) = -n \log(\sqrt{2\pi}b) - \frac{1}{2b^2} \sum_{i=1}^n (y_i - a)^2. \qquad (2.102)$$

Substituting this transformed objective function into (2.101), we obtain:

$$(\hat{\mu}_{ML}, \hat{\sigma}_{ML}) = \arg \max_{a,b \in \mathbb{R}} \underbrace{-n \log(\sqrt{2\pi}b) - \frac{1}{2b^2} \sum_{i=1}^{n} (y_i - a)^2}_{=:\mathcal{L}(a,b)} \qquad (2.103)$$

where $\mathcal{L}(a, b)$ is called the *log likelihood function*.

How do we identify the maximizer in (2.103)? If you have studied calculus or encountered this concept elsewhere, you might have heard or learned that a stationary point, where the derivative is zero, is often the maximizer. It turns out that this is indeed the case: the maximizer corresponds to the stationary point. In practice, a more rigorous understanding would involve knowledge of convex optimization, a subject you may have come across but may not be intimately familiar with. For now, we will rely on the statement that the maximizer is found at the stationary point. If you wish to delve deeper into this topic, you might consider taking a course (and/or reading a book) on convex optimization.

Relying upon the statement, we search for the stationary point by taking the derivative w.r.t. a and b:

$$\frac{d}{da}\mathcal{L}(a, b) = -\frac{1}{b^2} \sum_{i=1}^{n} (y_i - a) \cdot (-1);$$

$$\frac{d}{db}\mathcal{L}(a, b) = -\frac{n}{\sqrt{2\pi b}} \cdot \sqrt{2\pi} + \frac{1}{b^3} \sum_{i=1}^{n} (y_i - a)^2. \qquad (2.104)$$

Equating them to 0, we obtain the stationary point $(a^\star, b^\star)$ as below:

$$\frac{d}{da}\mathcal{L}(a, b) \bigg|_{a=a^\star, b=b^\star} = \frac{1}{b^{\star 2}} \sum_{i=1}^{n} (y_i - a^\star) = 0 \quad \longrightarrow \quad a^\star = \frac{\sum_{i=1}^{n} y_i}{n};$$

$$\frac{d}{db}\mathcal{L}(a, b) \bigg|_{a=a^\star, b=b^\star} = -n + \frac{1}{b^{\star 2}} \sum_{i=1}^{n} (y_i - a^\star)^2 = 0 \longrightarrow b^{\star 2} = \frac{1}{n} \sum_{i=1}^{n} (y_i - a^\star)^2.$$

This solution pertains to a specific set of observations $\{y_i\}_{i=1}^{n}$. When the random process $\{Y_i\}_{i=1}^{n}$ is fed into the estimator, the ML estimator can be expressed as:

$$\hat{\mu}_{ML} = \frac{\sum_{i=1}^{n} Y_i}{n}, \qquad \hat{\sigma}_{ML}^2 = \frac{1}{n} \sum_{i=1}^{n} (Y_i - \hat{\mu}_{ML})^2. \qquad (2.105)$$

Observe that the first expression corresponds precisely to the sample mean, while the second calculates the average of $(Y_i - \hat{\mu}_{ML})^2$, which can be interpreted as the sample variance. This interpretation aligns well with our intuition.

MLE in the limit of n   Just as in the Bernoulli case, a natural question emerges: Do $\hat{\mu}_{ML}$ and $\hat{\sigma}_{ML}^2$ converge to the true values μ and σ^2, respectively, as the number of observations n increases? To explore this, let's once again make use of the LLN. Applying the LLN to $\hat{\mu}_{\text{ML}}$, we obtain:

$$\hat{\mu}_{\text{ML}} = \frac{\sum_{i=1}^{n} Y_i}{n} \xrightarrow{\text{in prob}} \mathbb{E}[Y_i] = \mu \qquad \text{as } n \to \infty. \qquad (2.106)$$

Hence, the same mean converges to the true mean.

To assess the convergence of $\hat{\sigma}_{\text{ML}}^2$, we begin by expressing it as:

$$\hat{\sigma}_{\text{ML}}^2 = \frac{1}{n} \sum_{i=1}^{n} (Y_i - \hat{\mu}_{\text{ML}})^2$$

$$= \frac{1}{n} \sum_{i=1}^{n} Y_i^2 - \hat{\mu}_{\text{ML}}^2 \qquad (2.107)$$

where the second equality is due to $\hat{\mu}_{\text{ML}} = \frac{\sum_{i=1}^{n} Y_i}{n}$. Let's consider the first term in the second line of the above equation. Since $\{Y_i\}_{i=1}^{n}$ are i.i.d. random variables, their squared counterpart $\{Y_i^2\}_{i=1}^{n}$ are also i.i.d. This can be understood by noting that for independent random variables, say X and Y, any functions, for example, $f(X)$ and $g(Y)$, are also independent. Now, let's proceed to apply the LLN to $\{Y_i^2\}_{i=1}^{n}$. To do so, we first compute:

$$\mathbb{E}[Y_i^2] = \sigma^2 + (\mathbb{E}[Y_i])^2 = \sigma^2 + \mu^2. \qquad (2.108)$$

We also need to ensure the finiteness of $\text{var}(Y_i^2)$. To investigate this, consider:

$$\text{var}(Y_i^2) = \mathbb{E}[Y_i^4] - (\sigma^2 + \mu^2)^2 \qquad (2.109)$$

In fact, the computation of $\mathbb{E}[Y_i^4]$ is complicated, but it is doable. It turns out $\mathbb{E}[Y_i^4] = 3\sigma^4 + 6\mu^2\sigma^2 + \mu^4$. You can also exercise the moment calculation in Problem 6.5. Substituting this into the equation above, we have:

$$\text{var}(Y_i^2) = \mathbb{E}[Y_i^4] - (\sigma^2 + \mu^2)^2 = 2\sigma^4 + 4\mu^2\sigma^2. \qquad (2.110)$$

Consequently, it is indeed finite. As you might guess, the precise calculation of the variance is not of paramount importance. What truly matters is confirming its finiteness. Therefore, you need not concern yourself with the exact computation of $\mathbb{E}[Y_i^4]$. Simply remember that it is finite.

Applying the LLN to $\{Y_i^2\}_{i=1}^{n}$ together with (2.108),

$$\frac{\sum_{i=1}^{n} Y_i^2}{n} \xrightarrow{\text{in prob}} \mathbb{E}[Y_i^2] = \sigma^2 + \mu^2. \qquad (2.111)$$

Recall what you have learned from the course on Calculus. Suppose $\{a_i\}_{i=1}^{\infty}$ and $\{b_i\}_{i=1}^{\infty}$ are *convergent* sequences. Then, the limit of the sum of the two sequences converges to the sum of the two individual limits:

$$\lim_{n \to \infty} (a_n + b_n) = \lim_{n \to \infty} a_n + \lim_{n \to \infty} b_n. \tag{2.112}$$

Moreover, for any well-behaved functions, including the square function, the limit of the function applied to a convergent sequence is the same as the function of the limit. For example:

$$\lim_{n \to \infty} a_n^2 = \left(\lim_{n \to \infty} a_n \right)^2. \tag{2.113}$$

These statements are quite intuitive. If you don't recall the proof of these statements, you could attempt to derive them using the definition of convergence. If you've forgotten the definition of convergence, you might revisit the relevant content in your calculus studies or refer to resources like Wikipedia.

On the other hand, if you feel comfortable accepting these statements without the need for a formal proof, that's perfectly fine. You don't have to delve into them unless you plan to pursue more mathematically intensive work in your future career. In practice, many facts like those presented in (2.112) and (2.113) align with our intuition, so you need not fret about mastering this kind of hardcore mathematics.

It turns out the convergence facts ((2.112) and (2.113)) also hold for the *convergence in probability*. Again, let us not worry about the proof. Applying these to (2.107) together with (2.106), we get:

$$\sigma_{\text{ML}}^2 = \frac{1}{n} \sum_{i=1}^{n} Y_i^2 - \hat{\mu}_{\text{ML}}^2 \xrightarrow{\text{in prob}} (\sigma^2 + \mu^2) - \mu^2 = \sigma^2. \tag{2.114}$$

As expected, σ_{ML}^2 converges to the true value σ^2.

Look ahead

We have explored two applications of the ML principle: (i) Bernoulli parameter estimation; and (ii) Gaussian parameter estimation. Although we derived the optimal ML estimators, we have not yet discussed their performances, especially with respect to the number of observations n. It is reasonable to anticipate that the performances will enhance as the sample size n increases. In the upcoming section, we will demonstrate how performance evolves with n through analysis, and provide empirical validation using Python.

2.7 ML: Performance and **Python** Simulation

Recap

In the preceding section, we derived the optimal estimator for the mean and the variance of Gaussian distribution. Assuming no prior knowledge about the statistics of the parameters, the optimal estimator has been proved to be the ML estimator:

$$(\hat{\mu}_{\text{opt}}, \hat{\sigma}^2_{\text{opt}}) = (\hat{\mu}_{\text{ML}}, \hat{\sigma}^2_{\text{ML}}) = \arg\max_{a,b\in\mathbb{R}} \underbrace{f(y_1, \ldots, y_n | \mu = a, \sigma = b)}_{\text{likelihood}}$$

$$= \arg\max_{a,b\in\mathbb{R}} \frac{1}{(\sqrt{2\pi}b)^n} e^{-\frac{1}{2b^2}\sum_{i=1}^n (y_i - a)^2} \qquad (2.115)$$

$$= \arg\max_{a,b\in\mathbb{R}} -n\log(\sqrt{2\pi}b) - \frac{1}{2b^2}\sum_{i=1}^n (y_i - a)^2.$$

Here the second-to-last equality follows from the fact that Y_i's are i.i.d. $\sim \mathcal{N}(a, b^2)$ assuming $(\mu, \sigma) = (a, b)$. The last equality holds because applying the increasing function $\log(\cdot)$ does not affect the optimal solution. By finding the stationary point through differentiation, we have demonstrated that the maximizer aligns with our intuition: the sample mean and sample variance:

$$\hat{\mu}_{\text{ML}} = \frac{\sum_{i=1}^n Y_i}{n}, \quad \hat{\sigma}^2_{\text{ML}} = \frac{1}{n}\sum_{i=1}^n (Y_i - \hat{\mu}_{\text{ML}})^2.$$

By applying the Law of Large Numbers (LLN), we have also established that the ML estimate converges to the true value (in probability) as the number n of observations approaches infinity. Specifically, as $n \to \infty$,

$$\hat{\mu}_{\text{ML}} = \frac{\sum_{i=1}^n Y_i}{n} \xrightarrow{\text{in prob}} \mathbb{E}[Y_i] = \mu,$$

$$\hat{\sigma}^2_{\text{ML}} = \frac{1}{n}\sum_{i=1}^n (Y_i - \hat{\mu}_{\text{ML}})^2 = \frac{1}{n}\sum_{i=1}^n Y_i^2 - \hat{\mu}^2_{\text{ML}} \xrightarrow{\text{in prob}} (\sigma^2 + \mu^2) - \mu^2 = \sigma^2.$$

While it is reasonable to anticipate an enhancement in the estimator's performance with increasing sample size n, we have not yet provided verification.

Outline
In this section, we will verify this through both analysis and simulation. This section comprises three parts. We will start by introducing a metric that quantifies the accuracy of the estimator. Subsequently, we will analyze the performance of the estimator and illustrate how its behavior changes with the sample size. Finally, we will validate the analysis through a simulation using Python.

A performance measure The performance metric that we introduced earlier in the context of MAP estimators is the probability of error. Applying the same metric into the problem setting at hand (e.g., the μ-estimation problem), the error probability reads:

$$P_e := \mathbb{P}(\hat{\mu} \neq \mu). \tag{2.116}$$

However, this metric is not suitable for the estimation setting. The issue arises from the fact that the error probability P_e is always zero, regardless of the quality of the estimate. Consequently, this metric does not enable us to differentiate between good and poor estimators. The primary reason for P_e consistently being zero is that the variable of interest μ is a continuous value.

In the context of estimation, particularly when one seeks to infer a continuous quantity, a more appropriate and well-established metric is the mean squared error (MSE). The MSEs for the mean and variance estimates are defined respectively as:

$$\mathsf{MSE}(\hat{\mu}) := \mathbb{E}\left[(\hat{\mu} - \mu)^2\right]; \qquad \mathsf{MSE}(\hat{\sigma}^2) := \mathbb{E}\left[(\hat{\sigma}^2 - \sigma^2)^2\right]. \tag{2.117}$$

In contrast to the error probability, the MSE is never zero unless the estimate perfectly matches the true value, which rarely happens in practice.

MSE of the mean estimate Let us derive the MSE of the ML estimate. First focus on the mean estimate $\hat{\mu}_{\mathsf{ML}}$:

$$\mathsf{MSE}(\hat{\mu}_{\mathsf{ML}}) = \mathbb{E}\left[(\hat{\mu}_{\mathsf{ML}} - \mu)^2\right]. \tag{2.118}$$

Putting $\hat{\mu}_{\mathsf{ML}} = \frac{\sum_{i=1}^{n} Y_i}{n}$ into the above and using the fact that Y_i's are i.i.d. $\sim \mathcal{N}(\mu, \sigma^2)$, we obtain:

$$\text{MSE}(\hat{\mu}_{\text{ML}}) = \mathbb{E}\left[\left(\frac{1}{n}\sum_{i=1}^{n}Y_i - \mu\right)^2\right]$$

$$= \mathbb{E}\left[\left(\frac{1}{n}\sum_{i=1}^{n}(Y_i - \mu)\right)^2\right]$$

$$= \frac{1}{n^2}\mathbb{E}\left[\left(\sum_{i=1}^{n}(Y_i - \mu)\right)^2\right] \qquad (2.119)$$

$$= \frac{1}{n^2}\sum_{i=1}^{n}\mathbb{E}\left[(Y_i - \mu)^2\right]$$

$$= \frac{1}{n}\sigma^2$$

where the third equality is due to the homogeneity property of expectation; the second-to-last equality follows from the fact that the variance of the sum of the independent random variables $(Y_i - \mu)$'s are the sum of their individual variances; and the last equality is because each Y_i has the variance of σ^2.

MSE of the variance estimate Next, we consider the MSE of the variance estimate $\hat{\sigma}^2_{\text{ML}}$:

$$\text{MSE}(\hat{\sigma}^2_{\text{ML}}) = \mathbb{E}\left[(\hat{\sigma}^2_{\text{ML}} - \sigma^2)^2\right]$$
$$= \mathbb{E}\left[\hat{\sigma}^4_{\text{ML}}\right] - 2\sigma^2\mathbb{E}\left[\hat{\sigma}^2_{\text{ML}}\right] + \sigma^4 \qquad (2.120)$$

where the second equality is due to the linearity of expectation. For computational convenience, without loss of generality,[1] we will assume that the mean is zero. The reason that we can assume $\mu = 0$ is the following. Suppose we have a non-zero mean $\mu \neq 0$. We can then always subtract μ from Y_i's so that the mean of the subtracted version $Y_i - \mu$ is zero. Since $(Y_i - \mu)$'s are shifted by a constant μ, the variance remains the same as that of Y_i's.

Under the mean-zero assumption, we first compute $\mathbb{E}[\hat{\sigma}^2_{\text{ML}}]$. Applying $\hat{\mu}_{\text{ML}} = \frac{\sum_{i=1}^{n}Y_i}{n}$ and $\hat{\sigma}^2_{\text{ML}} = \frac{1}{n}\sum_{i=1}^{n}(Y_i - \hat{\mu}_{\text{ML}})^2$, we get:

---

[1] The phrase "without loss of generality" indicates that we can simplify the general case into a simpler version with appropriate adjustments.

$$\mathbb{E}[\hat{\sigma}_{\mathsf{ML}}^2] = \mathbb{E}\left[\frac{1}{n}\sum_{i=1}^{n}(Y_i - \hat{\mu}_{\mathsf{ML}})^2\right]$$

$$= \frac{1}{n}\mathbb{E}\left[\sum_{i=1}^{n}Y_i^2 - 2\hat{\mu}_{\mathsf{ML}}\sum_{i=1}^{n}Y_i + n\hat{\mu}_{\mathsf{ML}}^2\right]$$

$$= \frac{1}{n}\mathbb{E}\left[\sum_{i=1}^{n}Y_i^2 - n\hat{\mu}_{\mathsf{ML}}^2\right] \qquad (2.121)$$

$$= \mathbb{E}\left[Y_1^2\right] - \mathbb{E}\left[\hat{\mu}_{\mathsf{ML}}^2\right]$$

$$= \sigma^2 - \frac{1}{n^2}\mathbb{E}\left[\sum_{i=1}^{n}Y_i^2 + \sum_{i \neq j}Y_i Y_j\right]$$

$$= \left(1 - \frac{1}{n}\right)\sigma^2$$

where the third equality is due to $\sum_{i=1}^{n}Y_i = n\hat{\mu}_{\mathsf{ML}}$; the fourth equality follows from the fact that Y_i's are i.i.d., and so are Y_i^2's; and the last two equalities come from the mean-zero assumption.

Let us now focus on $\mathbb{E}[\hat{\sigma}_{\mathsf{ML}}^4]$ in (2.120):

$$\mathbb{E}[\hat{\sigma}_{\mathsf{ML}}^4] = \frac{1}{n^2}\mathbb{E}\left[\left\{\sum_{i=1}^{n}(Y_i - \hat{\mu}_{\mathsf{ML}})^2\right\}^2\right]$$

$$= \frac{1}{n^2}\left[\mathrm{var}\left(\sum_{i=1}^{n}(Y_i - \hat{\mu}_{\mathsf{ML}})^2\right) + \left(\mathbb{E}\left[\sum_{i=1}^{n}(Y_i - \hat{\mu}_{\mathsf{ML}})^2\right]\right)^2\right] \qquad (2.122)$$

where the second equality is due to the variance property. Here, we focus on two key quantities of interest: the variance and the mean of $\sum_{i=1}^{n}(Y_i - \hat{\mu}_{\mathsf{ML}})^2$. Ideally, you might hope that $(Y_i - \hat{\mu}_{\mathsf{ML}})$'s are i.i.d., similar to Y_i's. If that were the case, it would simplify the computation significantly. Unfortunately, this is not the situation, making the derivation more complex. For instance, when $n = 2$, $Y_1 - \hat{\mu}_{\mathsf{ML}} = \frac{Y_1 - Y_2}{2}$ and $Y_2 - \hat{\mu}_{\mathsf{ML}} = -\frac{Y_1 - Y_2}{2}$. These are indeed dependent, actually being anti-correlated each other.

By employing non-trivial linear-algebra techniques, we can represent $\sum_{i=1}^{n}(Y_i - \hat{\mu}_{\mathsf{ML}})^2$ as the sum of another set of i.i.d. random variables, expressed as:

$$\sum_{i=1}^{n}(Y_i - \hat{\mu}_{\mathsf{ML}})^2 = X_2^2 + X_3^2 + \cdots + X_n^2 \qquad (2.123)$$

where X_i's are i.i.d. $\sim \mathcal{N}(0, \sigma^2)$. One noticeable observation to highlight is that the number of X_i's contributed in the summation is $n - 1$ instead of n. The derivation of (2.123) is difficult, as it requires non-trivial matrix manipulation and significant linear-algebra techniques such as eigenvalue decomposition (Strang 2022). As such, we will not provide a detailed proof here. In Problem 6.3, however, you will have the opportunity to prove (2.123) and figure out how X_i's are expressed in terms of Y_i's. Applying (2.123) to (2.122) yields:

$$
\begin{aligned}
\mathbb{E}[\hat{\sigma}_{\text{ML}}^4] &= \frac{1}{n^2} \left[\text{var} \left(\sum_{i=2}^{n} X_i^2 \right) + \left(\mathbb{E} \left[\sum_{i=2}^{n} X_i^2 \right] \right)^2 \right] \\
&= \frac{1}{n^2} \left[(n-1)\text{var} \left(X_2^2 \right) + ((n-1)\mathbb{E} \left[X_2^2 \right])^2 \right] \\
&= \frac{1}{n^2} \left[(n-1) \left(\mathbb{E}[X_2^4] - (\mathbb{E}[X_2^2])^2 \right) + (n-1)^2\sigma^4 \right]
\end{aligned}
\tag{2.124}
$$

where the second equality follows from the fact that X_i's are i.i.d. $\sim \mathcal{N}(0, \sigma^2)$; and the last equality is a consequence of the variance property. Now, let's calculate the fourth moment of X_2: $\mathbb{E}[X_2^4]$. Utilizing an interesting moment generating function, which will be explored in depth in Problem 6.5, one can demonstrate that:

$$
\mathbb{E}[X_2^4] = 3\sigma^4.
\tag{2.125}
$$

Applying this into (2.124) then yields:

$$
\begin{aligned}
\mathbb{E}[\hat{\sigma}_{\text{ML}}^4] &= \frac{1}{n^2} \left[(n-1) \left(\mathbb{E}[X_2^4] - (\mathbb{E}[X_2^2])^2 \right) + (n-1)^2\sigma^4 \right] \\
&= \frac{1}{n^2} \left[2(n-1)\sigma^4 + (n-1)^2\sigma^4 \right] \\
&= \frac{n^2-1}{n^2}\sigma^4.
\end{aligned}
\tag{2.126}
$$

Substituting (2.121) and (2.126) into (2.120), we finally arrive at:

$$
\begin{aligned}
\text{MSE}(\hat{\sigma}_{\text{ML}}^2) &= \mathbb{E} \left[\hat{\sigma}_{\text{ML}}^4 \right] - 2\sigma^2 \mathbb{E} \left[\hat{\sigma}_{\text{ML}}^2 \right] + \sigma^4 \\
&= \frac{n^2-1}{n^2}\sigma^4 - 2\sigma^2 \cdot \left(1 - \frac{1}{n} \right) \sigma^2 + \sigma^4 \\
&= \frac{2n-1}{n^2}\sigma^4.
\end{aligned}
\tag{2.127}
$$

Python simulation We will empirically validate the MSE analysis for the mean (2.119) and variance (2.127) estimates. To begin, we below implement a Python script that calculates the MSE of the mean estimate and compares it with the corresponding analytical derivation (2.119).

```python
import numpy as np
import matplotlib.pyplot as plt

# Gaussian parameters
mu=0
sigma2=0.1
# number of independent trials
N=10000
# number of observations
n=np.arange(1,31,2) # [1,3,5,...,29]

def MSE_mu(n,N):
    # N: number of independent trials
    # n: number of observations
    SE_mu=np.zeros(N)
    for i in range(N):
        # Generate i.i.d. samples
        Y=mu + np.sqrt(sigma2)*np.random.randn(1,n)
        # ML estimate: sample mean
        YML=Y.mean()
        # Compute the square error
        SE_mu[i]= np.square(YML-mu)
    # Compute MSE
    return SE_mu.mean()

MSE_emp=np.zeros(len(n))
for i in range(len(n)): MSE_emp[i]=MSE_mu(n[i],N)
# MSE by theory
MSE_theory = sigma2/n

plt.figure(figsize=(4,4),dpi=500)
plt.plot(n,MSE_emp,'b*:',label='MSE (empirical)')
plt.plot(n,MSE_theory,color='red',label='MSE (theory)')
plt.xlabel('$n$ (number of observations)')
plt.ylabel('Mean Square Error')
plt.title('MSE of the mean estimate')
plt.legend()
plt.yscale('log')
plt.show()
```

Figure 2.16 illustrates the empirical MSE for $\hat{\mu}_{ML}$ alongside its analytical counterpart. It is evident that with a significantly large number of independent trials ($N = 10,000$), the empirical MSE precisely aligns with its analytical counterpart.

Next, we repeat the same process, this time focusing on the variance estimate. The code implementation for this is provided below.

```python
import numpy as np
import matplotlib.pyplot as plt

# Gaussian parameters
mu=0
sigma2=0.1
```

```
# number of independent trials
N=10000
# number of observations
n=np.arange(1,31,2) # [1,3,5,...,29]

def MSE_var(n,N):
    # N: number of independent trials
    # n: number of observations
    SE_var=np.zeros(N)
    for i in range(N):
        # Generate i.i.d. samples
        Y=mu + np.sqrt(sigma2)*np.random.randn(1,n)
        # sample mean
        YML=Y.mean()
        # ML estimate: sample variance
        s2ML= np.square(Y-YML).mean()
        # Compute the square error
        SE_var[i]= np.square(s2ML-sigma2)
    # Compute MSE
    return SE_var.mean()

MSE_emp=np.zeros(len(n))
for i in range(len(n)): MSE_emp[i]=MSE_var(n[i],N)
# MSE by theory
MSE_theory = (2*n-1)/n**2*sigma2**2
```

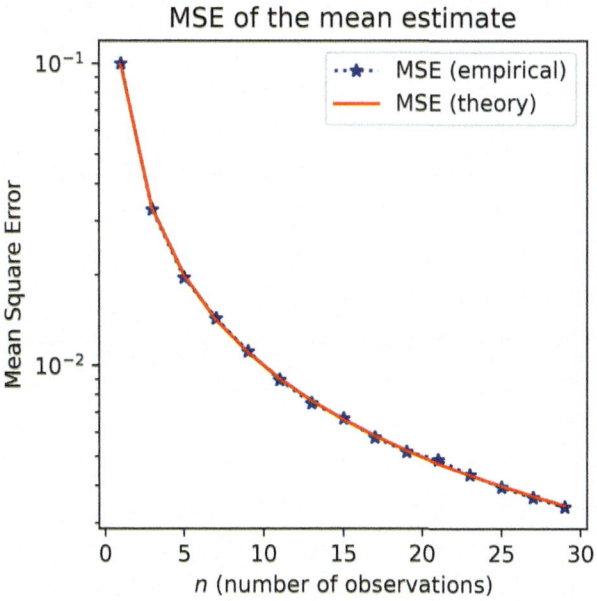

Fig. 2.16 The mean square error of the mean estimate as a function of n

```
plt.figure(figsize=(4,4),dpi=500)
plt.plot(n,MSE_emp,'b*:',label='MSE (empirical)')
plt.plot(n,MSE_theory,color='red',label='MSE (theory)')
plt.xlabel('$n$ (number of observations)')
plt.ylabel('Mean Square Error')
plt.title('MSE of the variance estimate')
plt.legend()
plt.yscale('log')
plt.show()
```

In Figure 2.17, we also verify that the empirical MSE closely aligns with the
theoretical MSE (2.120).

Look ahead

In Part II, we have thus far studied several concepts and key principles: (i)
random processes, Bernoulli process, Gaussian process, Markov process; (ii)
the MAP principle in the context of cancer prediction; (iii) the ML principle
and its application to Bernoulli and Gaussian parameter estimation; and (iv)
the Law of Large Numbers.

As mentioned earlier, prior to embarking on Part III (dedicated to applications), we need to study one more important theorem that plays a crucial role
for Gaussian noise modeling. That is, the *Central Limit Theorem (CLT)*. In the
upcoming section, we will investigate the CLT.

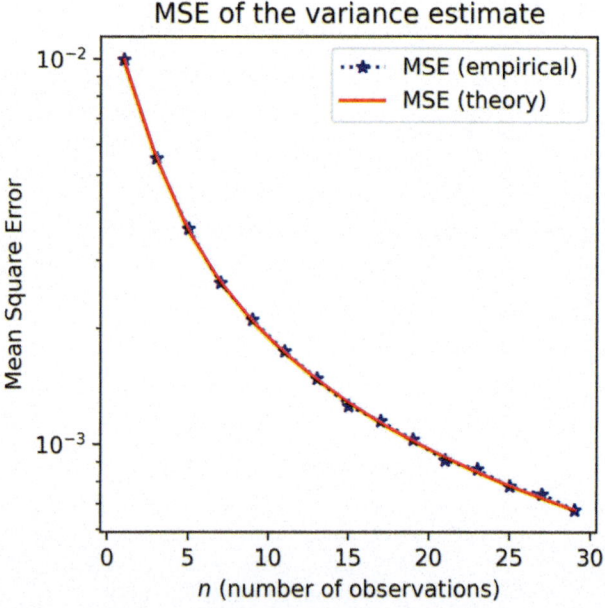

Fig. 2.17 The mean square error of the variance estimate as a function of n

2.8   Central Limit Theorem

Recap
Over the past two sections, we have studied the MLE for Gaussian parameter estimation. Given the i.i.d. Gaussian samples $\{Y_i\}_{i=1}^n \sim \mathcal{N}(\mu, \sigma^2)$, the optimal ML estimates were derived as:

$$\hat{\mu}_{\text{ML}} = \frac{\sum_{i=1}^n Y_i}{n}; \quad \hat{\sigma}^2_{\text{ML}} = \frac{1}{n}\sum_{i=1}^n (Y_i - \hat{\mu}_{\text{ML}})^2.$$

We next derived the MSEs of the ML estimates alongside their empirical verification based on **Python** simulation:

$$\begin{aligned} \text{MSE}(\hat{\mu}_{\text{ML}}) &= \mathbb{E}\left[(\hat{\mu}_{\text{ML}} - \mu)^2\right] = \frac{1}{n}\sigma^2; \\ \text{MSE}(\hat{\sigma}^2_{\text{ML}}) &= \mathbb{E}\left[(\hat{\sigma}^2_{\text{ML}} - \sigma^2)^2\right] = \frac{2n-1}{n^2}\sigma^4. \end{aligned} \quad (2.128)$$

We previously noted that, in preparation for Part III, which is dedicated to applications, we must delve into one more critical theorem that is essential for modeling Gaussian noise, a common occurrence in various contexts. That is, the *Central Limit Theorem (CLT)*.

Outline
In this section, we will investigate the CLT in depth. The section consists of three parts. First, we will present the precise statement of the CLT. Next, we will delve into two critical claims that hold significant importance in proving the theorem. Claim 1 pertains to the sum of multiple independent continuous random variables, and Claim 2 facilitates a simpler expression of a seemingly complex probability density function through the Laplace transform. Finally, we will employ these two claims to prove the theorem.

A rough statement of the CLT Consider the sequence $\{X_i\}_{i=1}^\infty$, which is an i.i.d. process with $\mathbb{E}[X_i] = \mu$ and a finite variance $\text{var}(X_i) = \sigma^2 < \infty$. A rough statement of the CLT is that in the limit of $n \to \infty$,

$$X_1 + X_2 + \cdots + X_n \text{ converges to a } \textit{Gaussian} \text{ random variable.}$$

Here the mean of the sum should read $\mathbb{E}[X_1 + \cdots + X_n] = n\mathbb{E}[X_1] = n\mu$ and its variance is:

$$\text{var}(X_1 + \cdots + X_n) = \text{var}(X_1) + \cdots + \text{var}(X_n)$$
$$= n\text{var}(X_1) = n\sigma^2 \tag{2.129}$$

where the first equality is due to the independence of $\{X_i\}_{i=1}^n$. A remarkable observation in the CLT is that the probability density function of the convergent random variable is *Gaussian*, regardless of the initial distribution of the original random process.

A precise statement of the CLT (Pólya 1920) The above statement is a bit rough in the following two aspects. Firstly, the converged random variable has an infinite mean and variance. Secondly, the notion of convergence to another random variable lacks a rigorous definition. To render the converged random variable finite in both mean and variance, a common approach is to employ a process called *normalization*, which involves subtracting the mean and dividing by the standard deviation:

$$Z_n := \frac{X_1 + X_2 + \cdots + X_n - n\mu}{\sqrt{n\sigma^2}}. \tag{2.130}$$

By applying this normalization process, we obtain $\mathbb{E}[Z_n] = 0$ and $\text{var}(Z_n) = 1$. With this in mind, the statement of the CLT becomes:

$$Z_n = \frac{X_1 + X_2 + \cdots + X_n - n\mu}{\sqrt{n\sigma^2}} \xrightarrow{\text{in distribution}} Z \sim \mathcal{N}(0, 1). \tag{2.131}$$

Here what it means by "convergence in distribution" is that the cumulative density distribution of Z_n is the same as that of Z in the limit:

$$\lim_{n \to \infty} F_{Z_n}(z) = \int_{-\infty}^{z} \frac{1}{\sqrt{2\pi}} e^{-\frac{1}{2}t^2} dt \quad \forall z \in \mathbb{R}. \tag{2.132}$$

A side note: The choice of defining convergence in distribution with respect to the cumulative distribution function (cdf) rather than the probability density function (pdf) is due to some peculiar and uncommon situations in which the cdf converges while the pdf does not. However, for the sake of simplicity in our illustration, we will use the pdf instead of the cdf. If you desire a more detailed understanding of this topic, it may be necessary to take advanced mathematics courses such as "measure theory".

Recall the notion of convergence in probability, which pertains to the probabilistic comparison between a random variable and a deterministic converged quantity. In contrast, convergence in distribution deals with the comparison of distribution between two random variables (like Z_n and Z in the above case). You may now understand why we have multiple definitions of convergence when working with random processes (not deterministic sequences).

An equivalent statement of the CLT There is a simpler and equivalent representation of (2.131) that we find more preferable. This expression is grounded in the following observation:

$$\underbrace{\frac{X_1 + X_2 + \cdots + X_n - n\mu}{\sqrt{n\sigma^2}}}_{} = \underbrace{\frac{X_1 - \mu}{\sqrt{n\sigma^2}}}_{=:X_1'} + \underbrace{\frac{X_2 - \mu}{\sqrt{n\sigma^2}}}_{=:X_2'} + \cdots + \underbrace{\frac{X_n - \mu}{\sqrt{n\sigma^2}}}_{=:X_n'}. \quad (2.133)$$

Here X_i' is a shifted version of X_i. Consequently, the sequence $\{X_i'\}$ remains i.i.d., but with a mean of $\mathbb{E}[X_i'] = 0$ and a variance of:

$$\begin{aligned}
\text{var}(X_i') &= \text{var}\left(\frac{X_1 - \mu}{\sqrt{n\sigma^2}}\right) \\
&= \frac{1}{n\sigma^2}\text{var}\,(X_1 - \mu) \\
&= \frac{\sigma^2}{n\sigma^2} = \frac{1}{n}
\end{aligned}$$

where the second equality is due to $\text{var}(cX) = \mathbb{E}[(cX)^2] - (\mathbb{E}[cX])^2 = c^2\text{var}(X)$ for any constant c and a random variable X; and the third equality is because $\text{var}(X_1 - \mu) = \text{var}(X_1) = \sigma^2$ (a constant shift does not alter the variance). Hence, in terms of $\{X_i'\}_{i=1}^\infty$, the CLT can be expressed as:

$$X_1' + X_2' + \cdots + X_n' \xrightarrow{\text{in dist}} Z \sim \mathcal{N}(0, 1). \quad (2.134)$$

For notational simplicity, let us re-use the simpler non-prime notation: $\{X_i\}_{i=1}^\infty$. With the simpler notation, the CLT then says:

$$Z_n := X_1 + X_2 + \cdots + X_n \xrightarrow{\text{in dist}} Z \sim \mathcal{N}(0, 1). \quad (2.135)$$

where $\{X_i\}_{i=1}^\infty$ is i.i.d. with $\mathbb{E}[X_i] = 0$ and $\text{var}(X_i) = \frac{1}{n}$. Here we will prove the equivalent statement (2.135), i.e.,

$$\lim_{n \to \infty} f_{Z_n}(z) = \frac{1}{\sqrt{2\pi}}e^{-\frac{1}{2}z^2} \quad \forall z \in \mathbb{R}. \quad (2.136)$$

There are two important claims that help streamlining the proof. So we will first investigate the two claims and then utilize them to complete the proof (2.136).

Claim 1 Observe in (2.135) that there are multiple summations involved in the relationship between Z_n and $\{X_i\}_{i=1}^n$. Ultimately, our goal is to determine the pdf of Z_n. The first claim pertains to the pdf of the sum of several independent random variables. To illustrate, let us consider just two random variables, say X and Y. Define $Z = X + Y$. Recall the case of *discrete* random variables where the pmf of the sum of two independent discrete random variables is expressed as their *convolution*. Claim 1 asserts that this concept also applies to *continuous* random variables:

$$\text{(Claim 1): } f_Z(z) = (f_X * f_Y)(z) := \int_{-\infty}^{\infty} f_X(x) f_Y(z-x) dx. \qquad (2.137)$$

The proof of this claim is not particularly challenging, so we will skip it here. Instead, you will have an opportunity to prove it in Problem 6.7.

Claim 2 Recall the interested random variable Z_n:

$$Z_n = X_1 + X_2 + \cdots + X_n.$$

Applying Claim 1 (2.137) many times, we obtain:

$$f_{Z_n}(z) = (f_{X_1} * f_{X_2} * \cdots * f_{X_n})(z). \qquad (2.138)$$

We are dealing with numerous convolutions. The convolution formula is complicated as the term "convoluted" implies. Consequently, the expression for $f_{Z_n}(z)$ can become highly complex. This is where Claim 2 comes into play. Claim 2 states that convolutions can be significantly simplified when considered in the context of the Laplace transform. To gain a more detailed understanding of this, let us examine the scenario where there are only two independent random variables, say X and Y, and we have $Z = X + Y$. Claim 2 suggests that the Laplace transform $F_Z(s)$ of $f_Z(z)$ can be expressed as the product of the individual Laplace transforms ($F_X(s)$, $F_Y(s)$) w.r.t. ($f_X(x)$, $f_Y(y)$):

$$\text{(Claim 2): } F_Z(s) = F_X(s) F_Y(s) \qquad (2.139)$$

where the Laplace transform is defined as:

$$F_Z(s) := \int_{-\infty}^{+\infty} e^{-sz} f_Z(z) dz. \qquad (2.140)$$

The proof of this claim is straightforward and relies on the "change of variable" technique, which we have used several times before. Hence, we will omit the proof here, but you can check it in Problem 6.7.

Setup for the proof of the CLT We are now ready to prove the CLT (2.135). Applying Claim 2 into (2.138) many times (more precisely, $n - 1$ times), we get:

$$F_{Z_n}(s) = F_{X_1}(s) F_{X_2}(s) \cdots F_{X_n}(s)$$
$$= \left[F_{X_1}(s) \right]^n \tag{2.141}$$

where the second equality is because each of $\{X_i\}_{i=1}^{\infty}$ is identically distributed. For the sake of simplicity, we will not concern ourselves with an exceptionally rare and practically irrelevant scenario where the Laplace transform does not exist.

Now, let's consider how to proceed with (2.141). More precisely, how can we compute $F_{Z_n}(s)$ given what we know: $\mathbb{E}[X_1] = 0$ and $\mathbb{E}[X_1^2] = \frac{1}{n}$? These moment information (first and second moments) play a crucial role in the computation. A key observation is that these moments appear as coefficients in the *Taylor series expansion* of $F_{X_1}(s)$. It turns out that this expansion provides an explicit expression for $F_{Z_n}(s)$, thereby shedding light on the pdf of Z_n. To understand this, let's attempt to derive the Taylor series expansion of $F_{X_1}(s)$.

Taylor series expansion (Stewart 2015) We first need to compute the kth derivative of $F_{X_1}(s)$:

$$\begin{aligned}
F_{X_1}^{(k)}(s) &:= \frac{d^k F_{X_1}(s)}{ds^k} \\
&= \frac{d^k}{ds^k} \left\{ \int_{-\infty}^{\infty} e^{-sx} f_{X_1}(x) dx \right\} \\
&= \int_{-\infty}^{\infty} (-1)^k x^k e^{-sx} f_{X_1}(x) dx
\end{aligned} \tag{2.142}$$

where the second equality is due to the definition of the Laplace transform: $F_{X_1}(s) := \int_{-\infty}^{+\infty} e^{-sx} f_{X_1}(x) dx$; and the third equality is because of the interchangeability of integration and differentiation (let's not worry about rare situations where this interchangeability does not hold). By applying the Taylor expansion at $s = 0$, we obtain:

$$F_{X_1}(s) = \sum_{k=0}^{\infty} \frac{F_{X_1}^{(k)}(0)}{k!} s^k. \tag{2.143}$$

Plugging $s = 0$ into (2.142), we get:

$$F_{X_1}^{(k)}(0) = \int_{-\infty}^{\infty} (-1)^k x^k f_{X_1}(x) dx = (-1)^k \mathbb{E}[X_1^k].$$

This together with (2.143) yields:

$$F_{X_1}(s) = \sum_{k=0}^{\infty} \frac{(-1)^k \mathbb{E}[X_1^k]}{k!} s^k. \tag{2.144}$$

Applying $\mathbb{E}[X_1] = 0$ and $\mathbb{E}[X_1^2] = \frac{1}{n}$ to the above yields:

$$F_{X_1}(s) = 1 + \frac{1}{2n} s^2 + \sum_{k=3}^{\infty} \frac{(-1)^k \mathbb{E}[X_1^k]}{k!} s^k. \tag{2.145}$$

What can we say about $\mathbb{E}[X_1^k]$ for $k \geq 3$ in the above? Observe that $\mathbb{E}[X_1^k] = \mathbb{E}[(X_1^2)^{\frac{k}{2}}]$ and $\mathbb{E}[X_1^2] = \frac{1}{n}$ scale like $\frac{1}{n}$. It turns out this leads to the fact that $\mathbb{E}[(X_1^2)^{\frac{k}{2}}]$ decays at a rate of $\frac{1}{n^{k/2}}$, which exhibits a faster rate of decay for $k \geq 3$ compared to $\frac{1}{n}$. This scaling leads to:

$$\lim_{n \to \infty} F_{Z_n}(s) = \lim_{n \to \infty} \left(1 + \frac{1}{2n} s^2 + \sum_{k=3}^{\infty} \frac{(-1)^k \mathbb{E}[X_1^k]}{k!} s^k \right)^n$$
$$= \lim_{n \to \infty} \left(1 + \frac{1}{2n} s^2 \right)^n \tag{2.146}$$
$$= e^{\frac{s^2}{2}}$$

where the last equality is due to the fact that

$$e^x = \lim_{n \to \infty} \left(1 + \frac{x}{n} \right)^n. \tag{2.147}$$

The rigorous proof for the second equality in (2.146) has been omitted. Instead, we provide the intuition behind it: $\mathbb{E}[(X_1^2)^{\frac{k}{2}}]$ decays at a rate of $\frac{1}{n^{k/2}}$, which exhibits a much faster decay than $\frac{1}{n}$ for $k \geq 3$. The rigorous proof would involve advanced mathematics, which might introduce unnecessary complexity without adding significant insights.

Proof of the CLT From (2.146), what can we say about $f_{Z_n}(z)$ in the limit of n? To figure this out, we need to do the Laplace *inverse* transform:

$$f_{Z_n}(z) = \text{InverseLaplace}(F_{Z_n}(s))(z) := \frac{1}{2\pi i} \lim_{T \to \infty} \int_{\text{Re}(s)-iT}^{\text{Re}(s)+iT} F_{Z_n}(s)e^{sz}ds$$

(2.148)

where $\text{Re}(s)$ denotes the real-component value in s. But you may feel headache because the inverse transform looks very complicated. Therefore, we will adopt a more intuitive approach: guess-and-check. One advantageous aspect of the Laplace transform is that it is a one-to-one mapping. Moreover, you can easily calculate the following:

$$\text{Laplace Transform} \left(\frac{1}{\sqrt{2\pi}} e^{-\frac{z^2}{2}} \right) = e^{\frac{s^2}{2}}.$$

This together with (2.146) gives:

$$\lim_{n \to \infty} f_{Z_n}(z) = \frac{1}{\sqrt{2\pi}} e^{-\frac{z^2}{2}}, z \in \mathbb{R}.$$

(2.149)

This is indeed the Gaussian distribution with mean 0 and variance 1. So this completes the proof of the CLT.

Look ahead

In Part I, we delved into several fundamental concepts in probability: the sample space, events, conditional probability, total probability law, independence, and random variables. Part II expanded our understanding with deeper concepts and key principles: random processes, the MAP principle, the ML principle, the Law of Large Numbers, and the Central Limit Theorem.

The overarching objective of Part III is to showcase practical applications of these concepts and principles within the context of three significant domains: (i) communication, (ii) community detection in social networks; (iii) machine learning, and (iv) speech recognition. In the next section, we will first focus on the communication application.

Problem Set 6

Problem 6.1 (*Linear algebra basics*) Consider the following square matrices:

$$A = \begin{bmatrix} a_{11} & a_{12} \\ a_{21} & a_{22} \end{bmatrix}, B = \begin{bmatrix} b_{11} & b_{12} \\ b_{21} & b_{22} \end{bmatrix}, C = \begin{bmatrix} c_{11} & c_{12} \\ c_{21} & c_{22} \end{bmatrix}. \tag{2.150}$$

Let (λ_i, v_i) be the ith eigenvalue and eigenvector of A, i.e., $Av_i = \lambda_i v_i$ for $i \in \{1, 2\}$.

(a) Show that $\det(A) = \det(A^T)$ where $(\cdot)^T$ indicates the transpose operation.
(b) Show that $\det(cA) = c^2\det(A)$ for $c \in \mathbb{R}$.
(c) Show that $\det(A^{-1}) = \frac{1}{\det(A)}$.
(d) Show that $\det(A) = \lambda_1 \lambda_2$.
(e) Show that $\text{trace}(A) = \lambda_1 + \lambda_2$ where the trace operation is defined as the sum of the diagonal entries, i.e., $\text{trace}(A) := a_{11} + a_{22}$.
(f) Let $V = [v_1, v_2] \in \mathbb{R}^{2 \times 2}$ and $\Lambda = \text{diag}(\lambda_1, \lambda_2) := \begin{bmatrix} \lambda_1 & 0 \\ 0 & \lambda_2 \end{bmatrix}$. Show that

$$A = V\Lambda V^{-1}. \tag{2.151}$$

Now suppose A is a symmetric matrix, i.e., $A = A^T$. Then, show that

$$A = V\Lambda V^T. \tag{2.152}$$

(g) Show that $\det(AB) = \det(A)\det(B)$.
(h) Show that

$$\det\left(\begin{bmatrix} A & 0 \\ C & B \end{bmatrix}\right) = \det(A)\det(B). \tag{2.153}$$

(i) Show that

$$\det(I + AB) = \det(I + BA). \tag{2.154}$$

Remark: All the mentioned properties are applicable to arbitrary n-by-n matrices (A, B, C). Proving these properties is not straightforward, but you are welcome to attempt verification if you wish.

Problem 6.2 (*Sylvester determinant theorem*) The property mentioned in Problem 6.1(i) is a special case of *Sylvester determinant theorem*, stated below: for $A \in \mathbb{R}^{n \times m}$ and $B \in \mathbb{R}^{m \times n}$,

$$\det(I_n + AB) = \det(I_m + BA) \tag{2.155}$$

where I_n (or I_m) denotes an n-by-n (or m-by-m) identity matrix. This problem explores the proof of this theorem.

(a) Show that

$$\det\left(\begin{bmatrix} I_m & -B \\ A & I_n \end{bmatrix}\right) \det\left(\begin{bmatrix} I_m & B \\ 0 & I_n \end{bmatrix}\right) = \det\left(\begin{bmatrix} I_m & B \\ 0 & I_n \end{bmatrix}\right) \det\left(\begin{bmatrix} I_m & -B \\ A & I_n \end{bmatrix}\right)$$
(2.156)

(b) Using the property in Problem 6.1(h), show that

$$\det\left(\begin{bmatrix} I_m & 0 \\ A & I_n + AB \end{bmatrix}\right) = \det(I_n + AB).$$
(2.157)

(c) Using part (b) and the property in Problem 6.1(g), show that

$$\det\left(\begin{bmatrix} I_m & -B \\ A & I_n \end{bmatrix}\right) \det\left(\begin{bmatrix} I_m & B \\ 0 & I_n \end{bmatrix}\right) = \det(I_n + AB),$$

$$\det\left(\begin{bmatrix} I_m & B \\ 0 & I_n \end{bmatrix}\right) \det\left(\begin{bmatrix} I_m & -B \\ A & I_n \end{bmatrix}\right) = \det(I_m + BA).$$
(2.158)

(d) Using parts (a) and (c), prove the Sylvester determinant theorem (2.155).

Problem 6.3 (*A key fact used in the derivation of* $\mathrm{MSE}(\hat{\sigma}^2_{\mathrm{ML}})$) This problem delves into the proof of the non-trivial fact (2.123) that we posited during the derivation of $\mathrm{MSE}(\hat{\sigma}^2_{\mathrm{ML}})$. Recalling the claim, consider i.i.d. samples Y_i's $\sim \mathcal{N}(0, \sigma^2)$ and $\hat{\mu}_{\mathrm{ML}} = \frac{\sum_{i=1}^n Y_i}{n}$. Then, there exists i.i.d. $\{X_i\}_{i=1}^n$ with $\mathcal{N}(0, \sigma^2)$ such that

$$\sum_{i=1}^n (Y_i - \hat{\mu}_{\mathrm{ML}})^2 = \sum_{i=2}^n X_i^2.$$
(2.159)

(a) Let $Y = [Y_1, \ldots, Y_n]^T \in \mathbb{R}^n$. Show that

$$\sum_{i=1}^n (Y_i - \hat{\mu}_{\mathrm{ML}})^2 = Y^T \left(I_n - \frac{1}{n} uu^T \right) Y$$
(2.160)

where $u = [1, 1, \ldots, 1]^T \in \mathbb{R}^n$.

(b) Let $M = I_n - \frac{1}{n} uu^T$. Show that the eigenvalues of M are $(\lambda_1, \lambda_2, \ldots, \lambda_n) = (0, 1, \ldots, 1)$ and the first eigenvector is u, i.e., M can be decomposed as:

$$M = V \Lambda V^T$$
(2.161)

where $V = [u, v_2, \ldots, v_n] \in \mathbb{R}^{n \times n}$ and $\Lambda = \mathrm{diag}(0, 1, \ldots, 1)$.

(c) Let $X := [X_1, \ldots, X_2]^T = V^T Y$. Show that X_i's are i.i.d. $\sim \mathcal{N}(0, \sigma^2)$ and

$$Y^T M Y = X^T \Lambda X = \sum_{i=2}^{n} X_i^2. \tag{2.162}$$

Note: Applying (2.162) into (2.160), we prove the claim (2.159).

Problem 6.4 (*Moment generating function of Gaussian distribution*) Let $X \sim \mathcal{N}(\mu, \sigma^2)$ and $f_X(x)$ be the pdf of X:

$$f_X(x) = \frac{1}{\sqrt{2\pi\sigma^2}} e^{-\frac{1}{2\sigma^2}(x-\mu)^2}, \ \forall x \in \mathbb{R}. \tag{2.163}$$

Let $F_X(t)$ be the moment generating function of $f_X(x)$:

$$F_X(t) := \mathbb{E}[e^{tX}] = \int_{-\infty}^{\infty} e^{tx} f_X(x) dx. \tag{2.164}$$

Show that

$$F_X(t) = e^{\mu t + \frac{1}{2}\sigma^2 t^2}. \tag{2.165}$$

Problem 6.5 (*Moments of Gaussian distribution*) Let $X \sim \mathcal{N}(0, \sigma^2)$ and $F_X(t) := \mathbb{E}[e^{tX}]$ be the moment generating function of X.

(a) Show that

$$\mathbb{E}[X^k] = F_X^{(k)}(t)\Big|_{t=0} \tag{2.166}$$

where $F_X^{(k)}(t)$ indicates the kth derivative of $F_X(t)$.

(b) Show that

$$\mathbb{E}[X^3] = 0, \ \mathbb{E}[X^4] = 3\sigma^2. \tag{2.167}$$

(c) Show that

$$\mathbb{E}[X^{2k-1}] = 0, \ \mathbb{E}[X^{2k}] = (2k-1)!!\sigma^2, \ k \in \mathbb{N} \tag{2.168}$$

where $(2k-1)!!$ denotes the double factorial:

$$(2k-1)!! := \prod_{i=1}^{k}(2i-1) = (2k-1)(2k-3)\cdots 3 \cdot 1. \tag{2.169}$$

Problem 6.6 (*Weak Law of Large Numbers*) Let $Y_1, Y_2, \ldots$ be an i.i.d. discrete random process with mean μ and variance $\sigma^2 < \infty$. Let

$$S_n := \frac{Y_1 + Y_2 + \cdots + Y_n}{n}. \tag{2.170}$$

For any $\epsilon > 0$, show that

$$\mathbb{P}\left(|S_n - \mathbb{E}[Y]| \leq \epsilon\right) \longrightarrow 1 \qquad \text{as } n \to \infty, \tag{2.171}$$

i.e., the sample mean S_n is within $\mathbb{E}[Y] \pm \epsilon$ with high probability as n tends to infinity. *Remark:* The above convergence w.r.t. the empirical sample mean S_n is named the Weak Law of Large Numbers, WLLN for short.

Problem 6.7 (*Sum of independent random variables*) Suppose that continuous random variables X and Y are independent. Let $Z = X + Y$.

(a) Express the pdf $f_Z(z)$ in terms of the two pdfs $f_X(x)$ and $f_Y(y)$.
(b) Express the Laplace transform of $f_Z(z)$ in terms of the Laplace transforms of $f_X(x)$ and $f_Y(y)$. Here the Laplace transform is defined as:

$$F_Z(s) := \int_{-\infty}^{\infty} e^{-sz} f_Z(z) dz. \tag{2.172}$$

(c) Assume that X and Y are i.i.d. Gaussian with mean 0 and variance σ^2. Using the fact that the Laplace transform is one-to-one mapping, show that Z is a Gaussian random variable with mean 0 and $2\sigma^2$.
Note: You don't need to prove the one-to-one mapping property of the Laplace transform.

Problem 6.8 (*MLE*) Let $(X_1, X_2, \ldots, X_n)$ be i.i.d. Gaussian random variables with mean μ and variance σ^2. Assume that $(X_1, X_2, \ldots, X_n)$ are observed and fed into the estimation block.

(a) Suppose the variance σ^2 is known. Find the MLE $\hat{\mu}_{ML}$ for the mean μ. Also compute the mean square error (MSE) of the estimator: $\mathbb{E}[(\hat{\mu}_{ML} - \mu)^2]$. How does the MSE depend on n?
Hint: You can use the fact that the maximizer in the associated optimization occurs at the stationary point.
(b) Suppose that the mean μ is known. Find the MLE $\hat{\sigma}^2_{ML}$ for the variance σ^2. Also compute the MSE of the estimator: $\mathbb{E}[(\hat{\sigma}^2_{ML} - \sigma^2)^2]$. How does the error depend on n?
Hint: You can use the fact that the maximizer in the associated optimization occurs at the stationary point.

Problem 6.9 (*MLE versus other estimator*) Consider the same setting as in Problem 6.8.

(*a*) Suppose both the mean and the variance are *unknown*. Find the MLE for the mean and variance simultaneously: $\hat{\mu}_{ML}$ and $\hat{\sigma}^2_{ML}$.
    Hint: You can use the fact that the maximizer in the associated optimization occurs at the stationary point.
(*b*) Compute $\mathbb{E}[\hat{\mu}_{ML}]$ and compare it with the true value μ. Also compute $\mathbb{E}[\hat{\sigma}^2_{ML}]$ and compare it with the true value σ^2.
(*c*) Instead of using $\hat{\sigma}^2_{ML}$ in part (*c*), consider another estimator:

$$\hat{\sigma}^2_{\text{other}} = \frac{1}{n-1} \sum_{i=1}^{n} (X_i - \hat{\mu}_{ML})^2. \tag{2.173}$$

Compute $\mathbb{E}[\hat{\sigma}^2_{\text{other}}]$.
(*d*) In fact, people prefer using the estimator in (2.173). Discuss why the above estimator (2.173) is more preferred.
    Hint: Ponder on how $\mathbb{E}[\hat{\sigma}^2_{\text{other}}]$ looks. Or search wikipedia with key words like the biased and unbiased estimators.

Problem 6.10 (*MSE for the unbiased estimator*) Consider the same setting as in Problem 6.9. In fact, the second estimator for the variance introduced therein is a very well-known estimator, named the *unbiased* estimator:

$$\hat{\sigma}^2_{\text{unbiased}} = \frac{1}{n-1} \sum_{i=1}^{n} (X_i - \hat{\mu}_{ML})^2. \tag{2.174}$$

Show that the mean squared error of the unbiased estimator is:

$$\text{MSE}(\hat{\sigma}^2_{\text{unbiased}}) := \mathbb{E}\left[(\hat{\sigma}^2_{\text{unbiased}} - \sigma^2)\right] = \frac{2}{n-1} \sigma^4. \tag{2.175}$$

Hint: Think about the derivation of $\text{MSE}(\hat{\sigma}^2_{ML})$.

Problem 6.11 (*MLE of Laplace distribution*) Suppose $\{X_i\}_{i=1}^n$ are i.i.d., each being according to the following pdf (named the *Laplace distribution*):

$$f(x) = \frac{1}{2} e^{-|x-\theta|}, \qquad x \in \mathbb{R} \tag{2.176}$$

where the model parameter $\theta \in \mathbb{R}$. Since the density decays slower than the Gaussian distribution, the Laplace distribution can be a better model when there are many *outliers* in the data. In this problem, we wish to find the maximum likelihood estimate (MLE) of θ.

(a) Show that the Laplace distribution is indeed a valid pdf, i.e., satisfying the sum-up-to-one constraint.
(b) Draw the Laplace distribution when $\theta = 1.5$.
(c) Given three samples $(X_1, X_2, X_3) = (0, 1, 3)$, find the MLE of θ.
(d) Given five samples $(X_1, X_2, X_3, X_4, X_5) = (0, 1, 3, 5, 1000)$, find the MLE of θ.

Problem 6.12 (*Role of Chebyshev's inequality*) Let X_i be a binary random variable that indicates the answer of the ith person when asked if she/he advocates the Democratic Party: 1 (yes); 0 (no), where $i \in \{1, 2, \ldots, n\}$. Let p be the fraction of people who advocate the Democratic Party. Assume that $\{X_i\}$ is the Bernoulli process with $\mathsf{Bern}(p)$. Also assume that $(X_1, X_2, \ldots, X_n)$ are observed and fed into the estimation block.

(a) Find the MLE for p: $\hat{p}_{\mathsf{ML}}$.
(b) Using Chebyshev's inequality, compute a bound on the number n of samples that we need to guarantee that the probability of exceeding a margin of ± 0.03 error is 0.05:

$$\mathbb{P}(|\hat{p}_{\mathsf{ML}} - p| > 0.03) \leq 0.05.$$

Hint: You may want to use the fact that $p(1 - p) \leq 0.25$.

Problem 6.13 (*Role of the central limit theorem*) Consider the same setting as in Problem 6.12.

(a) Using the central limit theorem, compute an estimate on the number of samples so as to guarantee that the probability of exceeding a margin of ± 0.03 error is 0.05:

$$\mathbb{P}(|\hat{p}_{\mathsf{ML}} - p| > 0.03) \leq 0.05.$$

Hint: You may want to use the table for the cdf of the standard normal and the fact that $p(1 - p) \leq 0.25$.
(b) Is the estimated number in part (a) more or less conservative than the number you obtained in part (b) of Problem 6.12?

Problem 6.14 (*Central limit theorem:* Python *simulation*) Remember that the central limit theorem holds for an arbitrary distribution of X_i's:

$$Z_n := X_1 + X_2 + \cdots + X_n \overset{\text{in dist}}{\longrightarrow} Z \sim \mathcal{N}(0, 1) \qquad (2.177)$$

where $\{X_i\}_{i=1}^{\infty}$ is i.i.d. with $\mathbb{E}[X_i] = 0$ and variance $\mathsf{var}(X_i) = \frac{1}{n}$. In this problem, we will empirically verify this for a simple binary case where

$$\mathbb{P}\left(X_i = +\frac{1}{\sqrt{n}}\right) = \mathbb{P}\left(X_i = -\frac{1}{\sqrt{n}}\right) = \frac{1}{2}. \tag{2.178}$$

(a) Show that $\mathbb{E}[X_i] = 0$ and $\text{var}(X_i) = \frac{1}{n}$ for all $i \in \{1, \ldots, n\}$.

(b) Write a **Python** script for generating X_i's when $n = 100$. Also print the sample mean and the sample variance of the generated X_i's.

   Hint: You may want to use a built-in function `bernoulli` in the `scipy.stats` package.

(c) Write a **Python** script for constructing $Z_n = X_1 + \cdots + X_n$. We repeat this construction 1,000 times, generating a variable named `Z_copies`, of size 1,000. Print the sample mean and the sample variance of `Z_copies`. Also plot the empirical density of `Z_copies` using the following code:

```
import matplotlib.pyplot as plt
import seaborn as sns
from scipy.stats import norm

Z = np.arange(-4., 4., 0.01)

plt.figure(figsize=(4,4), dpi=150)
plt.plot(Z,norm.pdf(Z),color='red',label='Gaussian pdf')
sns.histplot(Z_copies,stat='density',bins=10,label='emp')
plt.xlabel('Z')
plt.title('n=100')
plt.legend()
plt.show()
```

Here `norm.pdf(Z)` implements the standard Gaussian distribution (with mean 0 and variance 1), and `sns.histplot` draws the empirical density of `Z_copies`.

(d) Repeat part (c) for an increased sample size $n = 10,000$. Explain how the empirical density of `Z_copies` looks compared to the true standard Gaussian distribution.

Problem 6.15 (*True or False?*)

(a) Consider the problem of inferring X given (Y_1, Y_2):

$$\begin{align} Y_1 &= X + Z_1; \\ Y_2 &= X + Z_1 + Z_2. \end{align} \tag{2.179}$$

Here X is equally likely to be $+1$ or -1 and Z_i's are i.i.d. $\sim \mathcal{N}(0, \sigma^2)$. Then the ML estimate of X using Y_1 only is the same as the ML estimate based on (Y_1, Y_2).

(b) Let $(X_1, X_2, \ldots, X_n)$ be i.i.d. Gaussian random variables with mean μ and variance σ^2. Assume that $(X_1, X_2, \ldots, X_n)$ are observed and fed into the estimation block. Consider two estimators for the variance:

$$\hat{\sigma}^2_{\text{ML}} = \frac{1}{n} \sum_{i=1}^{n} (X_i - \hat{\mu}_{\text{ML}})^2; \tag{2.180}$$

$$\hat{\sigma}^2_{\text{unbiased}} = \frac{1}{n-1} \sum_{i=1}^{n} (X_i - \hat{\mu}_{\text{ML}})^2 \tag{2.181}$$

where $\hat{\mu}_{\text{ML}} = \frac{1}{n} \sum_{i=1}^{n} X_i$. Then, $\hat{\sigma}^2_{\text{unbiased}}$ outperforms $\hat{\sigma}^2_{\text{ML}}$, i.e.,

$$\mathbb{E}[(\hat{\sigma}^2_{\text{unbiased}} - \sigma^2)^2] \leq \mathbb{E}[(\hat{\sigma}^2_{\text{ML}} - \sigma^2)^2]. \tag{2.182}$$

Chapter 3
Information Technology Applications

The MAP and ML principles are prevalent in IT applications.

3.1 Communication: Probabilistic Channel Modeling

Recap

In the previous section, we explored an important theorem that underpins the Gaussian modeling of numerous random quantities of interest: the Central Limit Theorem (CLT). The CLT pertains to an i.i.d. random process $\{X_i\}_{i=1}^n$. For the sake of illustration, we considered a simple yet generalizable setup in which $\mathbb{E}[X_i] = 0$ and $\text{var}(X_i) = \frac{1}{n}$. In this context, the CLT can be stated as:

$$Z_n := X_1 + X_2 + \cdots + X_n \xrightarrow{\text{in dist}} Z \sim \mathcal{N}(0, 1) \qquad (3.1)$$

where the notation $\xrightarrow{\text{in dist}}$ signifies the convergence in distribution:

$$\lim_{n \to \infty} f_{Z_n}(z) = \frac{1}{\sqrt{2\pi}} e^{-\frac{1}{2}z^2} \qquad \forall z \in \mathbb{R}. \qquad (3.2)$$

To prove the CLT, we relied on two key claims. We utilized the first claim to express the pdf of Z_n as the convolution of the pdfs w.r.t. X_i's: $f_{Z_n}(z) = (f_{X_1} * f_{X_2} * \cdots * f_{X_n})(z)$. Employing the second claim, which allowed for a concise representation using Laplace transforms, we obtained: $F_{Z_n}(s) = [F_{X_1}(s)]^n$. Subsequently, by using a Taylor series expansion, we demonstrated that:

© The Author(s), under exclusive license to Springer Nature Singapore Pte Ltd. 2025
C. Suh, *Probability for Information Technology*,
https://doi.org/10.1007/978-981-97-4032-1_3

$$\lim_{n\to\infty} F_{Z_n}(s) = \lim_{n\to\infty} \left(1 + \frac{1}{2n}s^2\right)^n = e^{\frac{s^2}{2}}. \qquad (3.3)$$

Finally, making use of the one-to-one mapping property of Laplace transforms, together with the fact that

$$\mathrm{LaplaceTransform}\left(\frac{1}{\sqrt{2\pi}}e^{-\frac{z^2}{2}}\right) = e^{\frac{s^2}{2}},$$

we proved the CLT (3.2).

Now, let us recap the knowledge we've accumulated thus far. In Part I, we delved into various probability concepts: the sample space, events, conditional probability, independence, random variables, and the Gaussian distribution. In Part II, we took a deeper dive into essential concepts and key principles, covering: (i) random processes and three notable examples; (ii) the MAP and ML principles; (iii) the Law of Large Numbers; and (iv) the Central Limit Theorem. The objective of Part III is to demonstrate how these concepts and principles play a pivotal role in the context of four significant applications: (i) communication; (ii) community detection in social networks; (iii) machine learning; and (iv) speech recognition.[1]

Outline
In this section, our focus will be on the first application, communication, where we aim to elucidate its connection with probability. It comprises five parts. We will begin by revisiting the definition of communication and then introduce the architecture of *digital* communication, which we will put a special emphasis on. Second, we will investigate an *uncertain* entity that arises in communication and therefore the one that is intimately related to probability. The entity is *noise*. As mentioned earlier several times, noise can be modeled as a well-known *Gaussian* random variable. Third, we will delve into the physical characteristics of noise. Following our exploration of the physical properties, we will develop a mathematical model for noise. Lastly, we will build upon this mathematical framework to demonstrate that the probability distribution of noise conforms to the Gaussian distribution.

Digital communication (Shannon 2001) Let's recall the definition of communication. Communication is the process of transmitting information from one point to another. In this context, the originating point is referred to as the *transmitter*, while

---

[1] The contents to be presented in Part III are adapted from the author's previous publications: (i) (Suh 2023a) on communication and speech recognition; (ii) (Suh 2023b) on community detection in social networks; and (iii) (Suh 2022) on machine learning. However, the logical flow and narrative have been uniquely tailored to align with the specific theme of this book.

the destination is termed the *receiver*. The conduit through which this information traverses is known as the *channel*.

In broad terms, there are two primary forms of communication, categorized based on the nature of the information source that we intend to convey: (i) *analog* communication in which the information source is in its raw state, such as sound waveforms, images, and text; (ii) *digital* communication in which the information source is represented as a binary string, often referred to as *bits* which consists of a sequence of binary digits. In this book, our emphasis will be on digital communication, as it serves as the cornerstone for the majority of contemporary communication systems.

The dominance of digital communication can be attributed to a pivotal discovery made by a brilliant scientist, *Claude E. Shannon*, in the mid 1900s. Shannon's breakthrough revelation was that any form of information source could be accurately represented using binary bits without any loss of the underlying meaning. To illustrate this concept, consider the example of English text composed of English letters. A key observation here is that each letter can take only a *finite* number of possibilities, specifically the total number of English alphabets, which amounts to 26. Here we exclude special characters such as spaces. As a result, it becomes evident that mere 5 bits ($\lceil \log_2 26 \rceil = 5$) suffice to represent each letter. This reasoning applies similarly to other types of information sources.

Figure 3.1 illustrates the architecture of digital communication that Shannon introduced. Bits serve as the foundational information units we aim to transmit. Within the communication system, two key components play pivotal roles. The transmitter incorporates an *encoder*, symbolized as a white box in the figure. The encoder receives the input bits and transforms them into a signal ready for transmission over the channel. Correspondingly, at the receiver's end, another white box, referred to as a *decoder*, comes into play. The decoder's primary objective is to take the received signal and endeavor to reconstruct the original bits as accurately as possible. This process is essential for retrieving the transmitted information intact.

Modem We introduce a terminology you might encounter in the context of communication. The physical world inherently operates in an analog manner, meaning that a signal sent into the channel must manifest as a physical quantity, typically an electromagnetic signal, like an electrical voltage signal. Consequently, the encoder's task is to *modulate* the digital information (bits) into an electrical voltage signal. Conversely, when we receive a signal (the output from the channel), it is in the form of a voltage signal. Here, the decoder's role is to *demodulate* the analog signal in order to reconstruct the original bits. As a result, it is common to refer to the encoder

Fig. 3.1 The architecture of digital communication

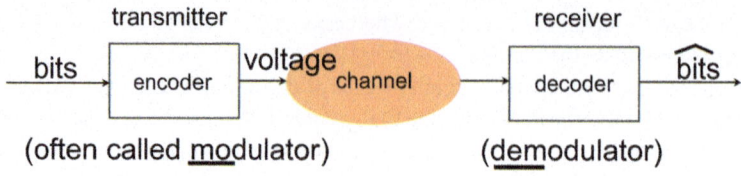

Fig. 3.2 MODEM: MOdulator & DEMoluator

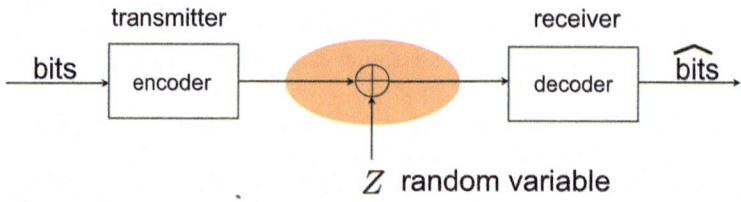

Fig. 3.3 An additive noise channel

and decoder collectively as a *modem*, which highlights the "mo" from modulator and the "dem" from demodulator. You can see an illustration of this in Figure 3.2.

A probabilistic model of the channel An essential characteristic of the channel is its inherent uncertainty. The received voltage signal does not follow a deterministic relationship with the transmitted signal. In most practical scenarios, the channel can be described as an additive process, where the received signal is the result of combining the transmitted signal with an additional source of noise, as illustrated in Figure 3.3. The additional noise, say Z, can be described as a random quantity. In the language of probability, such a random quantity is commonly referred to as a *random variable*. To be more precise, it is a *continuous* random variable, characterized by a probability density function. As mentioned at the outset, researchers have discovered that Z can be accurately modeled as the well-known *Gaussian* random variable. In the remainder of this section, we will provide the rationale behind this modeling choice by examining the physical properties of the noise.

Physics Let's begin by delving into the physics behind the occurrence of noise in communication systems. By communication systems, we are specifically referring to those involving electronic circuits, which we will simply term as electronic communication systems. In these electronic communication systems, the level of a signal is contingent upon its voltage. As you might learn from an introductory circuit course, voltage levels are intricately linked to the movement of electrons. In essence, the fewer electrons present, the higher the voltage level.

In the early 1900s, a physicist, John B. Johnson, made a significant discovery: the behavior of electrons contributes to the emergence of uncontrollable noise (Johnson 1928). This discovery stemmed from an interesting observation: electrons exhibit random agitation due to thermal effects. As a result, this random agitation causes voltage levels to become unpredictable, as they may rely on an uncontrollable factor like temperature. In practice, achieving precise control over the movement of electrons is nearly impossible. Johnson interpreted this unwanted random fluctuation as a primary source of *noise*.

During that period, Johnson aimed to comprehend the statistical nature of this noise phenomenon. However, he encountered a challenge since his expertise lay in experimental physics rather than mathematics. Fortunately, he had a brilliant and mathematically adept colleague at Bell Labs, Harry Nyquist, who played a pivotal role in this journey. Johnson shared his experimental findings with Nyquist, and Nyquist was able to formulate a mathematical theory that unraveled the statistical behavior of this noise (Nyquist 1928). This theory laid the foundation for the Gaussian noise model that we will explore in the sequel.

At the time, this noise was named *thermal noise* due to its reliance on temperature as a determining factor. In reality, noise can also be influenced by factors such as device imperfections and measurement inaccuracies. However, the primary source of randomness is attributed to thermal noise. Therefore, our primary focus will be on Nyquist's mathematical theory concerning thermal noise.

Assumptions regarding the thermal noise The mathematical theory starts with establishing specific assumptions regarding the physical properties that thermal noise adheres to. These assumptions are four folded.

1. *(Infinite additive sub-noises)*: The thermal noise, as a result of the random motion of electrons, is considered to be an aggregate outcome of numerous additive "sub-noises". It is assumed that the number of these sub-noises is infinite, reflecting the abundance of electrons in an electrical signal.
2. *(Mutually independent sub-noises)*: Electrons in the signal are generally uncorrelated with each other. Consequently, the second assumption posits that these sub-noises are mutually independent.
3. *(Equal energy contribution)*: No individual electron has a significantly dominant effect on the additive noise. Hence, a simplified version of this observation is that each sub-noise contributes an equal amount of energy to the total energy within the additive noise.
4. *(Finite noise energy)*: The energy of the noise, in relation to the energy of the voltage signal of interest, is not excessively large; it doesn't grow without bounds. Therefore, the fourth reasonable assumption is that the noise energy is finite.

A mathematical model grounded in the four assumptions Let us represent the four assumptions in mathematical terms. To do this, we will introduce some mathematical notations, expressing the total additive noise Z as the infinite sum of sub-noises, say $X_1, X_2, \ldots, X_n, \ldots$:

$$Z = \lim_{n \to \infty} \underbrace{X_1 + X_2 + \cdots + X_n}_{=:Z_n}. \tag{3.4}$$

The mathematical expressions for the first and second assumptions are as follows: $n \to \infty$ and $(X_1, \ldots, X_n)$ are *mutually independent*. A mathematical assumption that we will take w.r.t. the third assumption is: $(X_1, \ldots, X_n)$ are *identically distributed*. So it is the i.i.d. assumption.

Without loss of generality, one can assume that Z has a zero mean. Here what it means by "without loss of generality" is that the general case can be readily covered with some proper modification to the ground assumption that follows after the phrase. You may wonder why the zero-mean assumption can serve as the foundational assumption. To see this, consider a general scenario wherein there exists a bias in Z: $\mathbb{E}[Z] = \mu \neq 0$. In such a case, we can always counteract this bias by subtracting the bias value μ from the received signal. This ensures that the mean of the *effective noise* becomes zero. More precisely, we subtract the bias μ from the received signal $Y = X + Z$ (where X represents the transmitted signal) to obtain:

$$Y - \mu = X + (Z - \mu). \tag{3.5}$$

In this context, the term $Z - \mu$ can be considered as the *effective noise*, which has indeed a zero mean. You might be curious about how we can figure out the value of μ. An established approach for estimating μ is through the maximum likelihood (ML) principle that we have already learned. So the idea involves setting the transmitted signal X to zero, obtaining the received signals that contain only noise, and subsequently calculating their sample mean:

$$\hat{\mu}_{\mathsf{ML}} = \frac{\sum_{i=1}^{n} Y_n}{n} = \frac{\sum_{i=1}^{n} Z_n}{n} \tag{3.6}$$

where the second equality is due to our setting $X = 0$.

Due to the i.i.d. assumption, each sub-noise possesses an equal amount of energy:

$$\mathbb{E}[X_i^2] = \mathbb{E}[X_j^2], \quad \forall i, j \in \{1, \ldots, n\}. \tag{3.7}$$

Here energy is simply quantified as the square of the voltage, adhering to the principle that energy is proportional to the square of voltage. We are disregarding other factors such as "resistance". The energy of the aggregated noise Z_n is as follows:

$$\begin{aligned}
\mathbb{E}[Z_n^2] &= \mathsf{var}(Z_n) \\
&= \mathsf{var}(X_1) + \mathsf{var}(X_2) + \cdots + \mathsf{var}(X_n) \\
&= n\mathsf{var}(X_1) =: \sigma^2
\end{aligned} \tag{3.8}$$

where the first equality is due to $\mathbb{E}[Z_n] = 0$ (the zero-mean assumption); and the second and third equalities come from the i.i.d. assumption.

Lastly, let us examine the finite energy assumption. Denoted by σ^2, this finite energy implies that, as illustrated in (3.8), the energy of each sub-noise must decrease with the addition of more and more sub-noises:

$$\mathbb{E}[X_i^2] = \frac{\sigma^2}{n}, \quad i \in \{1, \ldots, n\}. \tag{3.9}$$

The statistical behavior of the thermal noise in the limit of n We aim to analyze the pdf of the random variable Z_n as the number of sub-noises n approaches infinity. Recall that $Z_n = X_1 + X_2 + \cdots + X_n$ and the aforementioned assumptions are summarized as follows:

$$
\begin{aligned}
&(X_1, \ldots, X_n) \text{ i.i.d.;} \\
&\mathbb{E}[X_i] = 0 \quad \forall i; \\
&\text{var}(X_i) = \frac{\sigma^2}{n} \quad \forall i.
\end{aligned}
\tag{3.10}
$$

What does this remind you of? Yes, it is the *Central Limit Theorem* (Pólya 1920; Bertsekas & Tsitsiklis 2008). The only difference in this context is that we have $\text{var}(X_i) = \frac{\sigma^2}{n}$ instead of $\text{var}(X_i) = \frac{1}{n}$. This variation results in $\text{var}(Z) = \sigma^2$ instead of $\text{var}(Z) = 1$. Therefore, applying the CLT, we obtain:

$$Z = \lim_{n \to \infty} Z_n \sim \mathcal{N}(0, \sigma^2). \tag{3.11}$$

Look ahead

We have introduced some physics-inspired assumptions regarding additive thermal noise to illustrate that it can accurately be represented by a Gaussian random variable. In the upcoming section, we will delve into the design of transmission and reception strategies within the context of the Gaussian channel. Through this exploration, we will demonstrate the pivotal role played by the MAP principle.

3.2   Communication: The MAP Principle

Recap

In the previous section, we explored a probabilistic model for the noise, which serves as a source of uncertainty in communication. This noise is typically superimposed on a transmitted signal, as depicted in Figure 3.4. Because of the inherent uncertainty in noise, it is described in terms of a random variable, say Z. Based on several physical characteristics of the noise, we demonstrated that Z can be modeled as a Gaussian random variable. Specifically, the physical characteristics are: (i) the noise arises from the combination of numerous "subnoises"; (ii) there is almost no correlation among these sub-noises; (iii) no single sub-noise dominates; and (iv) it possesses finite energy. We subsequently translated these characteristics into the following mathematical model:

$$Z = \lim_{n \to \infty} \underbrace{X_1 + X_2 + \cdots + X_n}_{=:Z_n} \tag{3.12}$$

where $\{X_i\}_{i=1}^n$ are i.i.d. with $\mathbb{E}[X_i] = 0$ and $\mathsf{var}(X_i) = \frac{\sigma^2}{n}$ for all i. Lastly applying the CLT into the above, we proved:

$$Z = \lim_{n \to \infty} Z_n \sim \mathcal{N}(0, \sigma^2). \tag{3.13}$$

Outline

In this section, we will investigate transmission and reception strategies in the context of the additive Gaussian noise channel. One critical aspect we would like to emphasize throughout the process is that the optimal receiver relies on the MAP and ML principles, which have been highlighted previously. Specifically this section comprises four parts. We will begin by examining a straightforward yet widely-employed transmission scheme. Under this simple transmission scheme, we will then propose an intuitive reception scheme. Subsequently, we will derive the optimal receiver by applying the MAP principle.

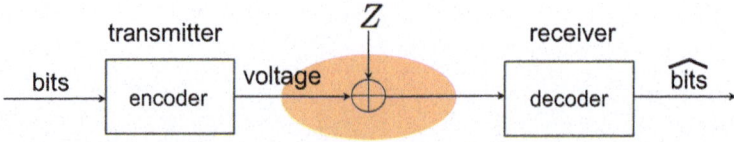

Fig. 3.4   The additive Gaussian noise channel

Lastly, we will demonstrate that, under reasonable assumptions, the optimal
decision rule simplifies to the ML estimation, and interestingly it coincides
with the intuitive reception scheme.

A simple transmission scheme for sending one bit As a simple setting, we consider
the transmission of a single bit, say B. Since the specific value of the transmitted
bit is unknown to the receiver, the bit of interest B can be regarded as a binary
random variable. In electronic communication systems, transmitted signals are typically represented as analog voltage signals. Therefore, we need to modulate B into
a voltage signal. To this end, we employ a straightforward binary mapping scheme,
as illustrated in Figure 3.5.

For this binary mapping, we associate the case $B = 0$ with a voltage level v_0,
and similarly, v_1 corresponds to $B = 1$. The selection of v_0 and v_1 is influenced by
a communication budget constraint in the system. From a physical perspective, the
transmitted voltage corresponds to an expenditure of energy. According to circuit
theory, the energy is proportional to the square of the voltage. To keep things simple,
let us assume that the energy spent on transmitting a voltage v in Volts is precisely
v^2 in Joules.

In light of this assumption, a sound choice for (v_0, v_1) could be the one that
maximizes the distance $|v_0 - v_1|$, as a larger distance makes it easier to distinguish
between the two voltage levels. However, this choice should adhere to the energy
budget constraint $v_i^2 \leq E$ for $i \in \{0, 1\}$. This simple optimization problem yields a
solution at the boundary points: $v_0 = -\sqrt{E}$ and $v_1 = +\sqrt{E}$. Consequently, we opt
for this choice, as depicted in Figure 3.5. This is a well-known transmission scheme
in the literature, called *Pulse Amplitude Modulation (PAM)*. It's worth noting that the
bit information is conveyed by modulating the amplitude of a voltage pulse which
reflects the binary signal.

Guess a reasonably good reception scheme Given the above transmission strategy,
what would constitute an effective reception scheme? To determine this, let us first
examine two key factors that influence a reception rule. The first factor is the *a priori*

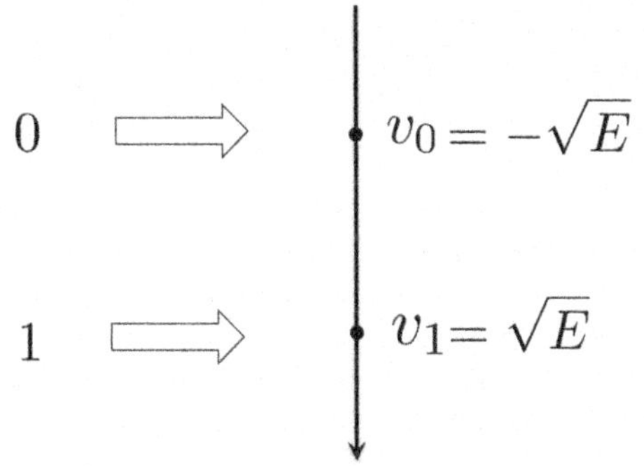

Fig. 3.5 Pulse Amplitude
Modulation (PAM):
Mapping the value of one bit
into one of the two possible
voltage levels

$$0 \implies \quad v_0 = -\sqrt{E}$$

$$1 \implies \quad v_1 = \sqrt{E}$$

probability: $\mathbb{P}(B = 0)$ and $\mathbb{P}(B = 1)$. Why does this matter? It becomes evident when you think about an extreme scenario where we possess prior knowledge (prior to the communication process) that the information bit is guaranteed to be 0. In such a case, there would be no need to take a look at the received voltage; we could simply assert that the information bit is always 0, regardless of the received signal. In practice, however, we often lack access to such prior information. In such cases, what we can do is assume that the information bit is *equally likely* to be 1 or 0. Therefore, we make the following assumption:

$$\text{(Assumption): } \mathbb{P}(B = 0) = \mathbb{P}(B = 1) = \frac{1}{2}. \qquad (3.14)$$

The second factor is *noise statistics*. Understanding the statistical properties of the noise is essential for the receiver to make a decision. To see this, consider the Gaussian noise. The Gaussian noise is more likely to cluster around zero, as depicted in Figure 3.6.   Here the height of a point on a bell-shaped curve represents the frequency density of occurrences at that point. Given this, your intuition suggests that an effective receiver should choose the voltage value closer to the received voltage out of the two possible transmitted voltages. In other words, if the received voltage is above the midpoint (0 in this case), we declare $\hat{B} = 1$; otherwise, $\hat{B} = 0$. This is an intuitive and well-known rule called the *Nearest Neighbor (NN)* rule. It is reasonable to guess that the optimal decision rule in the case of Gaussian noise might be the NN rule, and it turns out that this is indeed the case. For the remainder of this section, we will prove this.

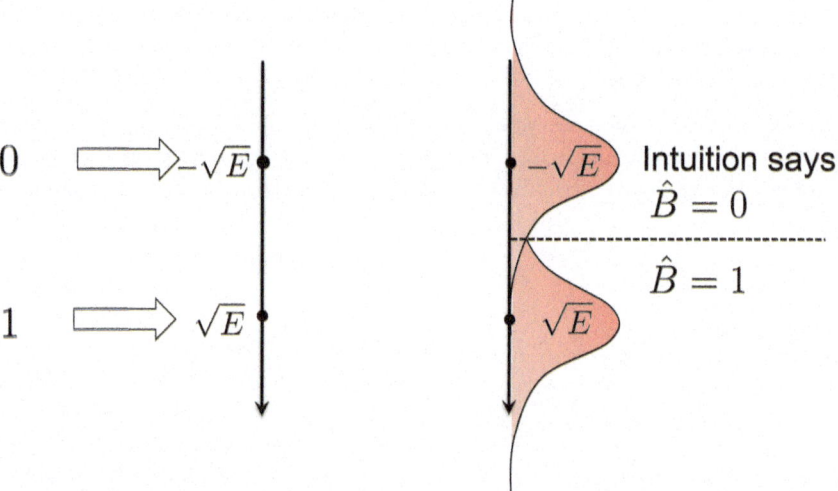

Fig. 3.6 A guess about the optimal receiver

The optimal decision rule From Part II, we have learned that the optimal decision rule is the one that maximizes the probability of making a correct decision. Therefore, our focus is on this specific probability. For a given realization of the received signal $Y = y$, the correct decision probability is expressed as follows:

$$\mathbb{P}(B = \hat{B}|Y = y) \tag{3.15}$$

where $\hat{B}$ is the estimate of B; see Figure 3.7. Consequently, the optimal decision rule is the MAP rule:

$$\hat{B}_{\text{opt}} = \hat{B}_{\text{MAP}} = \arg \max_{\hat{b} \in \{0,1\}} \mathbb{P}(B = \hat{b}|Y = y). \tag{3.16}$$

The MAP solution Let us massage the optimization problem (3.16) to obtain an explicit rule. As we did a couple of times earlier, using Bayes' law, we can express $\mathbb{P}(B = \hat{b}|Y = y)$ as:

$$\mathbb{P}(B = \hat{b}|Y = y) = \frac{\mathbb{P}(B = \hat{b})f_Y(y|B = \hat{b})}{f_Y(y)}. \tag{3.17}$$

Here, we introduce the probability density function $f_Y(y)$ to effectively handle the problematic probability-zero event $Y = y$. Since $f_Y(y)$ is not dependent on $B = \hat{b}$, the MAP rule simplifies to:

$$\hat{B}_{\text{MAP}} = \arg \max_{\hat{b} \in \{0,1\}} \mathbb{P}(B = \hat{b})f_Y(y|B = \hat{b}). \tag{3.18}$$

Under the equal probability assumption (3.14): $\mathbb{P}(B = 0) = \mathbb{P}(B = 1) = \frac{1}{2}$, this simplifies to the ML rule:

$$\hat{B}_{\text{MAP}} = \hat{B}_{\text{ML}} = \arg \max_{\hat{b} \in \{0,1\}} \underbrace{f_Y(y|B = \hat{b})}_{\text{likelihood}}. \tag{3.19}$$

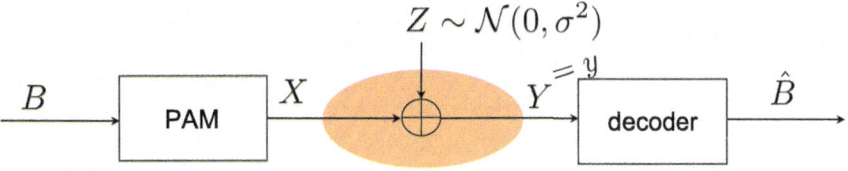

Fig. 3.7 The optimal receiver is the one that maximizes $\mathbb{P}(B = \hat{B}|Y = y)$

Relationship with the NN rule Let us prove that the optimal ML rule coincides with our initial guess: the NN rule. We first see that for the Gaussian channel, $f_Y(y|B = 1)$ can be rewritten as:

$$
\begin{aligned}
f_Y\left(y|B = 1\right) &= f_Y\left(y|X = +\sqrt{E}\right) \\
&= f_Z\left(y - \sqrt{E}|X = +\sqrt{E}\right) \quad\quad (3.20) \\
&= f_Z\left(y - \sqrt{E}\right)
\end{aligned}
$$

where the first equality comes from the fact that the event $B = 1$ is equivalent to the event $X = +\sqrt{E}$ due to our encoding rule, PAM; the second is due to the fact that the event $Y = y$ is equivalent to the event $Z = y - X = y - \sqrt{E}$; and the last is because of the independence between Z and X.

Here $f_Z(\cdot) = \frac{1}{\sqrt{2\pi\sigma^2}}e^{-\frac{(\cdot)^2}{2\sigma^2}}$. So, the ML rule for the Gaussian channel becomes the following: Decide $\hat{B} = 1$ if $f_Z(y - \sqrt{E}) \geq f_Z(y + \sqrt{E})$; $\hat{B} = 0$ otherwise. In situations where $f_Z(y - \sqrt{E}) = f_Z(y + \sqrt{E})$, we might consider a different approach by flipping a fair coin and making a random decision. For the sake of simplicity, in this case, we opt for the decision $\hat{B} = 1$, which still respects the ML rule.

Using the Gaussian pdf, we can simplify the deciding condition $f_Z(y - \sqrt{E}) \geq f_Z(y + \sqrt{E})$. It is equivalent to:

$$
(y - \sqrt{E})^2 \leq (y + \sqrt{E})^2. \quad\quad (3.21)
$$

Simplifying the above further, we arrive at the following equivalent condition:

$$
y \geq 0. \quad\quad (3.22)
$$

In summary, the ML decision rule takes the received voltage $Y = y$ and decides $\hat{B}$ as follows:

$$
y \geq 0 \quad\Longrightarrow\quad \hat{B} = 1 \text{ was sent;}
$$

$$
\text{Otherwise} \Longrightarrow \hat{B} = 0 \text{ was sent.}
$$

Figure 3.8 illustrates the ML decision rule. This rule selects the transmitted voltage level that is closest to the received voltage ("closest" in the standard sense of Euclidean distance). Therefore, the ML rule precisely corresponds to the NN rule.

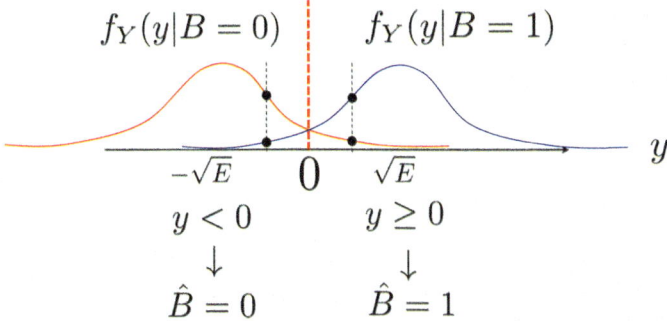

Fig. 3.8 The ML decision rule is equivalent to the NN rule in the additive Gaussian noise channel

Look ahead

We have derived the optimal MAP receiver for transmitting a single bit using PAM and established that, under the equal chance assumption for B (3.14), the optimal receiver is equivalent to the ML and NN rule. This naturally leads to the question about the optimal receiver's performance. In the context of communication, one popular metric for performance assessment is the *error probability*:

$$P_e := \mathbb{P}(\hat{B}_{\text{MAP}} \neq B). \qquad (3.23)$$

In typical communication systems, it is imperative to have a very low error probability, typically in the range of around 10^{-6} to 10^{-10} to ensure reliable communication. However, one can verify that under the PAM transmission scheme that we employed, the error probability P_e is far from the desired range. Not to worry; we will delve into this in the next section. Then, a natural follow-up question arises: Is there a way to reduce P_e to arbitrarily close to 0, ensuring we achieve the desired range? As it turns out, there exists another, yet still simple, transmission scheme that can achieve this. In the upcoming section, we will explore this scheme and demonstrate that P_e can indeed be made arbitrarily small under the scheme.

3.3 Communication: The MAP Rule Under Multi-shot Transmission

Recap

In the previous section, we examined a simple transmission scheme for sending a single bit and figured out the role of the MAP and ML principles in developing the optimal receiver under the additive Gaussian noise channel. See Figure 3.9. The employed encoding rule (PAM) assigned $B = 1$ to $X = +\sqrt{E}$ and $B = 0$ to $X = -\sqrt{E}$. In this setup, we showed that the optimal receiver is the MAP rule, and under the reasonable assumption $\mathbb{P}(B = 0) = \mathbb{P}(B = 1) = \frac{1}{2}$, it simplifies to the ML rule, and can be further simplified to the intuitive NN rule.

Finally, we posed a question: What about the error probability performance of the optimal receiver?

$$P_e := \mathbb{P}(\hat{B}_{\mathrm{MAP}} \neq B). \tag{3.24}$$

We claimed that the error probability P_e does not reach the desired range of 10^{-6} to 10^{-10}. However, the good news is that there exists another approach to reduce P_e to values arbitrarily close to zero, enabling us to meet the specified error probability requirement.

Outline

In this section, we will delve into a method for achieving reliable communication. The section consists of four parts. To begin, we will introduce another readily available communication resource. That is, *time*. We will then investigate the statistical characteristics of noise signals spanning multiple time slots. Following this, we will derive the optimal receiver for a straightforward multi-shot transmission scheme that will be introduced shortly. Lastly, we will analyze the error probability performance of the multi-shot communication scheme equipped with the optimal receiver, thereby demonstrating that P_e can be made small enough.

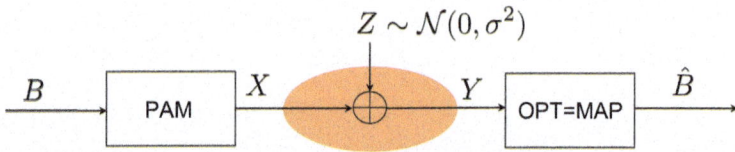

Fig. 3.9 A single-bit transmission via PAM over the additive Gaussian noise channel

Time is another communication resource One crucial observation in the transmission scheme introduced earlier is that we use *only one time slot*, even though there are potentially multiple time slots available for communication. In other words, we have not fully utilized another very natural communication resource: *time*. Therefore, one natural alternative is to make use of *multiple time slots*. In fact, in Sect. 2.3, we took a similar approach in the context of an inference problem for cancer prediction. In testing, we have the option of conducting *multiple experiments*, and a straightforward method to enhance performance is to employ multiple tests.

Sending one bit over n time slots We employ a multi-shot transmission scheme. Suppose we send still one bit B but through n time slots. We assume that the energy budget per time slot is fixed at E. A natural question that arises is: What is the optimal multi-shot communication scheme that minimizes the error probability given the constraint? In fact, answering this question is quite challenging. It concerns the central topic of a foundational field named *Information Theory* (Shannon 2001; Cover 1999). Claude E. Shannon, who was mentioned in Sect. 3.1, is the Father of Information Theory. Instead of delving into the intricate details of communication schemes inspired by information theory, here we will take one naive trial based on a very simple idea that people often employ for our daily-life conversation: *repeating what we said*.

A transmission scheme based on the repetition idea might involve sending a voltage level of $\pm\sqrt{E}$ at time slot 1 and then re-transmitting the identical voltage at the subsequent time slots. See Figure 3.10. A fancy term for this type of scheme is *repetition coding*. In this scheme, the received signals can be described as follows: for $i \in \{1, 2, \ldots, n\}$,

$$Y_i = X_i + Z_i = \begin{cases} -\sqrt{E} + Z_i, & \text{if } B = 0; \\ +\sqrt{E} + Z_i, & \text{if } B = 1. \end{cases} \tag{3.25}$$

A probabilistic model for $\{Z_i\}_{i=1}^n$ Our earlier discussion in Sect. 3.1 allows us to assert that the statistics of the additive noise at each time slot follow a Gaussian

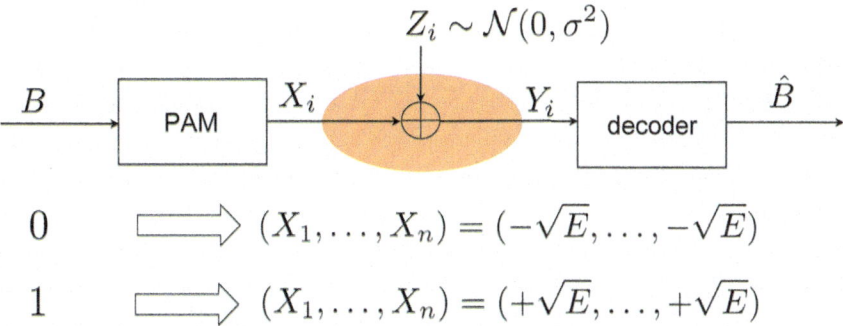

Fig. 3.10 Repetition coding: Sending the same voltage signal over n time slots

distribution. In practice, these statistics tend to remain constant over time. Therefore, it is reasonable to assume that the Gaussian parameters (the mean and the variance) remain unchanged over time. Additionally, the noises exhibit minimal correlation across different time slots. Consequently, another reasonable assumption is that the random variables Z_i's are *mutually independent*. In simpler terms, we assume that the noises are i.i.d. There is a specific terminology that describes such a noise, and it is known as *white* noise. Hence, we refer to this noise as Additive White Gaussian Noise, or simply AWGN. You might be curious about the term *white*. It stems from the fact that AWGN encompasses frequency components that span the entire spectrum, akin to the way *white light* contains all frequencies.

The optimal decision rule Given $(Y_1, \ldots, Y_n) = (y_1, \ldots, y_n)$, the optimal receiver follows the MAP rule by definition:

$$\hat{B}_{\mathsf{MAP}} = \arg\max_{\hat{b} \in \{0,1\}} \mathbb{P}(B = \hat{b} | Y_1 = y_1, \ldots, Y_n = y_n). \tag{3.26}$$

As we have demonstrated a couple of times earlier, assuming $\mathbb{P}(B = 0) = \mathbb{P}(B = 1) = \frac{1}{2}$, the MAP decision rule simplifies to the ML rule:

$$\hat{B}_{\mathsf{MAP}} = \hat{B}_{\mathsf{ML}} = \arg\max_{\hat{b} \in \{0,1\}} f_{Y_1, \ldots, Y_n}(y_1, \ldots, y_n | B = \hat{b}). \tag{3.27}$$

Now consider the likelihood function of interest:

$$f_{Y_1, \ldots, Y_n}\left(y_1, \ldots, y_n | B = \hat{b}\right)$$
$$\stackrel{(a)}{=} f_{Z_1, \ldots, Z_n}\left(y_1 - X_1, \ldots, y_n - X_n | B = \hat{b}\right)$$
$$\stackrel{(b)}{=} f_{Z_1, \ldots, Z_n}\left(y_1 - (2\hat{b} - 1)\sqrt{E}, \ldots, y_n - (2\hat{b} - 1)\sqrt{E} | B = \hat{b}\right)$$
$$\stackrel{(c)}{=} f_{Z_1, \ldots, Z_n}\left(y_1 - (2\hat{b} - 1)\sqrt{E}, \ldots, y_n - (2\hat{b} - 1)\sqrt{E}\right)$$
$$\stackrel{(d)}{=} \prod_{i=1}^{n} f_{Z_i}\left(y_i - (2\hat{b} - 1)\sqrt{E}\right)$$
$$\stackrel{(e)}{=} \prod_{i=1}^{n} \frac{1}{\sqrt{2\pi}\sigma} \exp\left(-\frac{1}{2\sigma^2}\left(y_i - (2\hat{b} - 1)\sqrt{E}\right)^2\right)$$
$$= \left(\frac{1}{\sqrt{2\pi}\sigma}\right)^n \exp\left(-\frac{1}{2\sigma^2}\sum_{i=1}^{n}\left(y_i^2 + (2\hat{b} - 1)^2 E - 2(2\hat{b} - 1)\sqrt{E}y_i\right)\right)$$
$$\stackrel{(f)}{=} \left(\frac{1}{\sqrt{2\pi}\sigma}\right)^n \exp\left(-\frac{1}{2\sigma^2}\sum_{i=1}^{n}(y_i^2 + E)\right) \times \exp\left(\frac{1}{\sigma^2}(2\hat{b} - 1)\sqrt{E}\sum_{i=1}^{n}y_i\right)$$

where (a) follows from the fact that the event $Y_i = y_i$ is equivalent to the event $Z_i = y_i - X_i$; (b) is because the simplest expression of our encoding rule reads

$X_i = (2B - 1)\sqrt{E}$; (c) is due to the independence between $\{Z_i\}_{i=1}^n$ and B; (d) is due to the independence of the additive noises at different time slots; (e) comes from the pdf of the Gaussian noise; and (f) comes from the fact that $(2\hat{b} - 1)^2 = 1$ for any $\hat{b} \in \{0, 1\}$.

One important observation here is that the first and second terms in the last equality $(\frac{1}{\sqrt{2\pi}\sigma})^n \exp\left(-\frac{1}{2\sigma^2} \sum_{i=1}^n (y_i^2 + E)\right)$ are *irrelevant* of $\hat{b}$. Hence, by applying a logarithmic function (an increasing function) to the remaining part, we obtain:

$$\hat{B}_{\mathsf{ML}} = \arg \max_{\hat{b} \in \{0,1\}} f_{Y_1, \dots, Y_n}(y_1, \dots, y_n | B = \hat{b})$$

$$= \arg \max_{\hat{b} \in \{0,1\}} \exp\left(\frac{(2\hat{b} - 1)\sqrt{E}}{\sigma^2} \sum_{i=1}^n y_i\right) \qquad (3.28)$$

$$= \arg \max_{\hat{b} \in \{0,1\}} (2\hat{b} - 1) \sum_{i=1}^n y_i.$$

Notice that the sum $\sum_{i=1}^n y_i$ plays a key role in the decision:

$\sum_{i=1}^n y_i \geq 0 \Longrightarrow \hat{B}_{\mathsf{ML}} = 1$;

$\sum_{i=1}^n y_i < 0 \Longrightarrow \hat{B}_{\mathsf{ML}} = 0$.

Again we declare $\hat{B}_{\mathsf{ML}} = 1$ when the equality occurs.

We can make an interesting interpretation for this rule. Essentially, we are consolidating all the received signals and creating a kind of "collective consensus". If the overall sum is positive, we conclude that the bit must correspond to a positive signal (and vice versa). This can be seen as a form of majority voting. More precisely, a soft version of majority voting. Why? Here what we mean by majority voting is that: we first make a binary hard decision for each received signal ($y_i \geq 0 \to \hat{B}_i = 1$; otherwise, $\hat{B}_i = 0$), and then choose the bit that garners the most votes ($\sum_{i=1}^n \hat{B}_i \geq \frac{n}{2} \to \hat{B}_{\mathsf{voting}} = 1$). In contrast, the ML rule involves aggregating y_i's before making a decision. So, it arrives at a single decision based on a sort of soft version of all the individual opinions. This rule coincides with our natural guess: the NN decision rule, but in this case, applied to the sum $\sum_{i=1}^n Y_i$.

Analysis of the error probability Now let us analyze the error probability:

$$P_e = \mathbb{P}(\hat{B}_{\mathsf{ML}} \neq B). \qquad (3.29)$$

There are two random variables involved in the error event: B and $\hat{B}_{\mathsf{ML}}$. How can we handle this? The answer lies in employing the *total probability law*. Applying this law, we obtain:

$$\mathbb{P}(\hat{B}_{\mathsf{ML}} \neq B) = \mathbb{P}(B = 0, \hat{B}_{\mathsf{ML}} \neq B) + \mathbb{P}(B = 1, \hat{B}_{\mathsf{ML}} \neq B)$$

$$= \mathbb{P}(B = 0)\mathbb{P}(\hat{B}_{\mathsf{ML}} = 1 | B = 0) + \mathbb{P}(B = 1)\mathbb{P}(\hat{B}_{\mathsf{ML}} = 0 | B = 1) \qquad (3.30)$$

where the second equality comes from the definition of conditional probability. As mentioned earlier, we assume that the a priori probabilities are equal (to 0.5 each). Focus on one of the error events by assuming that the information bit was actually 0. Then with the NN rule, we get:

$$\mathbb{P}(\hat{B}_{\mathrm{ML}} = 1 | B = 0) \stackrel{(a)}{=} \mathbb{P}\left(\sum_{i=1}^{n} Y_i \geq 0 | B = 0\right)$$

$$= \mathbb{P}\left(-n\sqrt{E} + \sum_{i=1}^{n} Z_i \geq 0 | B = 0\right)$$

$$\stackrel{(b)}{=} \mathbb{P}\left(\sum_{i=1}^{n} Z_i \geq n\sqrt{E}\right) \tag{3.31}$$

$$= \mathbb{P}\left(\frac{\sum_{i=1}^{n} Z_i}{\sqrt{n}\sigma} \geq \frac{\sqrt{n}E}{\sigma}\right)$$

$$\stackrel{(c)}{=} \int_{\frac{\sqrt{n}E}{\sigma}}^{\infty} \frac{1}{\sqrt{2\pi}} e^{-\frac{z^2}{2}} dz.$$

where (a) is due to the NN rule[2]; (b) is because of the independence between $\{Z_i\}_{i=1}^{n}$ and B; and (c) follows from the fact that $\frac{\sum_{i=1}^{n} Z_i}{\sqrt{n}\sigma} \sim \mathcal{N}(0, 1)$. In step (c), we utilize the property that the sum of Gaussian random variables remains Gaussian. The expression in the last equation in (3.31) denotes the area in the right tail of the standard Gaussian pdf. This quantity frequently arises in the performance analysis of various communication schemes. There is a specific term used to describe this expression, known as the Q-function:

$$Q(z) := \int_{z}^{\infty} \frac{1}{\sqrt{2\pi}} e^{-\frac{t^2}{2}} dt. \tag{3.32}$$

Applying the Q-function, we can then obtain:

$$\mathbb{P}(\hat{B}_{\mathrm{ML}} = 1 | B = 0) = Q\left(\frac{\sqrt{n}E}{\sigma}\right). \tag{3.33}$$

The Q-function is readily available in probability textbooks and on wikipedia. It can also be computed numerically by using a command "erfc" in Python (defined in the scipy.special package):

$$\mathrm{erfc}(x) := \int_{x}^{\infty} \frac{2}{\sqrt{\pi}} e^{-t^2} dt.$$

---

[2] For simplicity of analysis, we assume that for the event $Y = 0$, $\hat{B}$ is decided to be 1. Since it is the probability-zero event, the error probability analysis remains the same.

The relation between $Q(a)$ and $\texttt{erfc}(x)$ given by:

$$Q(a) := \int_a^\infty \frac{1}{\sqrt{2\pi}} e^{-\frac{z^2}{2}} dz$$

$$\overset{(a)}{=} \int_{\frac{a}{\sqrt{2}}}^\infty \frac{1}{\sqrt{\pi}} e^{-t^2} dt$$

$$= \frac{1}{2} \cdot \texttt{erfc}\left(\frac{a}{\sqrt{2}}\right)$$

where the second step comes from the change of variable $t := \frac{z}{\sqrt{2}}$ $(dz = \sqrt{2}dt)$.

Now consider the other error probability $\mathbb{P}(\hat{B}_{\text{ML}} = 0 | B = 1)$. Actually the error does not depend on which information bit is transmitted. The complete *symmetry* of the mapping from the bit values to the voltage levels and the NN decision rule strongly suggests that the two error probabilities should be equal. For the sake of completeness, we will go through the calculation for $\mathbb{P}(\hat{B}_{\text{ML}} = 0 | B = 1)$ and will confirm that it is indeed the case:

$$
\begin{aligned}
\mathbb{P}(\hat{B}_{\text{ML}} = 0 | B = 1) &= \mathbb{P}\left(\sum_{i=1}^n Y_i < 0 | B = 1\right) \\
&= \mathbb{P}\left(n\sqrt{E} + \sum_{i=1}^n Z_i < 0 | B = 1\right) \\
&= \mathbb{P}\left(\frac{\sum_{i=1}^n Z_i}{\sqrt{n}\sigma} < -\frac{\sqrt{nE}}{\sigma}\right) \\
&= Q\left(\frac{\sqrt{nE}}{\sigma}\right)
\end{aligned}
\tag{3.34}
$$

where the last equality is due to the symmetry of the Gaussian pdf. Applying this (together with (3.33)) into (3.30), we obtain:

$$P_e = Q\left(\frac{\sqrt{nE}}{\sigma}\right). \tag{3.35}$$

Signal-to-Noise energy Ratio (SNR) The expression (3.35) contains a crucial metric that frequently arises in communication systems. That is, the ratio between the energy budget E and the noise variance σ^2:

$$\text{SNR} := \frac{E}{\sigma^2}. \tag{3.36}$$

This ratio is called SNR, which stands for Signal-to-Noise energy Ratio. It serves as an intuitive metric that gauges the quality of a given channel: the larger the SNR, the better the channel. In many practically relevant channels, a typical SNR value falls in the range of 0 to 20 dB. The dB scale is often used for SNR, where SNR dB $= 10 \log_{10}$ SNR. So the typical range can be translated to SNR $= 1 \sim 100$.

Achieving the desired P_e level We can anticipate a decrease in the error probability P_e as n increases. Using Python, we can demonstrate that for a typical SNR value, say 10 dB, we can achieve the desired range of $P_e = 10^{-6} \sim 10^{-10}$ with only a few time slots, typically around 4 and 5.

Look ahead

In this section, we have analyzed the probability of error when employing the optimal ML receiver. One natural next step is to investigate how fast the error probability decays in the number of time slots n. However, a challenge arises as we lack an analytical expression for P_e, making it difficult to understand its behavior with respect to n. The good news is that there exists an approximation technique that allows us to derive an analytical expression that closely approximates the exact P_e. According to this approximation, we can demonstrate that the error probability diminishes exponentially as the number of time slots increases. In the next section, we will delve into this topic and provide empirical validation using Python.

3.4 Communication: Error Probability and **Python** Simulation

Recap

In the previous section, we have introduced the repetition coding scheme and derived the probability of error under the optimal ML receiver:

$$P_e = Q\left(\frac{\sqrt{nE}}{\sigma}\right) = Q\left(\sqrt{n\mathsf{SNR}}\right) \tag{3.37}$$

where $Q(z) := \int_z^\infty \frac{1}{\sqrt{2\pi}} e^{-\frac{t^2}{2}} dt$ and n denotes the number of time slots employed. However, this Q-function-based expression doesn't provide insights into the relationship between P_e and the number of time slots, as it is not in analytical form. As previously mentioned, there is an approximation technique that enables us to derive an analytical expression that closely approximates the exact P_e.

Outline

In this section, we will employ the approximation technique to demonstrate that the error probability decays exponentially as the number of time slots increases. This section comprises three parts. First, we will introduce an approximation for the Q-function to obtain an approximate analytical expression for the error probability. Next, we will conduct a Python simulation to discuss the accuracy of this approximation. We will also empirically confirm the analysis (3.37) via Python.

Approximate analytical expression for P_e Recall the error probability (3.37):

$$P_e = Q\left(\sqrt{n\mathsf{SNR}}\right). \tag{3.38}$$

While the Q-function can be found in standard statistical tables, it is useful for communication engineers to have a rule of thumb for how sensitive this SNR "knob" is in terms of P_e each setting offers. For instance, it would be useful to know by how much P_e decreases if we double SNR. To do this, it helps to use the following approximation:

$$Q(a) \approx \frac{1}{2} e^{-\frac{a^2}{2}}. \tag{3.39}$$

Here we use a rough approximation that we denote by "≈". This approxi-mation means that the exponent of $Q(a)$ is very close to that of $\frac{1}{2}e^{-\frac{a^2}{2}}$, i.e., $\log Q(a) \approx \log(\frac{1}{2}e^{-\frac{a^2}{2}})$. The reason that we consider this rough approximation is that the exponent plays an enough role as a proper measure when we deal with the probability of error. For example, the following probabilities of error (10^{-7} and $2 \cdot 10^{-7}$) are considered to provide almost the same performance although those dif-fer by two times, since they have roughly the same exponent. In fact, $\frac{1}{2}e^{-\frac{a^2}{2}}$ is an upper bound of the Q-function, and you will have a chance to check this rigorously in Problem 7.3.

Applying this approximation into (3.38), we obtain:

$$Q\left(\sqrt{n\text{SNR}}\right) \approx \frac{1}{2}e^{-\frac{n\text{SNR}}{2}}. \tag{3.40}$$

From this, we can observe that the error probability exhibits an *exponential* decay as the number of time slots increases. We can also verify this exponential relation-ship by conducting a **Python** simulation.   Since the logarithmic curve of P_e in Figure 3.11 displays a linear decrease, this indicates that the actual P_e follows an exponential decay pattern.

Likewise, **SNR** has an exponential impact on the probability of error. Hence, we can anticipate the extent to which P_e drops when we double the energy, i.e., $\tilde{E} = 2E$. For instance, in the case of $n = 1$, when we double the energy, we have:

$$Q\left(\sqrt{2\text{SNR}}\right) \approx \frac{1}{2}e^{-\text{SNR}} \approx \left[Q\left(\sqrt{\text{SNR}}\right)\right]^2. \tag{3.41}$$

Fig. 3.11 P_e decays exponentially in the number of time slots n

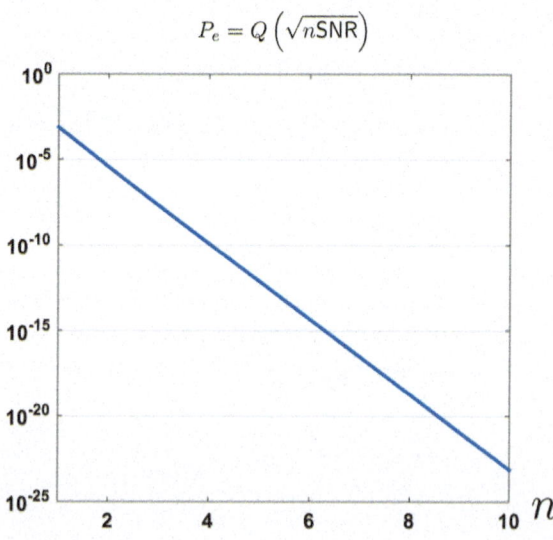

$$P_e = Q\left(\sqrt{n\text{SNR}}\right)$$

Doubling the energy has thus squared the probability of error and squaring a number less than 1 means that we have significantly reduced the probability of error.

Python simulation: Accuracy of the approximation To assess the accuracy of the approximation, we perform a Python simulation. See below a Python code for plotting P_e (3.38) and its approximation (3.40) in the case of $n = 1$, over a typical range of SNR from 0 to 20 dB.

```python
import numpy as np
from scipy.special import erfc
import matplotlib.pyplot as plt

SNRdB = np.arange(0,21,1)  # interval of SNR in dB scale
SNR = 10**(SNRdB/10)       # SNRdB = 10*log10(SNR)

# Q-function
Qfunc = 1/2*erfc(np.sqrt(SNR/2))
# An upper bound of the Q-function
Approx = 1/2*np.exp(-SNR/2)

plt.figure(figsize=(5,5), dpi=150)
plt.plot(SNRdB, Approx, 'r*:', label='0.5*exp(-SNR/2)')
plt.plot(SNRdB, Qfunc, label='Q(sqrt(SNR))')
plt.yscale('log')
plt.xlabel('SNR (dB)')
plt.grid()
plt.title('Q function vs. its approximation')
plt.legend()
plt.show()
```

Figure 3.12 illustrates the Q-function and its corresponding approximation. Observe that these two curves closely align on an exponential scale, with the approximation serving as an upper bound.

Python simulation: Empirical verification of the analysis (3.38) We aim to empirically validate the analysis (3.38) using Python. To keep it simple, we will focus on the case of $n = 1$. For empirical verification, we do multiple independent trials, say m, for sending one bit, using the optimal MAP rule under PAM transmission. By counting the number of error events out of m independent transmissions, we will compute an empirical error rate for different values of SNR, thereby demonstrating that the empirical error rate approaches the analysis result (3.38) with an increase in m. This empirical process for convergence verification is called the *Monte Carlo simulation* (Kroese et al. 2014).

First we construct m independent bits. We employ the built-in function `bernoulli`, defined in the `scipy.stats` package.

```python
from scipy.stats import bernoulli
import numpy as np

m = 10 # number of independent bits
```

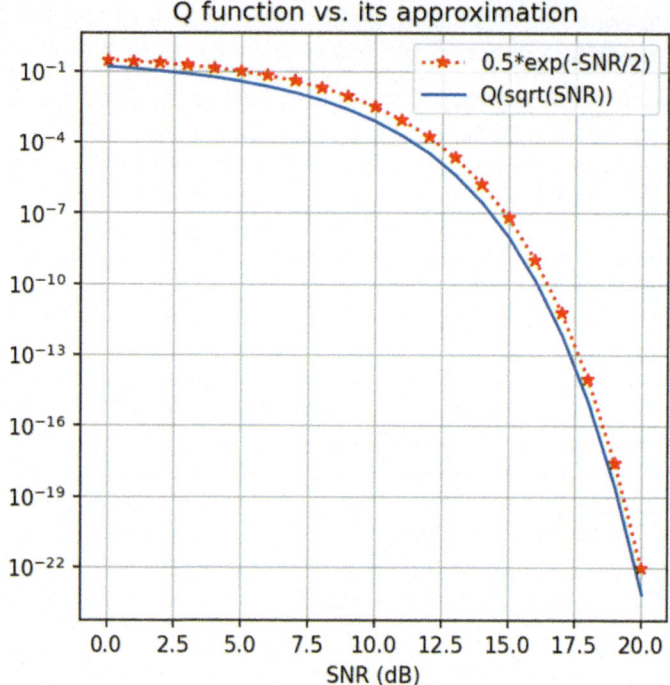

Fig. 3.12 Comparison of $Q(\sqrt{\mathrm{SNR}})$ and its upper bound $\frac{1}{2}e^{-\frac{\mathrm{SNR}}{2}}$

```
X = bernoulli(0.5)

# Generate m independent bits
b_samples = X.rvs(m)
print(b_samples)
```

```
[0 1 1 0 1 0 0 0 1 0]
```

Without loss of generality, assume the unit noise variance ($\sigma^2 = 1$). Then, the energy budget E is the same as SNR. Using the energy level, we can generate m transmitted signals. These signals are then sent over the additive Gaussian channel with an additive noise $\sim \mathcal{N}(0, 1)$.

```
SNRdB = 0
SNR = 10**(SNRdB/10)

# Use PAM to construct m transmitted signals
x_samples = np.sqrt(SNR)*(2*b_samples - 1)

# Passes through the Gaussian channel ~ N(0,1)
y_samples = x_samples + np.random.randn(m)
```

```
print(x_samples)
print(y_samples)
```

```
[-1.   1.   1.  -1.   1.  -1.  -1.  -1.   1.  -1.]
[-0.85448476   2.24874648 -0.06785438 -1.40327708   1.5978796
 -1.76768827 -0.20014252 -1.64453905   0.67536843   0.83061596]
```

Next, we employ the MAP rule to estimate $\hat{b}(y)$. In our case where we assume equal a priori probabilities and symmetric Gaussian noise centered around a mean of 0, the MAP rule simplifies to the ML rule, which coincides with the NN rule.

```
# MAP rule (=ML=NN in this case)
bhat = (y_samples > 0 )*1
print(bhat)
print(b_samples)
```

```
[1 0 0 0 0 0 0 0 1 0]
[1 0 0 0 0 0 0 1 1 0]
```

By comparing $\hat{b}(y)$ to b, we finally compute an empirical error rate.

```
# Compute an empirical error rate
num_errors = sum(b_samples != bhat)
error_rate = num_errors/m
print(error_rate)
```

```
0.1
```

Iterating the above procedures for different values of SNR, we plot the empirical error rate as in Figure 3.13. See below for the entire code. We set the number of independent trials as a large enough number, say $m = 200000$, in order to ensure an accurate error rate.

```
from scipy.stats import bernoulli
import numpy as np

m = 200000
X = bernoulli(0.5)
# SNR range
SNRdB = np.arange(0,13,1)
SNR = 10**(SNRdB/10)
error_rate = np.zeros(len(SNR))

for i, snr in enumerate(SNR):
    # Generate m independent bits
    b_samples = X.rvs(m)
    # Use the PAM to construct m transmitted signals
    x_samples = np.sqrt(snr)*(2*b_samples - 1)
    # Passes through the Gaussian channel ~ N(0,1)
    y_samples = x_samples + np.random.randn(m)
    # MAP rule (=ML=NN in this case)
    bhat = (y_samples > 0 )*1
    # Compute an empirical error rate
```

Fig. 3.13 Empirical error rate when the number of independent trials $m = 200000$

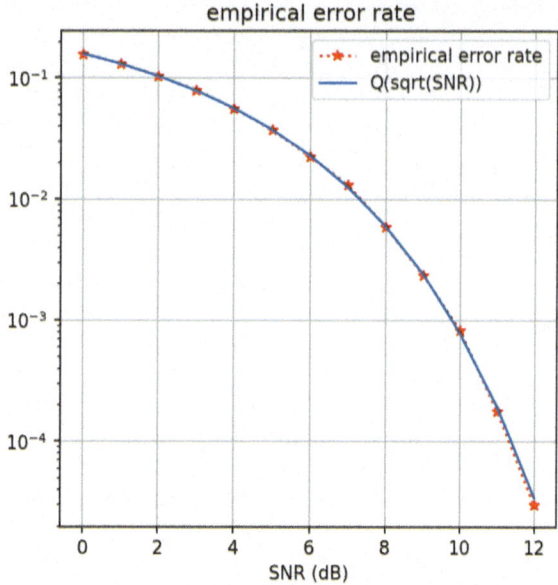

```
    num_errors = sum(b_samples != bhat)
    error_rate[i] = num_errors/m

# Q-function
Qfunc = 1/2*erfc(np.sqrt(SNR/2))

plt.figure(figsize=(5,5), dpi=150)
plt.plot(SNRdB,error_rate,'r*:',label='empirical error rate')
plt.plot(SNRdB,Qfunc,label='Q(sqrt(SNR))')
plt.yscale('log')
plt.xlabel('SNR (dB)')
plt.grid()
plt.title('empirical error rate')
plt.legend()
plt.show()
```

Look ahead

Over the past sections, we have demonstrated the role of probabilistic modeling and the MAP/ML principles for communication. In the next section, we will move onto the second application: Community detection in social networks.

Problem Set 7

Problem 7.1 (*Bits*) In Sect. 3.1, we learned that *bits* is a common currency of information that can represent any type of information sources. Consider an image signal that consists of many pixels. Each pixel is represented by three real values which indicate the intensity of red, green and blue color components respectively. Here we assume that each value is *quantized*, taking one of 256 equal-spaced values in $[0, 1)$, i.e., $\frac{0}{256}, \frac{1}{256}, \frac{2}{256}, \ldots, \frac{254}{256}, \frac{255}{256}$. Explain how bits can represent such an image source.

Problem 7.2 (*A simple noise example*) Consider a simple additive noise channel:

$$Y = X + Z \tag{3.42}$$

where X indicates a transmitted signal with E energy budget and Z denotes a noise strictly within $\pm\sigma$. Suppose we accept a 10% error rate and know some statistics on the noise: $\mathbb{P}(-\frac{\sigma}{2} \leq Z \leq \frac{\sigma}{2}) = 0.9$. Suppose we use only one time slot and have the same energy constraint as in the 0% error rate case (perfect transmission). How many bits (at least) can we transmit more? For simplicity, we assume that the number of bits can be any real value, not limited to an integer.

Problem 7.3 (*An upper bound of the Q-function*)

(a) Show that for $z \geq a$ and $a \geq 0$,

$$z^2 \geq (z - a)^2 + a^2. \tag{3.43}$$

(b) Using part (a), derive

$$Q(a) := \int_a^\infty \frac{1}{\sqrt{2\pi}} e^{-\frac{z^2}{2}} dz \leq \frac{1}{2} e^{-\frac{a^2}{2}}, \quad a > 0. \tag{3.44}$$

(c) Using a skeleton **Python** code provided in Sect. 3.4, plot the $Q(a)$ and the upper bound of $\frac{1}{2} e^{-\frac{a^2}{2}}$. Use a built-in function "yscale" (placed in `matplotlib.pyplot`) to exhibit y-axis in a logarithm (base 10) scale.

Problem 7.4 (*Useful bounds in probability*) Consider two events A_1 and A_2. Prove the following inequalities:

(a) $\mathbb{P}(A_1 \cap A_2) \leq \mathbb{P}(A_1)$.

(b) $\mathbb{P}(A_1) \leq \mathbb{P}(A_1 \cup A_2)$.

(c) $\mathbb{P}(A_1 \cup A_2) \leq \mathbb{P}(A_1) + \mathbb{P}(A_2)$.

Remark: The bound in part (c) is called the union bound.

Problem 7.5 (*Majority voting*) A transmitted signal X is equally likely to be $+1$ or -1 and is received at n antennas:

$$Y_i = X + Z_i, \quad i \in \{1, \ldots, n\}$$

where the additive noise Z_i at antenna i is Gaussian with mean zero and variance σ^2, and is independent across all antennas and of the transmitted signal X.

(a) Let $\hat{X}_i = \mathsf{ML}[X|Y_i]$ be the output of the ML receiver that decodes X based *solely* on Y_i.

(b) Derive the ML receiver that decodes X based on $(\hat{X}_1, \ldots, \hat{X}_n)$ (yet without access to $(Y_1, \ldots, Y_n)$):

$$\hat{X} = \mathsf{ML}[X|\hat{X}_1, \ldots, \hat{X}_n]. \tag{3.45}$$

(c) Derive the error probability when using the ML receiver derived in part (b). Use the Q-function notation if needed.

Problem 7.6 (*The optimal receiver principle*) Consider the optimal decision rule for X given (Y_1, Y_2):

$$Y_1 = X + Z_1;$$
$$Y_2 = Z_1 + Z_2.$$

Here X is equally likely to be $\sqrt{E}$ or $-\sqrt{E}$ and (Z_1, Z_2) are i.i.d. $\sim \mathcal{N}(0, \sigma^2)$, being independent of X.

(a) Is Y_1 the only information employed in the optimal decision rule? Give an intuitive (not necessarily rigorous) explanation to your answer.

(b) Find the optimal receiver for X given (Y_1, Y_2), and provide a rigorous explanation for your answer in part (a).

(c) Derive the error probability when using the optimal receiver derived in part (b).

(d) Consider the best receiver for X using Y_1 only. Find the additional energy, if any, needed to achieve the same error probability as that of the optimal receiver in part (b).

Problem 7.7 (*The MAP versus ML rules*) Consider an additive Gaussian noise channel with $Z \sim \mathcal{N}(0, \sigma^2)$. Consider a simple transmission scheme in which information bit 0 is mapped to $-\sqrt{E}$ and 1 to $+\sqrt{E}$. Suppose that the a priori probabilities are $\mathbb{P}(B = 0) = \frac{1}{4}$ and $\mathbb{P}(B = 1) = \frac{3}{4}$.

(a) Given the received signal $Y = y$, derive the MAP rule.

(b) Evaluate the error probability when using the MAP rule: $\mathbb{P}(\hat{B}_{\mathsf{MAP}} \neq B)$. Use the Q-function notation.

(c) Given the received signal $Y = y$, derive the ML rule.

(d) Evaluate the error probability when using the ML rule: $\mathbb{P}(\hat{B}_{\mathsf{ML}} \neq B)$. Use the Q-function notation.

(e) Compare the two error probabilities derived in parts (b) and (d) respectively. Which is smaller? Explain why.

Problem 7.8 (*The MAP versus ML versus NN rules*) A transmitter wishes to send an information bit $B \in \{0, 1\}$ to a receiver. The encoding strategy is the following simple mapping: $X = B$. Given X, the channel output Y respects a Gaussian distribution: $\mathcal{N}(0.1 + B, 0.1 + B)$. Assume that $B \sim \mathsf{Bern}(0.6)$.

(a) Given $Y = y$, derive the MAP rule $\hat{B}_{\mathsf{MAP}}$. You should describe how $\hat{B}_{\mathsf{MAP}}$ is decided as a function of y.

(b) What is the error probability when using the MAP rule? Use the Q-function notation.

(c) Given $Y = y$, derive the ML rule $\hat{B}_{\mathsf{ML}}$. Again you should describe how $\hat{B}_{\mathsf{ML}}$ is decided as a function of y.

(d) What is the error probability when using the ML rule?

(e) Given $Y = y$, derive the NN rule $\hat{B}_{\mathsf{NN}}$. Again you should describe how $\hat{B}_{\mathsf{NN}}$ is decided as a function of y.

(f) What is the error probability when using the NN rule?

Problem 7.9 (*The ML and MAP rules under multiple observations*) Given $X = x$, the random variables $\{Y_n, n \geq 1\}$ are i.i.d. according to an exponential distribution with mean x, i.e., $f_{Y_i}(y) = xe^{-xy}$, $\forall i$. Assume that $\mathbb{P}(X = 1) = 1 - \mathbb{P}(X = 2) = p \in (0, 1)$.

(a) Find the ML rule for decoding X given $Y^n := \{Y_1, \ldots, Y_n\}$.

(b) Find the MAP rule for decoding X given Y^n.

Problem 7.10 (*True or False?*)

(*a*) A transmitter wishes to send an information bit $B \in \{0, 1\}$ to a receiver. The transmitted signal X is simply B. Given X, the channel output Y follows a Gaussian distribution: $\mathcal{N}(B, 1 + B)$. The ML decision rule is the same as the NN rule.

(*b*) Suppose we send $X = \sqrt{E}$ or $-\sqrt{E}$, depending on an information bit, over an additive Gaussian noise channel in which the noise has zero mean and is independent of x. In Sect. 3.2, we learned that the MAP decision rule is optimal in a sense of minimizing the probability of error. Suppose the a priori probabilities are equal, i.e., $\mathbb{P}(x = \sqrt{E}) = \mathbb{P}(x = -\sqrt{E}) = \frac{1}{2}$. Then the MAP rule reduces to the NN rule.

(*c*) Suppose X is a discrete random variable which takes -1 or $+1$ with the equal probabilities. Let Y be a Gaussian random variable independent of X. Then $Z = X + Y$ is Gaussian.

3.5 Social Networks: Probabilistic Modeling

Recap
Over the past four sections, we have explored the role of probability in communication. We first focused on an uncertain entity that arises in communication. That is, *noise*. By translating the physical properties of noise into a mathematical framework and leveraging the central limit theorem, we illustrated that the noise can be modeled as a *Gaussian* random variable. Next we demonstrated the role of the MAP and ML principles in the design of the optimal receiver, in the context of an additive Gaussian noise channel. Finally we conducted an in-depth analysis of the performance of the optimal receiver via a probabilistic metric, *probability of error*.

Outline
In this section, we will delve into our second application. The application of interest is one prominent problem that arises in social networks: *Community detection*. The section consists of four parts. We will begin by providing an explanation of community detection and its relevance to the concept of probability. Subsequently, we will formulate the corresponding mathematical problem and introduce a fundamental question centered around an interesting concept, called *phase transition*. Following this, we will establish a connection with a communication problem, shedding light on community detection through a probabilistic lens. Finally, we will elucidate the answer to the above question by leveraging the principles of probability.

Community detection (Girvan & Newman 2002; Fortunato 2010; Abbe 2017) A community is a group of individuals who share common interests and/or reside in the same geographic area. Community detection is the task of identifying groups of individuals who exhibit similarities. For a visual representation, refer to Figure 3.14. Consider a scenario in which users are represented as nodes within a graph (as depicted on the left side of Figure 3.14), and there are two communities, labeled as the blue and red communities. Here, the objective of the problem is to determine the affiliation of each user to either the blue or red community, as visualized on the right side of Figure 3.14. Another term often used interchangeably with community detection is *clustering* (Bansal et al. 2004; Jalali et al. 2011). Notice that the output nodes are clustered either into the blue or red community.

You may wonder why this problem is of importance. The reason lies in the fact that this problem arises in a multitude of applications in the IT field. These applications span various domains, including social networks (e.g., Facebook, LinkedIn, Twitter, etc.) and biological networks (Chen & Yuan 2006). In social networks, the

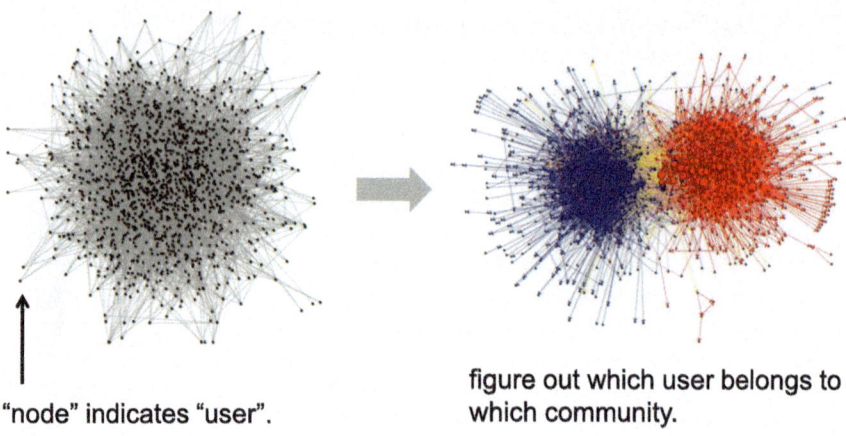

"node" indicates "user".

figure out which user belongs to
which community.

Fig. 3.14 Community detection in picture

knowledge of community structures can aid in identifying target groups for adver-
tising and recommending products. In biological networks, community detection
plays a pivotal role in DNA sequencing (Browning & Browning 2011; Das & Vikalo
2015; Chen et al. 2016; Si et al. 2014), which can be instrumental in cancer detection
and personalized medicine. It's worth noting that the utility of community detection
extends beyond these examples, encompassing a wide range of applications.

Problem formulation To perform community detection, we first need to consider
the type of information that is accessible. In many applications, one readily available
source of information is *relationship information*. This includes data on friendships
in Meta's network (formerly Facebook), connections on LinkedIn, or followers on
Twitter, among others. However, the actual community-type information is typically
not disclosed in practice. In Meta, only friendships information is accessible, while
the specific community to which a user belongs remains undisclosed. In situations
that concern privacy, such community-type information may be legally prohibited
from being made public, even if Meta possesses such information.

In order to give you a concrete feel as to what the relationships information
looks like, let us give you an example. Suppose that x_i indicates the community
membership of user i, and we associate x_i with node i. Then, one natural function
that one can think of is a relationship function operating on the values assigned to two
nodes, e.g., x_i and x_j. For instance, when $x_1 = x_2$, the relationship function would
yield $y_{12} = 0$; otherwise $y_{12} = 1$. To be specific, $y_{12} = x_1 \oplus x_2$ where $\oplus$ denotes the
bit-wise modulo sum.

Given a collection of parity values y_{ij}'s, the goal of the problem is to decode
$\mathbf{x} := [x_1, x_2, \ldots, x_n]$. However, aiming for an exact decoding of $\mathbf{x}$ is an unattainable
goal if we rely solely upon the parity information y_{ij}'s. To understand this, consider an
example illustrated in Figure 3.15. Suppose that $n = 4$ and (y_{12}, y_{13}, y_{14}) are given as
$(0, 1, 1)$. If $x_1 = 0$, then $\mathbf{x} = [0, 1, 1, 0]$. But an ambiguity arises regarding the value

Fig. 3.15 An example of community detection

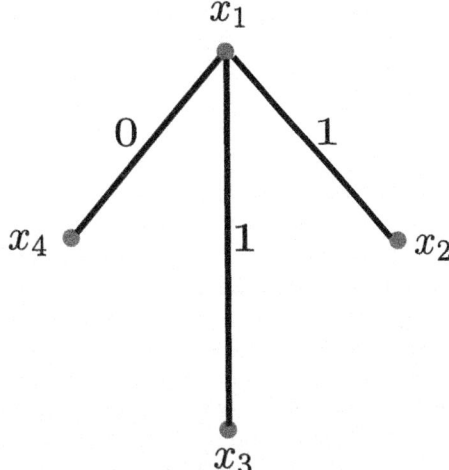

of x_1 because there is no way to uniquely infer x_1 only from the provided similarity information. The alternative solution $\mathbf{x} = [1, 0, 0, 1]$ is equally valid, meaning that there are always two valid solutions: (i) the correct one; and (ii) its flipped counterpart. To resolve this ambiguity, we should relax the goal as follows: decoding either $\mathbf{x}$ or $\mathbf{x} \oplus \mathbf{1}$ from y_{ij}'s.

Challenges From the example in Figure 3.15, you may guess that solving the problem is not that difficult. In the big data era, however, we face two significant challenges. The first challenge arises from the fact that the number of nodes (e.g., users in social networks), say n, is often exceedingly large in a wide variety of applications. For instance, as of April 2023, the number of Facebook users is 2.989 billion (the order of 10^9) (Meta 2023). This is around the one third of the world population. See Figure 3.16. Hence, it is reasonable to anticipate that the available information consists of only a fraction of the similarity parities. Note that the number of all possible pairs is astronomically large. In the above example, $\binom{n}{2} \approx 4.309 \times 10^{18}$.

In fact, this may not pose a significant challenge in certain scenarios where we have the flexibility to select any pair of two nodes as per our wish. In such cases, the task of community detection becomes straightforward. By choosing a set of consecutive pairs, like $(y_{12}, y_{23}, y_{34}, \ldots, y_{(n-1)n})$, one can easily decode $\mathbf{x}$ up to a global shift. It's worth noting that the number of pairwise measurements, $n - 1$, scales linearly with n, which is considerably fewer than the total number of pairs $\binom{n}{2}$. However, the situation can be challenging due to another factor. The challenge arises when, in many applications like Facebook, the similarity parities are provided *passively* within a given context. Here, "passively given" means that such information is not obtained at our discretion but is provided by the context. For example, the friendships information on Facebook is simply given by the context, and we cannot arbitrarily choose pairs of users to figure out if they are friends or not. In such cases, one might assume that the similarity information is offered in a *random* manner. This is where

Fig. 3.16 The number of facebook users is around 2.989 billion as of April 2023

the concept of *probability* comes into play. One natural probabilistic assumption made in the literature is that the information for any pair of users is provided with probability, say p, independently across all the other pairs.

A question While community detection seems very challenging due to the partial and random nature of the pairwise measurements, we have a good news. The good news is that as hinted from the example in Figure 3.15, decoding **x** does not necessarily require the entire observation of every measurement pair. Notice that the pairs are quite dependent with each other; hence, one can guess partial pairs might suffice to decode **x**. Then, one natural question arises.

> **? Question**
> Is there a *fundamental limit* on the number of measurement pairs required to make community detection possible?

An interesting phenomenon, called *phase transition*, occurs here. Phase transition is a fundamental occurrence frequently encountered in physics and information theory. In this context, it means that there exists a sharp threshold on the observation probability p above which reliable community detection is possible and below which it is impossible no matter what and whatsoever. The sharp threshold, also known as the fundamental limit, is solely dependent on the characteristics of the considered social network, regardless of any community detection strategies. Hence, it is like a scientific law dictated by nature. To gain a deeper understanding of this concept, it is crucial to clarify what is meant by *reliable* community detection. For the rest of this section, we will explore the precise definition of reliability and elucidate the sharp threshold, say p^*.

Translation to a communication problem Under the partial and random observation setting, y_{ij} is *statistically* related to x_i and x_j. Therefore, community detection can be viewed as an *inference problem*, indicating a close association with a communication problem. By transforming community detection into a communication problem, we can formulate a precise mathematical statement regarding the limit.

Figure 3.17 illustrates the translation to a communication problem. Consider **x** as the information source we aim to transmit, and $\hat{\mathbf{x}}$ as its estimate. It's important to note that we are provided with pairwise measurements. So one can think of a block diagram at the transmitter which converts **x** into pairwise measurements, say x_{ij}'s. Here $x_{ij} := x_i \oplus x_j$. Actually one can view this as a kind of *encoder*. Given that only a subset of these pairwise measurements is observed in a *random* fashion, we introduce another block that implements partial and random measurements to extract a subset of x_{ij}'s. This processing can be likened to a *channel* where the output y_{ij} admits:

$$y_{ij} = \begin{cases} x_{ij}, & \text{w.p. } p; \\ \text{erasure}, & \text{w.p. } 1 - p \end{cases}$$

where p denotes the observation probability and **erasure** indicates empty information. In the communication & information theory literature, this process is modeled as an *erasure channel* with erasure probability $1 - p$. Hence, we employ the terminology **erasure**. These y_{ij}'s are then fed into an algorithm block, thus yielding $\hat{\mathbf{x}}$. The algorithm block can be interpreted as a decoder.

Performance metrics & an optimization problem As in the communication setting, we can think of two performance metrics. The first is the one that we are interested in characterizing the limit on. That is, the number of pairwise measurements that are observed, also called *sample complexity*. In this problem context, the sample complexity would be concentrated around:

$$\text{sample complexity} \longrightarrow \binom{n}{2} p \quad \text{as } n \to \infty.$$

Why? Think about the Law of Large Numbers. The second performance metric is the one that is conventionally employed in the context of inference problems. That is, the probability of error defined as:

Fig. 3.17 Translation to a communication problem

$$P_e := \mathbb{P}\left(\hat{\mathbf{x}} \notin \{\mathbf{x}, \mathbf{x} \oplus \mathbf{1}\}\right).$$

An error occurs when $\hat{\mathbf{x}} \neq \mathbf{x}$ and $\hat{\mathbf{x}} \neq \mathbf{x} \oplus \mathbf{1}$.

There must be a tradeoff relationship between the sample complexity and P_e. The larger the sample complexity, the smaller P_e and vice versa. Hence, we can formulate the following optimization problem from which the tradeoff relationship can be characterized. Given p and n,

$$P_e^*(p, n) := \min_{\text{algorithm}} P_e. \tag{3.46}$$

Actually in the communication problem context, Shannon formulated a similar optimization problem in the course of characterizing the fundamental limit on the transmission rate, called *channel capacity*. The formulated problem was a notoriously difficult problem which has been open even so far. Here we see the same thing. It turns out the above problem (3.46) is also very difficult; hence, the exact error probability $P_e^*(p, n)$ is still open.

Phase transition But we have a good news. The good news is that *phase transition* occurs w.r.t. the sample complexity in the limit of n, and this allows us to characterize the sample complexity without deriving the exact $P_e^*(p, n)$. To have a deeper understanding of this, see Figure 3.18. If sample complexity is above a threshold, then P_e is made arbitrarily close to 0 as $n \to \infty$; otherwise (i.e., if it is below the threshold), $P_e \not\to 0$ no matter what we do and whatsoever. In other words, there exists a sharp

Fig. 3.18 Phase transition in community detection. If S is above the minimal sample complexity $S^* = \frac{n \log n}{n}$, then we can make P_e arbitrarily close to 0 as n tends to infinity. If $S < S^*$, then P_e cannot be made arbitrarily close to 0 no matter what and whatsoever

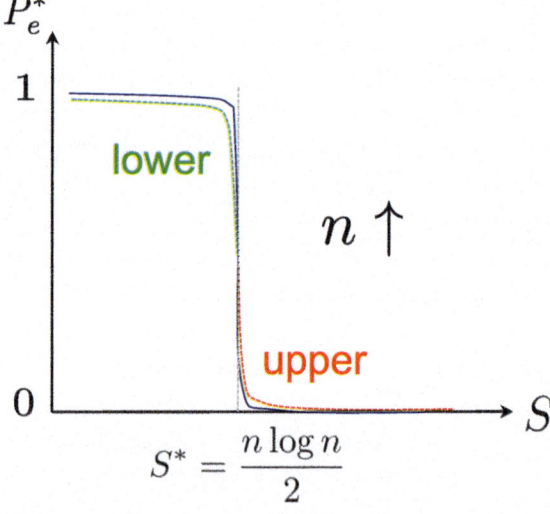

threshold on the sample complexity which determines the boundary between possible versus impossible detection. This sharp threshold is called the *minimal sample complexity* S^*.

A key point here is that in the process of characterizing the limit, *bounding techniques* were employed. Even in cases where we lack precise knowledge of P_e, we can assert that it can approach zero if its *upper* bound converges towards zero. This is due to the non-negativity of probability. Likewise, one can infer that P_e approaches 1 if its *lower* bound tends towards 1, as probabilities cannot exceed 1. Hence, it is possible to characterize the sample complexity without the need for exact calculations of P_e. This indirect yet smart approach was indeed employed in establishing the fundamental limit.

The minimal sample complexity is characterized as:

$$S^* = \frac{n \log n}{2}$$

where $\log(\cdot)$ indicates a natural logarithm. Notice that S^* is much smaller than the total number of possible pairs: $\binom{n}{2} \approx \frac{n^2}{2}$. This result implies that the limit on the observation probability is

$$p^* = \frac{S^*}{\binom{n}{2}} = \frac{\log n}{n}$$

where the second equality is because $\lim_{n \to \infty} \frac{n(n-1)}{n^2} = 1$. Notice that $p^* = \frac{\log n}{n}$ vanishes as $n \to \infty$, meaning that community detection requires only a negligible fraction of pairwise measurements for successful community detection.

Look ahead

It turns out the ML principle plays a significant role to prove the achievability of the limit p^*:

$$p > \frac{\log n}{n} \implies P_e \to 0 \text{ as } n \to \infty. \tag{3.47}$$

In the next section, we will invoke the ML principle to prove the above.

3.6   Social Networks: The ML Principle

Recap

In the previous section, we introduced the second application: *community detection*. Suppose there are two communities and $x_i \in \{0, 1\}$ denotes the community membership of user $i \in \{1, 2, \ldots, n\}$. Then, the goal of the problem is to decode $\mathbf{x} := [x_1, \ldots, x_n]$, i.e., figure out which user belongs to which group. As measurement information, we considered the one that can be easily accessed to in many applications: similarity parities which indicate whether two users belong to the same community. What we are given is assumed to be $y_{ij} = x_i \oplus x_j$. Since there is no way to distinguish $\mathbf{x}$ from $\mathbf{x} \oplus \mathbf{1}$ only from the pairwise measurements y_{ij}'s, we broaden a success event to admit "its flipped version $\mathbf{x} \oplus \mathbf{1}$" as the correct one. Motivated by big data applications such as Meta's social networks where the number n of users is often very large and hence gathering all the relationship information is almost impossible, we considered a setting in which only part of y_{ij}'s are accessible and the choice of pairs is not subject to our design. This motivated us to ask a question: Is there a fundamental limit on the number of randomly chosen pairwise measurements required to make community detection possible? We claimed that there indeed exists a limit, and we explored what this means precisely by translating it into a communication problem, illustrated in Figure 3.19.

Specifically, we introduced two performance metrics: (i) sample complexity (concentrated around $\binom{n}{2}p$); and (ii) the probability of error $P_e := \mathbb{P}(\hat{\mathbf{x}} \notin \{\mathbf{x}, \mathbf{x} \oplus \mathbf{1}\})$. We then claimed the minimal sample complexity (above which one can make $P_e \to 0$, under which one cannot make $P_e \to 0$ no matter what and whatsoever) is:

$$S^* = \frac{n \log n}{2}, \text{ i.e., } p^* = \frac{\log n}{n}.$$

Fig. 3.19 Translation of community detection into a communication problem

Outline

In this section, we will prove that p^* is achievable:

$$p > \frac{\log n}{n} \implies P_e \to 0 \text{ as } n \to \infty. \tag{3.48}$$

This section consists of three parts. First, we will invoke the ML principle to derive the optimal decoder. We will then analyze the decoding error probability under the ML decoding rule. Using a couple of bounding techniques, we will derive an upper bound of the error probability instead of directly attacking the exact probability. Lastly we will show that as long as $p > \frac{\log n}{n}$, the upper bound approaches 0 as the number n of users tends to infinity, thereby proving the achievability.

Encoder As illustrated in Figure 3.19, the encoder converts $\mathbf{x}$ (that we may want to interpret as the information source in light of the communication problem) into $x_{ij} := x_i \oplus x_j$. A distinctive feature of the encoder compared to the communication setting is that it is not subject to our design choice, but is given by the context. Let $\mathbf{X}$ be the output matrix of size n-by-n which contains x_{ij}'s as its entries. In the communication and information theory literature, there is a terminology which indicates the encoder output. It is called a *codeword* (Cover & Joy 2006). So we will adopt the same terminology. One obvious property of the codeword $\mathbf{X}$ is its symmetry: $x_{ij} = x_{ji}$. For instance, when $\mathbf{x} = (1000)$ and $n = 4$, the corresponding codeword reads:

$$\mathbf{X}(\mathbf{x}) = \begin{bmatrix} \underline{1} & 1 & 1 \\ & 0 & 0 \\ & & 0 \end{bmatrix}. \tag{3.49}$$

This expression omits diagonal components and symmetric counterparts as they can be inferred trivially. Let us call, a collection of codewords $\mathbf{X}(\mathbf{x})$'s, a *codebook*. The codebook is also a commonly used terminology in the information theory literature (Cover & Joy 2006).

The optimal decoder Let $\mathbf{Y} = [y_{ij}]$. The codebook is assumed to be known at the decoder. This assumption makes a trivial sense because the structure of pairwise measurements is revealed and therefore $\mathbf{X}(\mathbf{x})$ is determined by $\mathbf{x}$. We employ the optimal decision rule for decoding $\mathbf{x}$: the MAP rule. In this setting, we have no idea on the statistics of $\mathbf{x}$. So we consider a sort of the worst-case scenario in which $\mathbf{x}$ is uniformly distributed. In this case, there is no preference on some particular patterns of $\mathbf{x}$, i.e., every candidate is equally likely. As we exercised before, under the equi-probable setting, the MAP decoder simplifies to the ML counterpart:

$$\hat{\mathbf{x}}_{\mathrm{ML}} = \arg\max_{\mathbf{x}} \mathbb{P}\left(\mathbf{Y}|\mathbf{X}(\mathbf{x})\right).$$

The calculation of the likelihood function $\mathbb{P}(\mathbf{Y}|\mathbf{X}(\mathbf{x}))$ is straightforward, being the same as the one in the communication setting. For instance, suppose $n = 4$, $\mathbf{x} = (0000)$, and y_{ij}'s are all zeros except that $(y_{13}, y_{14}) = (\mathsf{e}, \mathsf{e})$:

$$\mathbf{Y}(\mathbf{x}) = \begin{bmatrix} \underline{0} \; \mathsf{e} \; \mathsf{e} \\ 0 \; 0 \\ 0 \end{bmatrix} \tag{3.50}$$

where e indicates the **erasure** (empty information). Then,

$$\mathbb{P}(\mathbf{Y}|\mathbf{X}(0000)) = (1-p)^2 p^4.$$

Here the number 2, the exponent of $1 - p$, denotes the number of erasures. The input pattern $\mathbf{x} = (0000)$ is a possible candidate for the ground truth, since the corresponding likelihood is not zero. We call such a pattern *compatible*. On the other hand, for $\mathbf{x} = (1000)$,

$$\mathbb{P}(\mathbf{Y}|\mathbf{X}(1000)) = 0.$$

Note that $x_{12} = 1$ (underscored in (3.49)) contradicts with $y_{12} = 0$ (underscored in (3.50)), thus forcing the likelihood to be 0. This implies that the pattern $\mathbf{x} = (1000)$ can never be a solution. In such a case, it is said to be *incompatible* with $\mathbf{Y}$. Then, what is the ML decoding rule? Here is how it works.

1. Eliminate all the input patterns which are *incompatible* with $\mathbf{Y}$.
2. If there is only one pattern that survives, declare it as the correct pattern.

However, this procedure is not sufficient to describe the ML decoding rule. The reason is that we may have a different erasure pattern that confuses the rule. To see this clearly, consider the following concrete example. Suppose that y_{ij}'s are all zeros except that $(y_{12}, y_{13}, y_{14}) = (\mathsf{e}, \mathsf{e}, \mathsf{e})$:

$$\mathbf{Y}(\mathbf{x}) = \begin{bmatrix} \mathsf{e} \; \mathsf{e} \; \mathsf{e} \\ 0 \; 0 \\ 0 \end{bmatrix}. \tag{3.51}$$

Then, we see that

$$\mathbb{P}(\mathbf{Y}|\mathbf{X}(0000)) = (1-p)^3 p^4;$$
$$\mathbb{P}(\mathbf{Y}|\mathbf{X}(0111)) = (1-p)^3 p^4.$$

The two patterns (0000, 0111) are both compatible and the likelihood functions are *equal*. In this case, what we can do for the best is to flip a coin, choosing one out of the two in a random manner. This forms the last step of the ML decoding rule.

3. If there are multiple survivals, choose one *randomly*.

A setup for analysis of the error probability For the achievability proof (3.48), let us analyze the probability of error when using the ML decoder. Starting with the definition of P_e and using the total probability law, we get:

$$P_e := \mathbb{P}\left(\hat{\mathbf{x}}_{\mathsf{ML}} \notin \{\mathbf{x}, \mathbf{x} \oplus \mathbf{1}\}\right)$$
$$= \sum_{\mathbf{a}} \mathbb{P}(\mathbf{x} = \mathbf{a})\mathbb{P}\left(\hat{\mathbf{x}}_{\mathsf{ML}} \notin \{\mathbf{a}, \mathbf{a} \oplus \mathbf{1}\}|\mathbf{x} = \mathbf{a}\right).$$

For a fixed $\mathbf{a}$, $\mathbb{P}\left(\hat{\mathbf{x}} \notin \{\mathbf{a}, \mathbf{a} \oplus \mathbf{1}\}|\mathbf{x} = \mathbf{a}\right)$ is a sole function of the *likelihood function* which depends only on erasure patterns for the $\binom{n}{2}$ independent channels. Also, the erasure patterns are independent of the channel input affected by $\mathbf{a}$. Therefore, this probability is irrelevant of what the value of $\mathbf{a}$ is. Applying this to the above, we get:

$$P_e = \mathbb{P}\left(\hat{\mathbf{x}}_{\mathsf{ML}} \notin \{\mathbf{0}, \mathbf{1}\}|\mathbf{x} = \mathbf{0}\right)$$
$$= \mathbb{P}\left(\bigcup_{\mathbf{a} \notin \{\mathbf{0},\mathbf{1}\}} \{\hat{\mathbf{x}}_{\mathsf{ML}} = \mathbf{a}\}|\mathbf{x} = \mathbf{0}\right) \tag{3.52}$$

where the second equality is because $\hat{\mathbf{x}}_{\mathsf{ML}} \notin \{\mathbf{0}, \mathbf{1}\}$ implies that $\hat{\mathbf{x}}_{\mathsf{ML}} = \mathbf{a}$ for some $\mathbf{a} \notin \{\mathbf{0}, \mathbf{1}\}$.

The union bound In general, the probability of the *union* of multiple events is not that simple to compute. Rather it is quite complicated especially when it involves a large number of multiple events. Even for the three-event case (A, B, C), the probability formula is complicated:

$$\mathbb{P}(A \cup B \cup C) = \mathbb{P}(A) + \mathbb{P}(B) + \mathbb{P}(C)$$
$$- \mathbb{P}(A \cap B) - \mathbb{P}(A \cap C) - \mathbb{P}(B \cap C) + \mathbb{P}(A \cap B \cap C).$$

Even worse, the number of associated multiple events in (3.52) amounts to $2^n - 2$, which significantly complicates the probability formula. Consequently, computing the probability becomes a non-trivial task. In order to make a progress, we will employ the bounding technique that we mentioned in Sect. 3.5. The technique leverages an upper bound and the following key insight: when an upper bound on the error probability tends to zero, the exact error probability also converges to zero. In this approach, a critical step is to establish a good enough upper bound. Fortunately, we have a widely recognized upper bound that serves this purpose well. It pertains to the union of multiple events and is termed the *union bound*. The union bound says that for events A and B,

$$\mathbb{P}(A \cup B) \leq \mathbb{P}(A) + \mathbb{P}(B).$$

The proof of the union bound is straightforward since $\mathbb{P}(A \cap B) \geq 0$. You can also check this in Problem 7.4. Applying the union bound to (3.52) yields:

$$P_e = \mathbb{P}\left(\bigcup_{\mathbf{a} \notin \{0,1\}} \{\hat{\mathbf{x}}_{\mathsf{ML}} = \mathbf{a}\} | \mathbf{x} = \mathbf{0}\right) \tag{3.53}$$
$$\leq \sum_{\mathbf{a} \notin \{0,1\}} \mathbb{P}\left(\hat{\mathbf{x}}_{\mathsf{ML}} = \mathbf{a} | \mathbf{x} = \mathbf{0}\right).$$

Further upper-bounding Let us focus on $\mathbb{P}(\hat{\mathbf{x}}_{\mathsf{ML}} = \mathbf{a} | \mathbf{x} = \mathbf{0})$. To gain insights into this term, consider an example where $n = 4$ and $\mathbf{a} = (1000)$. In this case, the error event implies that $\mathbf{X}(1000)$ must be *compatible*. A necessary condition for $\mathbf{X}(1000)$ being compatible under $\mathbf{x} = (0000)$ is: $(y_{12}, y_{13}, y_{14}) = (\mathsf{e}, \mathsf{e}, \mathsf{e})$. Notice that for all (i, j) entries whose values are different between the two codewords $\mathbf{X}(0000)$ and $\mathbf{X}(1000)$ (that we call *distinguishable positions*), erasures must occur; otherwise, (1000) cannot be compatible as its corresponding likelihood function would be 0. Using the fact that for two events A and B such that A implies B, $\mathbb{P}(A) \leq \mathbb{P}(B)$ (check in Problem 7.4), we get:

$$\begin{aligned}
&\mathbb{P}\left(\hat{\mathbf{x}}_{\mathsf{ML}} = (1000) | \mathbf{x} = \mathbf{0}\right) \\
&\leq \mathbb{P}\left(\mathbf{X}(1000) \text{ compatible} | \mathbf{x} = \mathbf{0}\right) \\
&\leq \mathbb{P}\left((y_{12}, y_{13}, y_{14}) = (\mathsf{e}, \mathsf{e}, \mathsf{e}) | \mathbf{x} = \mathbf{0}\right) \\
&= (1 - p)^3.
\end{aligned} \tag{3.54}$$

Here the number, the exponent placed above $1 - p$ in the last line, indicates the number of erasures that must occur in the *distinguishable positions*.

Observe that a key to determine the above upper bound (3.54) is the number of distinguishable positions between $\mathbf{X}(\mathbf{a})$ and $\mathbf{X}(\mathbf{0})$. Actually one can easily verify that the number of distinguishable positions (w.r.t. $\mathbf{X}(\mathbf{0})$) depends on the number of 1's in $\mathbf{a}$. To see this, consider an example of $\mathbf{a} = (\underbrace{11 \cdots 1}_{k} \underbrace{00 \cdots 0}_{n-k})$. In this case,

$$\mathbf{X}(\mathbf{a}) = \begin{bmatrix} 0\,0\,0\,1\,1 \\ 0\,0\,1\,1 \\ 0\,1\,1 \\ 0\,0 \\ 0 \end{bmatrix}.$$

Each of the first k rows contains $(n - k)$ ones; hence, the total number of distinguishable positions w.r.t. $\mathbf{X}(0)$ is:

$$\# \text{ of distinguishable positions} = k(n - k).$$

For a succinct description of the summation term in (3.53), let us classify the instance $\mathbf{a}$ depending on the number of 1's in $\mathbf{a}$. To this end, we introduce:

$$\mathcal{A}_k := \{\mathbf{a} | \|\mathbf{a}\|_1 = k\} \tag{3.55}$$

where $\|a\|_1 := |a_1| + |a_2| + \cdots + |a_n|$. Using this notation, we can then express (3.53) as:

$$
\begin{aligned}
P_e &\leq \sum_{k=1}^{n-1} \sum_{\mathbf{a} \in \mathcal{A}_k} (1 - p)^{k(n-k)} \\
&= \sum_{k=1}^{n-1} |\mathcal{A}_k| \, (1 - p)^{k(n-k)} \\
&= \sum_{k=1}^{n-1} \binom{n}{k} (1 - p)^{k(n-k)} \\
&= 2 \sum_{k=1}^{\frac{n}{2}} \binom{n}{k} (1 - p)^{k(n-k)} - \binom{n}{n} (1 - p)^{n \cdot 0} \\
&\leq 2 \sum_{k=1}^{\frac{n}{2}} \binom{n}{k} (1 - p)^{k(n-k)}
\end{aligned}
\tag{3.56}
$$

where the second-to-last step follows from the fact that $\binom{n}{k}(1 - p)^{k(n-k)}$ is symmetric around $k = n/2$.

The final step of the achievability proof Since we intend to prove the achievability when $p > \frac{\log n}{n}$, let's focus on the regime: $\lambda > 1$ where λ is defined such that $p := \lambda \frac{\log n}{n}$. Then, it suffices to show that in the regime of $\lambda > 1$,

$$\sum_{k=1}^{\frac{n}{2}} \binom{n}{k} (1 - p)^{k(n-k)} \longrightarrow 0 \text{ as } n \to \infty. \tag{3.57}$$

In this setting of $p = \lambda \frac{\log n}{n}$, p is arbitrarily close to 0 in the limit of n. This motivates us to employ the following upper bound on $1 - p$ (which is very tight in the regime):

$$1 - p \leq e^{-p}. \tag{3.58}$$

Check this in Problem 8.1. Also when n is large, the following bound is good enough to prove the achievability:

$$\binom{n}{k} = \frac{n(n-1)(n-2)\cdots(n-k+1)}{k!} \leq \frac{n^k}{k!} \leq n^k. \tag{3.59}$$

Applying the bounds (3.58) and (3.59) into (3.56), we get:

$$P_e \leq 2 \sum_{k=1}^{\frac{n}{2}} \binom{n}{k} (1-p)^{k(n-k)}$$

$$\leq 2 \sum_{k=1}^{\frac{n}{2}} n^k e^{-pk(n-k)}$$

$$\overset{(a)}{=} 2 \sum_{k=1}^{\frac{n}{2}} e^{k \log n - \lambda k \left(1-\frac{k}{n}\right) \log n} \tag{3.60}$$

$$= 2 \sum_{k=1}^{\frac{n}{2}} e^{-k\left(\lambda\left(1-\frac{k}{n}\right)-1\right) \log n}$$

where (a) follows from $n^k = e^{k \log n}$ and $p := \lambda \frac{\log n}{n}$.

Notice that for the range of $1 \leq k \leq \frac{n}{2}$, $1 - \frac{k}{n}$ is minimized at $\frac{1}{2}$. Applying this to the last line in (3.60), we can then get:

$$P_e \leq 2 \sum_{k=1}^{\frac{n}{2}} e^{-k\left(\frac{\lambda}{2}-1\right) \log n}.$$

Case I: $\lambda > 2$: A key observation is that if $\lambda > 2$, then we can apply the well-known summation formula w.r.t. the geometric series (where the common ratio is $e^{-\left(\frac{\lambda}{2}-1\right) \log n}$), thus obtaining:

$$P_e \leq 2 \sum_{k=1}^{\frac{n}{2}} \left(e^{-\left(\frac{\lambda}{2}-1\right) \log n}\right)^k \leq 2 \cdot \frac{e^{-\left(\frac{\lambda}{2}-1\right) \log n}}{1 - e^{-\left(\frac{\lambda}{2}-1\right) \log n}} \longrightarrow 0 \text{ as } n \to \infty.$$

Hence, we can make $P_e \to 0$ for the case of $\lambda > 2$.

Case II: $1 < \lambda \leq 2$: Observe in the last step in (3.60) that

$$\lambda\left(1 - \frac{k}{n}\right) - 1 = 0 \quad \Longleftrightarrow \quad k = \left(1 - \frac{1}{\lambda}\right) n,$$

and $\lambda(1 - \frac{k}{n}) - 1$ is a decreasing function in k. See Figure 3.20.

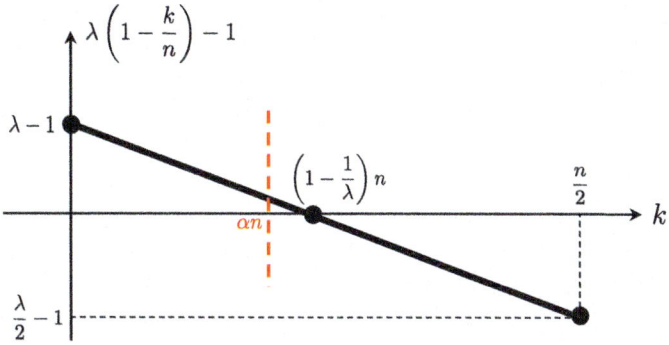

Fig. 3.20 Behavior of $\lambda\left(1 - \frac{k}{n}\right) - 1$ in k

This together with a choice of $\alpha < 1 - \frac{1}{\lambda}$ gives:

$$\lambda\left(1 - \frac{k}{n}\right) - 1 > 0 \quad \text{when } k \leq \alpha n.$$

Applying this to (3.60) yields:

$$P_e \leq 2 \sum_{k=1}^{\frac{n}{2}} \binom{n}{k}(1 - p)^{k(n-k)}$$

$$\leq 2 \sum_{k=1}^{\alpha n} e^{-k(\lambda(1-\frac{k}{n})-1)\log n} + \sum_{k=\alpha n+1}^{\frac{n}{2}} \binom{n}{k}(1 - p)^{k(n-k)}.$$

Again applying the summation formula of the geometric series to the first term in the last step of the above, we see that the first term vanishes as $n \to \infty$. Hence, we obtain:

$$P_e \lesssim \sum_{k=\alpha n+1}^{\frac{n}{2}} \binom{n}{k}(1 - p)^{k(n-k)}$$

$$\overset{(a)}{\leq} (1 - p)^{\alpha(1-\alpha)n^2} \sum_{k=\alpha n+1}^{\frac{n}{2}} \binom{n}{k}$$

$$\overset{(b)}{\leq} (1 - p)^{\alpha(1-\alpha)n^2} \cdot 2^n$$

$$\overset{(c)}{\leq} e^{-\lambda\alpha(1-\alpha)n\log n} \cdot e^{n\log 2}.$$

where (a) is due to the fact that $k(n - k) \geq \alpha(1 - \alpha)n^2$ (where the equality holds when $k = \alpha n + 1$); (b) comes from the binomial theorem $(\sum_{k=0}^{n} \binom{n}{k} = 2^n)$; and (c) comes from the fact that $1 - p \leq e^{-p}$ and $p := \lambda \frac{\log n}{n}$. Notice that the last term in the above tends to 0 as $n \to \infty$. This is because $\lambda \alpha (1 - \alpha)n \log n$ grows much faster than $n \log 2$. This implies $P_e \to 0$, which completes the achievability proof (3.48).

Look ahead

Using the ML principle, we proved the achievability of the community detection limit (3.48). It turns out the other way around holds as well: if $p < \frac{\log n}{n}$, then P_e cannot be made arbitrarily close to 0 even if any schemes are taken into consideration. In other words, the condition of $p > \frac{\log n}{n}$ is fundamentally *required* for reliable detection. However, due to the interest of this book, we will not cover this proof. The proof requires some sophisticated techniques that we do not want to emphasize in this book. Instead we will touch upon a particular yet efficient algorithm that ensures good performance when $p > \frac{\log n}{n}$. Remember the optimal ML decoding rule that we employed as an achievable scheme:

$$\hat{\mathbf{x}}_{\mathsf{ML}} = \arg \max_{\mathbf{x}} \mathbb{P}\left(\mathbf{Y}|\mathbf{X}(\mathbf{x})\right).$$

One challenge here is that this requires the number 2^n of likelihood computations that grows exponentially with the number n of users. Since n is typically very large in practice, the complexity is huge. It turns out there is another yet much efficient algorithm that yields almost the same performance as the ML rule. In the next section, we will study the efficient algorithm.

3.7 Social Networks: **Python Implementation**

Recap

In the preceding section, we applied the ML principle to establish the attainability of the limit $p^* = \frac{\log n}{n}$ for the task of community detection. Let's recall the optimal ML decoding rule:

$$\hat{\mathbf{x}}_{\mathsf{ML}} = \arg \max_{\mathbf{x}} \mathbb{P}\left(\mathbf{Y}|\mathbf{X}(\mathbf{x})\right)$$

where $\mathbf{x} = [x_1, \ldots, x_n]^T$ represents the vector indicating community memberships; $\mathbf{X}(\mathbf{x})$ refers to a codeword matrix constructed such that $x_{ij} = x_i \oplus x_j$ is the (i, j) entry; and $\mathbf{Y}$ is an observation matrix with y_{ij}'s ($y_{ij} = x_{ij}$ w.p. p and erasure otherwise). A significant challenge with the ML rule is its computational complexity. Since $\mathbf{x}$ takes one of the 2^n possible patterns, the ML rule requires the number 2^n of likelihood computations that grows exponentially with n. Hence, it is crucial to develop a computationally efficient algorithm that possibly yields the same performance as the ML rule. Indeed, efficient algorithms have been developed that achieve the optimal ML performance.

Outline

In this section, we will delve into an efficient algorithm. In fact, the efficient algorithm with the optimality is somewhat intricate. Hence, for the purpose of illustration, this section will focus on a simplified version of it. This simplified version captures the core concept of the algorithm, albeit with suboptimal performance. The section comprises four parts. First we will introduce an *adjacency matrix*, which plays a key role in the algorithm under consideration. We will then explain how the algorithm works. The algorithm, named the *spectral algorithm*, involves the process of identifying the principal eigenvector of the adjacency matrix. Subsequently, we will investigate an efficient method for computing the principal eigenvector: the *power method*. Finally we will implement the spectral algorithm using **Python**.

Adjacency matrix Given that the number of users n is typically quite large, pairwise measurements often constitute a significant volume of big data, even though only a portion of these measurements are observed. There exists a valuable entity that efficiently encapsulates such large datasets. That is, the *adjacency matrix* (Chartrand 1977).

The adjacency matrix, say $\mathbf{A}$, serves as an equivalent representation of the pairwise data, with each column and row corresponding to individual users. In this matrix, each entry a_{ij} conveys whether users i and j belong to the same community. Specifically, a value of $a_{ij} = +1$ indicates that the two users are part of the same community, while

$a_{ij} = -1$ implies the opposite. Lastly, $a_{ij} = 0$ indicates that there is no recorded measurement for the pair.

$$a_{ij} = \begin{cases} 1 - 2y_{ij}, & \text{w.p. } p; \\ 0, & \text{otherwise } (y_{ij} = \text{erasure}). \end{cases} \tag{3.61}$$

For instance, when $\mathbf{x} = [1, 0, 0, 1]^T$, we might have:

$$\mathbf{A} = \begin{bmatrix} +1 & -1 & 0 & 0 \\ -1 & +1 & +1 & -1 \\ 0 & +1 & +1 & -1 \\ 0 & -1 & -1 & +1 \end{bmatrix} \tag{3.62}$$

where we have erasures for (y_{13}, y_{14}).

In the scenario of a full measurement setting where $p = 1$, one can make a key observation that offers valuable insights into algorithms. Specifically, when $p = 1$, we arrive at the following:

$$\mathbf{A} = \begin{bmatrix} +1 & -1 & -1 & +1 \\ -1 & +1 & +1 & -1 \\ -1 & +1 & +1 & -1 \\ +1 & -1 & -1 & +1 \end{bmatrix}. \tag{3.63}$$

The key observation is that the rank of $\mathbf{A}$ is 1, implying that each row is a linear combination of the other rows. The question arises: can we derive the community pattern from this rank-1 matrix? Indeed, the answer is affirmative, and this forms the basis of the spectral algorithm that we will explore in the sequel.

Spectral algorithm (Shen et al. 2011) Using the fact that the rank of $\mathbf{A}$ (3.63) is 1, we can easily come up with its *eigenvector*.

$$\mathbf{v} = \begin{bmatrix} +1 \\ -1 \\ -1 \\ +1 \end{bmatrix}. \tag{3.64}$$

Let's confirm that $\mathbf{Av} = 4\mathbf{v}$ is indeed the case. The entries of $\mathbf{v}$, whether they are $+1$ or -1, precisely reveal the community memberships of the users. Consequently, in the ideal scenario where every pair is sampled, the principal eigenvector (which is the only eigenvector in this case) effectively recovers the community structure. This method, which involves taking the adjacency matrix and computing its principal eigenvector, is referred to as the *spectral algorithm*.

But what happens in the case of partial measurements where $p < 1$? In this scenario, it is not immediately clear whether the principal eigenvector can accurately determine the community memberships. Additionally, the components of the eigenvector may not strictly be limited to $+1$ or -1. To address the latter concern, we can introduce a thresholded eigenvector, say $\mathbf{v}_{\text{th}}$, where each entry simply takes on the sign of v_i:

$$v_{\text{th},i} = \begin{cases} +1, & v_i > 0; \\ -1, & \text{otherwise.} \end{cases} \tag{3.65}$$

It becomes evident that as long as p reaches a sufficiently large value, the thresholded eigenvector $\mathbf{v}_{\text{th}}$ can accurately recover the ground truth of communities as the number of users n tends to infinity. Given the scope of this book, we won't delve into an in-depth analysis of how large p needs to be for the spectral algorithm's successful detection. Instead, we will later provide empirical simulations using Python to demonstrate that for a substantial value of n, $\mathbf{v}_{\text{th}}$ indeed converges towards the ground truth as p increases.

Power method (Golub & Van Loan 2013) Another technical question arises when dealing with the computation of the principal eigenvector. What if the adjacency matrix is of considerable size? Keep in mind that in Meta's social networks, the order of n is on the order of 10^9. Using a naive approach to compute the eigenvector based on eigenvalue decomposition would entail a complexity of around n^3, which can be highly prohibitive. Fortunately, there exists a highly efficient and widely-used method for computing the principal eigenvector, known as the *power method*. This method is well-established and widely popular.

Before delving into the detailed explanation of how the power method works, let's first highlight some crucial observations that naturally pave the way for this method. Consider an adjacency matrix $\mathbf{A} \in \mathbb{R}^{n \times n}$, which possesses m eigenvalues λ_i's and their corresponding eigenvectors $\mathbf{v}_i$'s:

$$\mathbf{A} := \lambda_1 \mathbf{v}_1 \mathbf{v}_1^T + \lambda_2 \mathbf{v}_2 \mathbf{v}_2^T + \cdots + \lambda_m \mathbf{v}_m \mathbf{v}_m^T$$

where $\lambda_1 > \lambda_2 \geq \lambda_3 \geq \cdots \geq \lambda_m$ and $\mathbf{v}_i$'s are orthonormal: $\mathbf{v}_i^T \mathbf{v}_j = 1$ only if $i = j$; 0 otherwise. According to the definition, $(\lambda_1, \mathbf{v}_1)$ represent the principal eigenvalue and its corresponding eigenvector, respectively. Let's consider an arbitrary non-zero vector $\mathbf{v} \in \mathbb{R}^n$ such that $\mathbf{v}_1^T \mathbf{v} \neq 0$. Then, $\mathbf{A}\mathbf{v}$ can be expressed as:

$$\begin{aligned} \mathbf{A}\mathbf{v} &= \left(\sum_{i=1}^{m} \lambda_i \mathbf{v}_i \mathbf{v}_i^T \right) \mathbf{v} \\ &= \sum_{i=1}^{m} \lambda_i (\mathbf{v}_i^T \mathbf{v}) \mathbf{v}_i \end{aligned} \tag{3.66}$$

where the second equality is due to the fact that $\mathbf{v}_i^T \mathbf{v}$ is a scalar.

A crucial observation is that the first term $\lambda_1 (\mathbf{v}_1^T \mathbf{v}) \mathbf{v}_1$ in the summation above makes a significant contribution, primarily because λ_1 is the largest eigenvalue. This effect becomes even more pronounced when we apply the adjacency matrix to the resulting vector $\mathbf{Av}$. To gain a clearer understanding of this, consider the following:

$$\mathbf{A}^2 \mathbf{v} = \mathbf{A}(\mathbf{Av})$$

$$= \left(\sum_{i=1}^{m} \lambda_i \mathbf{v}_i \mathbf{v}_i^T \right) \left(\sum_{j=1}^{m} \lambda_j (\mathbf{v}_j^T \mathbf{v}) \mathbf{v}_j \right)$$

$$\overset{(a)}{=} \sum_{i=1}^{m} \lambda_i^2 (\mathbf{v}_i^T \mathbf{v})(\mathbf{v}_i^T \mathbf{v}_i)\mathbf{v}_i + \sum_{i \neq j} \lambda_i \lambda_j (\mathbf{v}_j^T \mathbf{v})(\mathbf{v}_i^T \mathbf{v}_j)\mathbf{v}_j \qquad (3.67)$$

$$\overset{(b)}{=} \sum_{i=1}^{m} \lambda_i^2 (\mathbf{v}_i^T \mathbf{v})\mathbf{v}_i$$

where (a) comes from the fact that $\mathbf{v}_i^T \mathbf{v}_j$ is a scalar; and (b) $\mathbf{v}_i$'s are orthonormal vectors, i.e., $\mathbf{v}_i^T \mathbf{v}_i = 1$ and $\mathbf{v}_i^T \mathbf{v}_j = 0$ for $i \neq j$. The difference in (3.67) when compared to $\mathbf{Av}$ in (3.66) lies in the presence of λ_i^2 instead of λ_i. The multiplication of $\mathbf{A}$ amplifies the contribution from the principal component in comparison to the other components. By iterating this process (multiplying $\mathbf{A}$ with the resulting vector repeatedly), we obtain the following:

$$\mathbf{A}^k \mathbf{v} = \sum_{i=1}^{m} \lambda_i^k (\mathbf{v}_i^T \mathbf{v})\mathbf{v}_i .$$

Observe that in the limit of k,

$$\frac{\mathbf{A}^k \mathbf{v}}{\lambda_1^k (\mathbf{v}_1^T \mathbf{v})} = \sum_{i=1}^{m} \left(\frac{\lambda_i}{\lambda_1} \right)^k \frac{\mathbf{v}_i^T \mathbf{v}}{\mathbf{v}_1^T \mathbf{v}} \mathbf{v}_i \longrightarrow \mathbf{v}_1 \quad \text{as } k \to \infty.$$

This implies that as we repeat the following process (multiplying $\mathbf{A}$ and subsequently normalizing the resulting vector), the normalized vector converges to the principal eigenvector. In precise terms, as k approaches infinity,

$$\frac{\mathbf{A}^k \mathbf{v}}{\sqrt{\|\mathbf{A}^k \mathbf{v}\|^2}} \longrightarrow \mathbf{v}_1 \text{ as } k \to \infty.$$

This key observation naturally paves the way for the *power method*:

1. Choose a random vector $\mathbf{v}$ and set $\mathbf{v}^{(0)} = \mathbf{v}$ and $t = 0$.
2. Compute $\mathbf{v}^{(t+1)} = \frac{\mathbf{Av}^{(t)}}{\sqrt{\|\mathbf{Av}^{(t)}\|^2}}$ and increase t by 1.
3. Iterate Step 2 until converged, e.g., $\|\mathbf{v}^{(t+1)} - \mathbf{v}^{(t)}\|^2 < \epsilon = 10^{-5}$.

The power method entails performing multiple (say k) matrix-vector multiplications. Each has a complexity of n^2 multiplications. Consequently, the overall computational complexity of the power method remains on the order of n^2, as long as the number of iterations k is not significantly large in relation to the size of n. This is often the case in practical applications. Importantly, this complexity is considerably smaller than the n^3 complexity associated with eigenvalue decomposition, especially when n is exceptionally large. Due to this significant computational advantage, the power method is widely adopted as an efficient algorithm for identifying the principal eigenvector in numerous applications.

Python implementation of the spectral algorithm Now, let's proceed to implement the spectral algorithm using Python. To begin, we generate the community memberships for a set of n users.

```python
from scipy.stats import bernoulli
import numpy as np

n = 8 # number of users
Bern = bernoulli(0.5)
# Generate n community memberships
x = Bern.rvs(n)
print(x)
```

```
[0 0 0 0 1 0 1 0]
```

We then generate random pairwise measurements.

```python
# Construct the codebook
X = np.zeros((n,n))
for i in range(len(x)):
    for j in range(i,len(x)):
        # Compute xij = xi + xj (modulo 2)
        X[i,j] = (x[i]+x[j]) % 2
        # Symmetric component
        X[j,i] = X[i,j]
print(X)
```

```
[[0. 0. 0. 0. 1. 0. 1. 0.]
 [0. 0. 0. 0. 1. 0. 1. 0.]
 [0. 0. 0. 0. 1. 0. 1. 0.]
 [0. 0. 0. 0. 1. 0. 1. 0.]
 [1. 1. 1. 1. 0. 1. 0. 1.]
 [0. 0. 0. 0. 1. 0. 1. 0.]
 [1. 1. 1. 1. 0. 1. 0. 1.]
 [0. 0. 0. 0. 1. 0. 1. 0.]]
```

Next we compute the adjacency matrix **A**.

```
# observation probability
p = 0.8
obs_bern = bernoulli(p)
# Construct an n-by-n mask matrix:
# entry = 1 (observed); 0 (otherwise)
mask_matrix = obs_bern.rvs((n,n))

# Construct the adjacency matrix
A = (1-2*X)*mask_matrix

print(1-2*X)
print(mask_matrix)
print(A)
```

```
[[ 1.  1.  1.  1. -1.  1. -1.  1.]
 [ 1.  1.  1.  1. -1.  1. -1.  1.]
 [ 1.  1.  1.  1. -1.  1. -1.  1.]
 [ 1.  1.  1.  1. -1.  1. -1.  1.]
 [-1. -1. -1. -1.  1. -1.  1. -1.]
 [ 1.  1.  1.  1. -1.  1. -1.  1.]
 [-1. -1. -1. -1.  1. -1.  1. -1.]
 [ 1.  1.  1.  1. -1.  1. -1.  1.]]
[[1 1 0 0 1 0 1 1]
 [1 1 1 0 0 1 1 1]
 [1 1 0 1 1 1 1 1]
 [1 1 1 1 0 1 1 1]
 [1 1 1 1 1 1 1 1]
 [1 1 1 1 1 0 1 1]
 [1 1 1 1 1 1 1 1]
 [1 0 1 1 1 1 1 1]]
[[ 1.  1.  0.  0. -1.  0. -1.  1.]
 [ 1.  1.  1.  0. -0.  1. -1.  1.]
 [ 1.  1.  0.  1. -1.  1. -1.  1.]
 [ 1.  1.  1.  1. -0.  1. -1.  1.]
 [-1. -1. -1. -1.  1. -1.  1. -1.]
 [ 1.  1.  1.  1. -1.  0. -1.  1.]
 [-1. -1. -1. -1.  1. -1.  1. -1.]
 [ 1.  0.  1.  1. -1.  1. -1.  1.]]
```

We run the power method to compute the principal eigenvector of **A**.

```
def power_method(A, eps=1e-5):
    # A computationally efficient algorithm
    # for finding the principal eigenvector
    # Choose a random vector
    v = np.random.randn(n)
    # normalization
    v = v/np.linalg.norm(v)

    prev_v = np.zeros(len(v))
    t = 0
```

```
    while np.linalg.norm(prev_v-v) > eps:
        prev_v = v
        v = np.array(np.dot(A,v)).reshape(-1)
        v = v/np.linalg.norm(v)
        t += 1
    print("Power method is terminated after %s iterations"%t)
    return v

v1 = power_method(A)
print(v1)
print(np.sign(v1))
print(1-2*x)
```

```
Power method is terminated after 8 iterations
[ 0.25389707  0.29847768  0.35720613  0.34957476 -0.40941936
  0.35720737 -0.40941936  0.36579104]
[ 1.   1.   1.   1.  -1.   1.  -1.   1.]
[ 1   1   1   1  -1   1  -1   1]
```

In the above, it's worth noting that the thresholded principal eigenvector np.sign(v1) aligns perfectly with the ground truth community vector 1-2*x.

Python: Performance of the spectral algorithm As claimed earlier, we will demonstrate through Python experiments that the principal eigenvector approaches the ground truth of community memberships as p increases. We will consider a realistic scenario with a large value of n, say $n = 4000$. To quantify the similarity between the principal eigenvector and the community vector, we will use a well-known correlation measure, namely the Pearson correlation (Freedman et al. 2007):

$$\rho_{X,Y} := \frac{\sigma_{X,Y}}{\sigma_X \sigma_Y} \qquad (3.68)$$

where $\sigma_{X,Y} = \mathbb{E}[XY] - \mathbb{E}[X]\mathbb{E}[Y]$, $\sigma_X = \sqrt{\mathbb{E}[X^2] - (\mathbb{E}[X])^2}$, and $\sigma_Y = \sqrt{\mathbb{E}[Y^2] - (\mathbb{E}[Y])^2}$. For the calculation of the Pearson correlation, we will utilize the built-in function pearsonr provided by the scipy.stats module.

```
from scipy.stats import bernoulli
from scipy.stats import pearsonr
import numpy as np
import matplotlib.pyplot as plt


n = 4000 # number of users
Bern = bernoulli(0.5)
# Generate n community memberships
x = Bern.rvs(n)

# Construct the codebook
X = np.zeros((n,n))
for i in range(len(x)):
    for j in range(i,len(x)):
        # Compute xij = xi + xj (modulo 2)
```

```python
        X[i,j] = (x[i]+x[j]) % 2
        # Symmetric component
        X[j,i] = X[i,j]

p = np.linspace(0.0003,0.0025,30)
limit = np.log(n)/n
p_norm = p/limit

def power_method(A, eps=1e-5):
    # A computionally efficient algorithm
    # for finding the principal eigenvector
    # Choose a random vector
    v = np.random.randn(n)
    # normalization
    v = v/np.linalg.norm(v)

    prev_v = np.zeros(len(v))
    t = 0
    while np.linalg.norm(prev_v-v) > eps:
        prev_v = v
        v = np.array(np.dot(A,v)).reshape(-1)
        v = v/np.linalg.norm(v)
        t += 1
    print("Power method is terminated after %s iterations"%t)
    return v

corr = np.zeros_like(p)

for i,val in enumerate(p):
    obs_bern = bernoulli(val)
    # Construct an n-by-n mask matrix:
    # entry = 1 (observed); 0 (otherwise)
    mask_matrix = obs_bern.rvs((n,n))

    # Construct the adjacency matrix
    A = (1-2*X)*mask_matrix
    # Power method
    v1 = power_method(A)
    # Threshold the principal eigenvector
    v1 = np.sign(v1)
    # Compute the ground truth
    ground_truth = 1-2*x
    # Compute Pearson correlation btw estimate & ground truth
    corr[i] = np.abs(pearsonr(ground_truth,v1)[0])
    print(p_norm[i], corr[i])

plt.figure(figsize=(5,5), dpi=200)
plt.plot(p_norm, corr)
plt.title('Pearson correlation btw estimate and ground truth')
plt.grid(linestyle=':', linewidth=0.5)
plt.show()
```

Fig. 3.21 Pearson correlation between the thresholded principal eigenvector and the ground-truth community vector as a function of $\lambda := \frac{p}{p^*} = \frac{p}{\log n/n}$

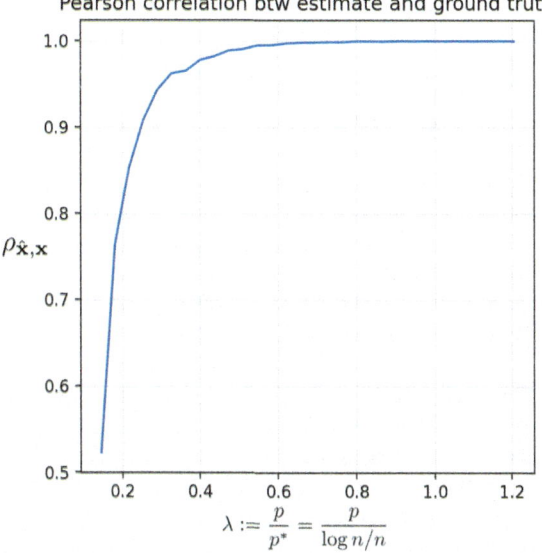

Pearson correlation btw estimate and ground truth

$$\lambda := \frac{p}{p^*} = \frac{p}{\log n/n}$$

Observe in Figure 3.21 that the thresholded principal eigenvector increasingly aligns with the ground-truth community vector (or its flip version) as p grows. This alignment is indicated by the high Pearson correlation for larger values of p. Particularly, when p approaches the limit $p^* = \frac{\log n}{n}$, the Pearson correlation nears 1, illustrating that the spectral algorithm nearly achieves the optimal performance as promised by the ML decoding rule. This is somewhat of a heuristic argument. To provide a more precise analysis, we should rely on the empirical error rate (instead of the Pearson correlation) computed across a sufficient number of random realizations of community vectors. However, for computational simplicity, we employ the Pearson correlation, which can be reliably calculated with only one random trial for each value of p.

Look ahead
Up to this point, we have showcased the role of probabilistic modeling and the MAP/ML principles in two applications: Communication and community detection. In the upcoming section, we will transition to our third application: *Machine learning*.

Problem Set 8

Problem 8.1 (*Bounds and combinatorics*)

(*a*) Let $p \geq 0$. Prove that

$$1 - p \leq e^{-p}.$$

Also specify the condition under which the equality holds.

(*b*) Show that for non-negative integers n and k ($k \leq n$):

$$\binom{n}{k} \leq e^{k \log n}.$$

(*c*) Show that for integers $n \geq 0$:

$$\sum_{k=0}^{n} \binom{n}{k} = 2^n.$$

Problem 8.2 (*The concept of reliable community detection*) Consider a scenario where there are n users clustered into two communities. Let $x_i \in \{0, 1\}$ denote the community membership for user $i \in \{1, 2, \ldots, n\}$. We make the assumption that we have access to only a portion of the pairwise measurements:

$$y_{ij} = \begin{cases} x_i \oplus x_j, & \text{w.p. } p; \\ \text{erasure}, & \text{w.p. } 1 - p, \end{cases}$$

for every pair $(i, j) \in \{(1, 2), (1, 3), \ldots, (1, n), (2, 3), \ldots, (n-1, n)\}$ and $p \in [0, 1]$. We also make the assumption that the values of y_{ij} are independent for all pairs (i, j). Given these y_{ij}'s, the goal is to decode the community membership vector $\mathbf{x} := [x_1, x_2, \ldots, x_n]$ or its complement $\mathbf{x} \oplus \mathbf{1} := [x_1 \oplus 1, x_2 \oplus 1, \ldots, x_n \oplus 1]$. Let $\hat{\mathbf{x}}$ represent an estimate. Define the probability of error as:

$$P_e = \mathbb{P}\left(\hat{\mathbf{x}} \notin \{\mathbf{x}, \mathbf{x} \oplus \mathbf{1}\}\right).$$

(*a*) Define the term *sample complexity* as the number of pairwise measurements that are not erased. Show that

$$\frac{\text{sample complexity}}{\binom{n}{2}} \longrightarrow p \text{ as } n \to \infty.$$

Hint: You may want to use the statement in Problem 6.6.

(b) Consider the following optimization problem. Given p and n,

$$P_e^*(p, n) := \min_{\text{algorithm}} P_e.$$

State the definition of *reliable detection*. Also state the definition of *minimum sample complexity* using the concept of *reliable detection*.

(c) Consider a slightly different optimization problem. Given p,

$$P_e^*(p) := \min_{\text{algorithm}, n} P_e.$$

The distinction here is that n is a design parameter. For $\epsilon > 0$, what are $P_e^*(\frac{\log n}{n} + \epsilon)$ and $P_e^*(\frac{\log n}{n} - \epsilon)$? Also explain why.

Problem 8.3 (*An upper bound*) Let $p = \lambda \frac{\log n}{n}$ where $\lambda > 1$. Show that

$$\sum_{k=1}^{\frac{n}{2}} \binom{n}{k} (1 - p)^{k(n-k)} \longrightarrow 0 \qquad \text{as } n \to \infty.$$

Hint: Use the bounds in Problem 8.1.

Problem 8.4 (*Erdős-Rényi random graph*) Consider a random graph $\mathcal{G}$ that two Hungarian mathematicians (named Paul Erdős and Alfréd Rényi) introduced in (Erdős, Rényi et al. 1960). The graph, named the Erdős-Rényi graph, includes n nodes and assumes that an edge appears w.p. $p \in [0, 1]$ for any pair of two nodes in an independent manner. We say that a graph is *connected* if for any two nodes, there exists a path (i.e., a sequence of edges) that connects one node to the other; otherwise, it is said to be *disconnected*. See Figure 3.22 for examples.

(a) Show that

$$\mathbb{P}(\mathcal{G} \text{ is disconnected}) \leq \sum_{k=1}^{n-1} \binom{n}{k} (1 - p)^{k(n-k)}.$$

Hint: Think about a necessary event for disconnectivity.

(b) Show that if $p > \frac{\log n}{n}$, then $\mathbb{P}(\mathcal{G} \text{ is disconnected}) \to 0$ as $n \to \infty$.
Hint: Recall the achievability proof for community detection that we did in Sect. 3.6.

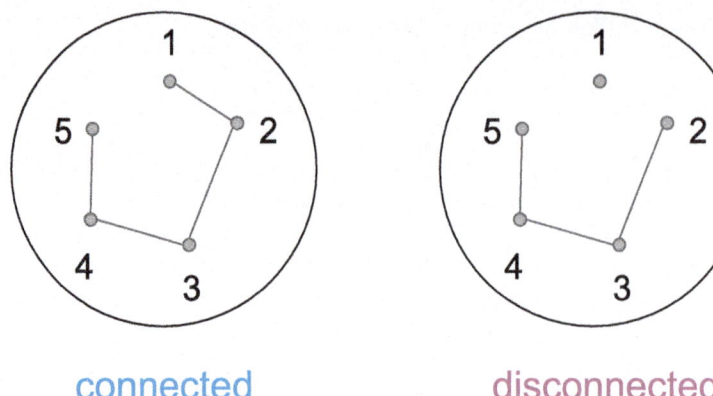

Fig. 3.22 Illustration of graph connectivity and disconnectivity

(c) It has been shown that if $p < \frac{\log n}{n}$, one can verify that $\mathbb{P}(\mathcal{G}$ is disconnected$) \to 1$ as $n \to \infty$. This together with the result in part (a) implies that the sharp threshold on p for graph *connectivity* is exactly the same as the one for community detection:

$$p^* = \frac{\log n}{n}.$$

Relate this to the fundamental limit on the observation probability in the community detection problem, i.e., explain why the limits are same.

Problem 8.5 (*True or False?*)

(a) Consider the two-community detection problem. Let $\mathbf{x} := [x_1, x_2, \ldots, x_n]$ be the community membership vector in which $x_i \in \{0, 1\}$ and n denotes the total number of users. Suppose we are given part of the comparison pairs with observation probability p. In Sect. 3.5, we formulated an optimization problem which aims to minimize the probability of error defined as $P_e := \mathbb{P}\left(\hat{\mathbf{x}} \notin \{\mathbf{x}, \mathbf{x} \oplus \mathbf{1}\}\right)$. Given p and n, denote by $P_e^*(p, n)$ the minimum probability of error. In Sect. 3.6, we did not intend to derive the exact $P_e^*(p, n)$; instead we developed a lower bound of $P_e^*(p, n)$ to demonstrate that for any $p > \frac{\log n}{n}$, the probability of error can be made arbitrarily close to 0 as n tends to infinity.

(b) Consider an *inference* problem in which we wish to decode $X \in \mathcal{X}$ from $Y \in \mathcal{Y}$ where $\mathcal{X}$ and $\mathcal{Y}$ indicate the ranges of X and Y, respectively. Given $Y = y$, the optimal decoder is:

$$\hat{X} = \arg\max_{x \in \mathcal{X}} \mathbb{P}\left(Y = y | X = x\right).$$

(c) Let $\mathbf{A} \in \mathbb{R}^{n \times n}$ be a matrix with m positive eigenvalues λ_i's and eigenvectors $\mathbf{v}_i$'s:

$$\mathbf{A} := \lambda_1 \mathbf{v}_1 \mathbf{v}_1^T + \lambda_2 \mathbf{v}_2 \mathbf{v}_2^T + \cdots + \lambda_m \mathbf{v}_m \mathbf{v}_m^T$$

where $\lambda_1 > \lambda_2 \geq \lambda_3 \geq \cdots \geq \lambda_m$ and $\mathbf{v}_i$'s are orthonormal: $\mathbf{v}_i^T \mathbf{v}_j = \mathbf{1}\{i = j\}$. Let $\mathbf{v} \in \mathbb{R}^n$ be some non-zero vector. Then,

$$\frac{\mathbf{A}^k \mathbf{v}}{\sqrt{\|\mathbf{A}^k \mathbf{v}\|^2}} \longrightarrow \mathbf{v}_1$$

as $k \rightarrow \infty$.

3.8   Machine Learning: Probabilistic Modeling

Recap
For the past sections, relying on the ML principle, we proved the achievability of the fundamental limit on observation probability p required for reliable community detection:

$$p > \frac{\log n}{n} \implies P_e \to 0 \quad \text{as } n \to \infty.$$

Through Python simulation, we also observed the phenomenon of phase transition at the limit $p^* = \frac{\log n}{n}$. In fact, the estimated vector through the spectral algorithm converges to the ground-truth community vector when p is close to $\frac{\log n}{n}$.

Outline
In this section, we will embark on our third application: *Machine learning*. The section consists of four parts. First, we will revisit the definition of machine learning, which we briefly introduced in Sect. 1.1. Subsequently, we will explore the connection between machine learning and probability, highlighting the *probabilistic* aspect in machine learning. Following that, we will formulate an optimization problem that plays a pivotal role in the design of machine learning models. Finally, we will examine a specific function that emerges within the context of optimization and emphasize that the choice of the function is intimately related to the Maximum Likelihood (ML) principle.

Review: Machine learning   Let's begin by revisiting the definition of machine learning. Machine learning is a methodology designed to enable machines to emulate human-like performance. It is the field that revolves around the study of *algorithms* (sets of instructions executable by a computer) to train a computer system in a way that the trained machine can effectively carry out a specific task of interest.

See Figure 3.23 for a visual representation. The focus here is on building a computer system that is definitely a machine. Being a system (essentially a function), it comprises both input and output components. The input, typically represented as x, signifies the information used to execute a specific task. The output, conventionally denoted as y, indicates the result of that task. For example, if a task of interest is the classification of cat versus dog image, x could represent the pixel values of an image and y would take on a binary value, indicating whether the provided image is of a cat (e.g., $y = 1$) or a dog (e.g., $y = 0$). One crucial aspect of machine learning lies in the utilization of *data*. The data often refer to input-output paired samples, denoted by:

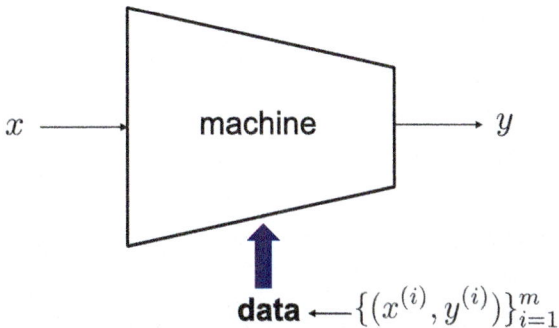

Fig. 3.23 Machine learning: A methodology for training a machine using data to enable it to perform tasks similar to human capabilities

$$\{(x^{(i)}, y^{(i)})\}_{i=1}^{m}, \tag{3.69}$$

where $(x^{(i)}, y^{(i)})$ represents the ith input-output sample and m denotes the total number of samples. It should be noted that people often use a different term, *example*, to refer to a sample.

A remark on the naming The reasoning behind the naming of machine learning becomes evident when we shift our perspective. Looking at it from the machine's standpoint, we can see that a machine acquires the ability to perform tasks by learning from data. Thus, the term *machine learning* is coined. This name was originally introduced in 1959 by Arthur Lee Samuel (Samuel 1967). See Figure 3.24.

Arthur Samuel is one of the pioneers in the field of Artificial Intelligence (AI), which includes machine learning as a subfield. AI is the domain dedicated to the exploration of creating intelligence in machines, distinct from natural intelligence demonstrated by intelligent beings such as humans and animals. One of his early accomplishments was the creation of a computer player for the board game checkers that exhibited human-like game play; see the right figure in Figure 3.24. During the development of computer checkers, he introduced numerous algorithms and concepts. Interestingly, these algorithms would go on to serve as the foundation for AlphaGo, a computer program designed for the Go board game (Silver et al. 2016). In 2016, AlphaGo achieved a groundbreaking feat by defeating one of the 9-dan professional players, Lee Sedol, with a score of 4 wins out of 5 games (News 2016).

The probabilistic aspect in machine learning The connection between machine learning and probability arises through data $\{(x^{(i)}, y^{(i)})\}_{i=1}^{m}$. This is because people often consider the data as one particular realization of a *random process*:

$$\{(X^{(i)}, Y^{(i)})\}_{i=1}^{m}. \tag{3.70}$$

Arthur Samuel '59   checkers

Fig. 3.24 (Left) Arthur Lee Samuel is an American pioneer in the field of artificial intelligence. Among his notable early achievements is the development of computer checkers, a significant precursor to AlphaGo; (Right) Checkers

One natural assumption made in practice is that the random process is i.i.d. across distinct examples, with each example being drawn from a joint distribution $\mathbb{P}_{X,Y}(x, y)$:

$$\{(X^{(i)}, Y^{(i)})\}_{i=1}^{m} \text{ i.i.d. } \sim \mathbb{P}_{X,Y}(x, y). \tag{3.71}$$

Training via optimization Another probabilistic aspect in machine learning pertains to the training process. To understand what this means, let us first examine a typical method for training a machine, which involves estimating a function of machine, say $f(\cdot)$. This estimation is typically done by solving an optimization problem. Then, you might be curious about the connection between an optimization problem and the training process.

Objective function To figure this out, let us consider what might serve as an *objective function* in the optimization problem for training. In the design of a machine learning model, our aim is to ensure that the predicted value $f(x^{(i)})$ closely approximates the ground-truth label $y^{(i)}$ for all examples:

$$y^{(i)} \approx f(x^{(i)}), \quad \forall i \in \{1, \ldots, m\}.$$

A natural question that follows is: How can we measure the degree of *closeness* indicated by the "$\approx$" symbol between $y^{(i)}$ and $f(x^{(i)})$? One prevalent approach in the field is to utilize a function known as a *loss* function, typically represented by:

$$\ell\left(y^{(i)}, f(x^{(i)})\right). \tag{3.72}$$

A fundamental requirement for the loss function $\ell(\cdot, \cdot)$ is that it should yield a small value when its two arguments are close and should equal zero when they are identical. Using such loss function (3.72), one can then formulate an optimization problem as:

$$\min_{f(\cdot)} \sum_{i=1}^{m} \ell(y^{(i)}, f(x^{(i)})). \qquad (3.73)$$

How to introduce optimization variable? Next, what is the best approach to minimize the objective function? To address this, we first need to identify a quantity, called the *optimization variable*, which influences the objective function. However, in this case, there is no traditional variable. Instead, we deal with a different entity that can be optimized: the function $f(\cdot)$. So, the issue becomes: How can we introduce an optimization variable? A common practice in the field involves representing the function $f(\cdot)$ using *parameters* (or *weights*), denoted as w, and then treating these weights as the optimization variable. With this approach, we can rewrite the problem (3.73) as follows:

$$\min_{w} \sum_{i=1}^{m} \ell(y^{(i)}, f_w(x^{(i)})) \qquad (3.74)$$

where $f_w(x^{(i)})$ denotes the function $f(x^{(i)})$ parameterized by w.

The optimization problem described above depends on how we define two functions: (i) $f_w(x^{(i)})$ w.r.t. the parameters w, and (ii) the loss function $\ell(\cdot, \cdot)$. In the field of machine learning, extensive research has been conducted to determine suitable choices for these functions.

A choice for $f_w(\cdot)$ Around the same period when the machine learning field was established, an architecture was proposed for the function $f_w(\cdot)$, particularly in the context of binary classifiers, where y can only assume one of two options, such as $y \in \{0, 1\}$. This architecture is known as the *Perceptron* and was introduced in 1957 by one of the early pioneers in AI, Frank Rosenblatt (Rosenblatt 1958). See Figure 3.25 for his portrait. Frank Rosenblatt was a psychologist who was intrigued by the workings of intelligent beings' brains. His research on the brain's functioning led him to develop the perceptron, providing valuable insights into neural networks that many of you might hear of.

How the brain works Here are the details of how the brain structure served as inspiration for the perceptron architecture. The brain contains numerous electrically excitable cells, namely *neurons*; see Figure 3.26. In the figure, a neuron is represented by a red circle, and the illustration displays a total of three neurons. These neurons possess three significant properties that influenced the development of the perceptron architecture.

The first is that neurons are electrical entities and possess a voltage. The second property is that neurons are interconnected through mediums, called *synapses*,

Fig. 3.25 Frank Rosenblatt (1928–1971) is an American psychologist notable as the inventor of perceptron. One sad story is that he died in 1971 on his 43rd birthday, in a boating accident

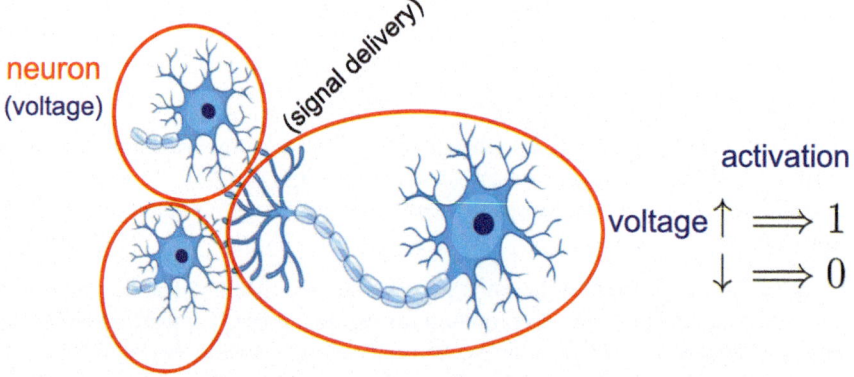

Fig. 3.26 Neurons are electrically excitable cells and are connected through synapses

responsible for transmitting electrical voltage signals between neurons. Depending on the strength of the synaptic connections, these signals can either increase or decrease as they pass from one neuron to another. The last property is that neurons take a specific action known as *activation*. This activation is contingent on the neuron's voltage level and results in the generation of an all-or-nothing pulse signal. For example, if the voltage level surpasses a certain threshold, the neuron produces an impulse signal with a specific magnitude, say 1; otherwise, it remains inactive.

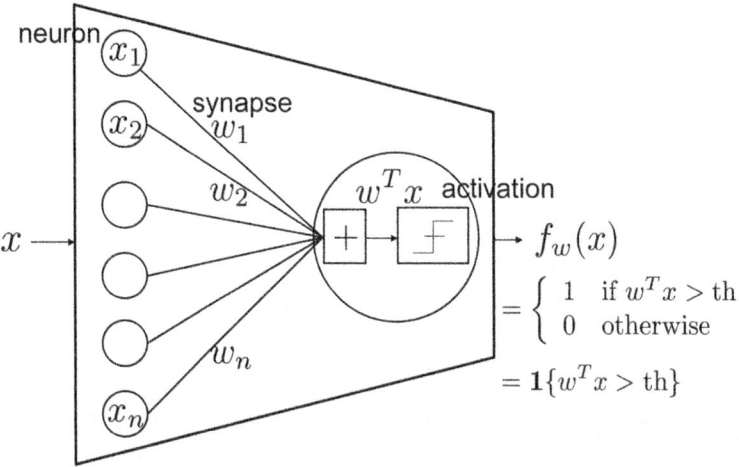

Fig. 3.27 The architecture of perceptron

Perceptron The above three properties inspired Frank Rosenblatt to introduce the perceptron architecture, as depicted in Figure 3.27.

An input $x := [x_1, x_2, \ldots, x_n]^T$ is an n-dimensional real-valued signal. Each component x_i corresponds to a voltage signal level for a respective neuron. These voltage signals x_i's are transmitted through synapses to another neuron located on the right side in the figure, which is depicted as a larger circle. The voltage level can either increase or decrease depending on the strength of synaptic connections. To account for this, a weight w_i is multiplied by x_i, resulting in $w_i x_i$ as the voltage signal delivered to the terminal neuron.

Based on an empirical observation that the voltage level at the terminal neuron increases with a larger number of connected neurons, Rosenblatt introduced an adder to sum all the voltage signals originating from multiple neurons. Consequently, he modeled the voltage signal at the terminal neuron as follows:

$$w_1 x_1 + w_2 x_2 + \cdots + w_n x_n = w^T x. \tag{3.75}$$

In an attempt to mimic *activation*, he modeled the output signal as:

$$f_w(x) = \begin{cases} 1 & \text{if } w^T x > \text{th}; \\ 0 & \text{otherwise,} \end{cases} \tag{3.76}$$

where "th" indicates a certain threshold level. It can also be simply denoted as

$$f_w(x) = 1\{w^T x > \text{th}\} \tag{3.77}$$

where $\mathbf{1}\{\cdot\}$ is the indicator function that returns 1 when $(\cdot)$ is true while returning 0 otherwise.

Activation functions Taking the percentron as a function class, one can formulate the optimization problem (3.74) as:

$$\min_{w} \sum_{i=1}^{m} \ell \left(y^{(i)}, \mathbf{1}\{w^T x^{(i)} > \text{th}\} \right). \tag{3.78}$$

This is an initial optimization problem that people came up with. However, there is a fundamental issue when attempting to solve this problem. The issue arises from the presence of an indicator function within the objective function, rendering it *non-differentiable*. As we have seen a couple of times earlier, a common approach to solving optimization problems involves "derivative computation". Consequently, the optimization problem expressed in (3.78), which incorporates a non-differentiable function, is considered undesirable. So, what are our options in this situation? One common approach adopted in the field is to *approximate* the activation function. There are several methods available for approximating it, and in the following discussion, we will explore one of these techniques.

Approximation via a logistic function A widely accepted approach involves using the following function, which facilitates a *smooth* transition from 0 to 1:

$$f_w(x) = \frac{1}{1 + e^{-w^T x}}. \tag{3.79}$$

Observe that when $w^T x$ is significantly small, the function $f_w(x)$ approximates to 0. As $w^T x$ increases, it experiences exponential growth, followed by logarithmic growth, and eventually reaches saturation at 1 when $w^T x$ becomes very large. Refer to Figure 3.28 for a visual representation of this behavior.

In fact, the function (3.79) is a well-known and widely used function in statistics, often referred to as the *logistic* function (Garnier & Quetelet 1838). The term *logistic* is derived from a Greek word, signifying slow growth similar to logarithmic growth. This function is also known by another name, the *sigmoid* function, which is named

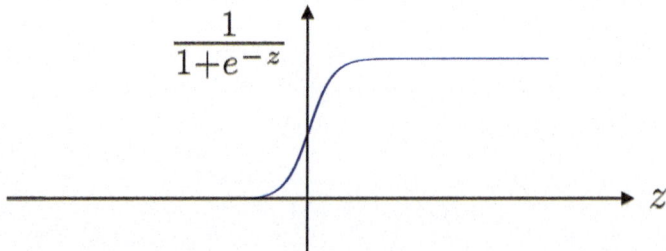

Fig. 3.28 Logistic function: $\sigma(z) = \frac{1}{1+e^{-z}}$

after its S-shaped curve resembling the lower-case Greek letter sigma. The logistic function possesses two advantageous characteristics. Firstly, it is differentiable, making it suitable for various optimization problems. Secondly, it can be interpreted as the *probability* of the output in a binary classifier, e.g., $\mathbb{P}(Y = 1)$ where Y represents the random variable for the ground-truth label in the binary classifier. This interpretability is a valuable feature of the logistic function.

Look ahead

When using the logistic activation, the selection of an appropriate *loss* function is a crucial consideration. The Maximum Likelihood (ML) principle plays a pivotal role in designing an *optimal* loss function. In the following section, we will delve into the nature of this optimality and explore how the ML principle informs the design of the optimal loss function.

3.9   Machine Learning: The Maximum Likelihood (ML) Principle

Recap
In the previous section, we established an optimization problem for designing a machine learning model based on the perceptron architecture:

$$\min_{w} \sum_{i=1}^{m} \ell\left(y^{(i)}, \hat{y}^{(i)}\right) \tag{3.80}$$

where $\{(x^{(i)}, y^{(i)})\}_{i=1}^{m}$ indicates a collection of input-output paired examples; $\ell(\cdot, \cdot)$ denotes a loss function; and $\hat{y}^{(i)} := f_{w}(x^{(i)})$ is the prediction function parameterized by the weights w. We introduced the logistic function as the activation function, a widely employed choice in the field:

$$f_{w}(x) = \frac{1}{1 + e^{-w^T x}}. \tag{3.81}$$

We claimed that the ML principle plays a crucial role in the design of the *optimal* loss function.

Outline
In this section, we will delve into the details of this claim. The section is organized into three parts. Firstly, we will explore the concept of the *optimal* loss function. Following that, we will examine how the ML principle guides the development of the optimal loss function. Next, we will study how to solve the optimization problem formulated. While there is no closed-form solution available, there is a prominent algorithm that allows us to obtain a *numerical* solution with the assistance of a computer. This algorithm is known as *gradient descent*. So, in the last part, we will explore how the gradient descent algorithm works.

Optimality in a sense of maximizing likelihood A binary classifier employing the logistic function (3.81) is called *logistic regression*. This terminology can be somewhat misleading since "regression" typically implies prediction rather than classification. The reason behind the use of this naming is that the classifier provides a continuous output, in contrast to a discrete output, say between 0 and 1. See Figure 3.29 for illustration.

Notice that the output $\hat{y}$ falls within the range of 0 and 1:

$$0 \leq \hat{y} \leq 1.$$

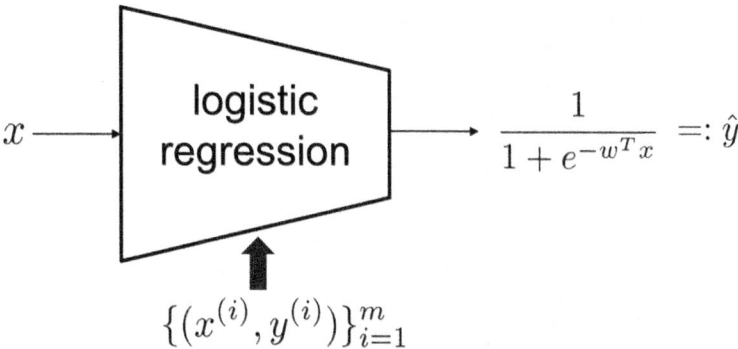

Fig. 3.29 Logistic regression

Consequently, the output can be interpreted as a *probabilistic* quantity. One natural assumption inspired by this interpretation is:

$$\text{Assumption}: \hat{y} = \mathbb{P}(Y = 1 | X = x) \tag{3.82}$$

where X and Y represent random variables for input and output, respectively. This assumption aligns intuitively with the idea that a high value of $\hat{y}$ (close to 1) naturally corresponds to a prediction of the ground-truth as 1, while a low value of $\hat{y}$ (close to 0) corresponds to a prediction of the ground-truth as 0.

The meaning of an optimal solution in logistic regression can be defined under the above assumption. To understand this, let's first examine the *likelihood* of the true classifier:

$$\mathbb{P}\left(Y^{(1)} = y^{(1)}, \ldots, Y^{(m)} = y^{(m)} | X^{(1)} = x^{(1)}, \ldots, X^{(m)} = x^{(m)}\right). \tag{3.83}$$

Notice that the classifier output $\hat{y}$ depends on the weights w, as the classifier is parameterized by w. Hence, given the assumption (3.82), the likelihood (3.83) is also a function of the weights w.

We are now ready to establish the notion of optimality for w. The optimal weight, say $w^{\star}$, is defined as the one that *maximizes the likelihood* (3.83):

$$w^{\star} := \arg\max_{w} \mathbb{P}\left(Y^{(1)} = y^{(1)}, \ldots, Y^{(m)} = y^{(m)} | X^{(1)} = x^{(1)}, \ldots, X^{(m)} = x^{(m)}\right). \tag{3.84}$$

Certainly, various definitions of optimality are possible, but here we employ the Maximum Likelihood (ML) principle, the most popular choice. This is precisely where the definition of the *optimal loss function*, say $\ell^{\star}(\cdot, \cdot)$, kicks in. We define $\ell^{\star}(\cdot, \cdot)$ as the one that satisfies the following condition:

$$\arg \min_{w} \sum_{i=1}^{m} \ell^{\star}(y^{(i)}, \hat{y}^{(i)})$$
$$= \arg \max_{w} \mathbb{P}\left(Y^{(1)} = y^{(1)}, \dots, Y^{(m)} = y^{(m)} | X^{(1)} = x^{(1)}, \dots, X^{(m)} = x^{(m)}\right).$$

(3.85)

It turns out that the optimal loss function $\ell^{\star}(\cdot, \cdot)$ due to the condition (3.85) is closely associated with a very well-known machine learning classifier: *logistic regression*, in which the loss function reads:

$$\ell^{\star}(y, \hat{y}) = \ell_{\text{logistic}}(y, \hat{y}) = -y \log \hat{y} - (1 - y) \log(1 - \hat{y}). \qquad (3.86)$$

This reveals that the ML principle plays a pivotal role in the design of the well-known classifier, logistic regression, which is the optimal classifier within the perceptron architecture. We will next provide the proof of (3.86).

Derivation of the optimal loss function $\ell^{\star}(\cdot, \cdot)$ Samples are typically drawn from diverse contexts. Therefore, it is reasonable to assume that these samples are independent of each other:

$$\{(X^{(i)}, Y^{(i)})\}_{i=1}^{m} \text{ are independent over } i. \qquad (3.87)$$

Under this assumption, we can then rewrite the likelihood (3.83) as:

$$\mathbb{P}\left(Y^{(1)} = y^{(1)}, \dots, Y^{(m)} = y^{(m)} | X^{(1)} = x^{(1)}, \dots, X^{(m)} = x^{(m)}\right)$$
$$\overset{(a)}{=} \frac{\mathbb{P}\left(X^{(1)} = x^{(1)}, Y^{(1)} = y^{(1)}, \dots, X^{(m)} = x^{(m)}, Y^{(m)} = y^{(m)}\right)}{\mathbb{P}\left(X^{(1)} = x^{(1)}, \dots, X^{(m)} = x^{(m)}\right)}$$
$$\overset{(b)}{=} \frac{\prod_{i=1}^{m} \mathbb{P}_{X,Y}\left(x^{(i)}, y^{(i)}\right)}{\prod_{i=1}^{m} \mathbb{P}_{X}(x^{(i)})} \qquad (3.88)$$
$$\overset{(c)}{=} \prod_{i=1}^{m} \mathbb{P}_{Y|X}\left(y^{(i)} | x^{(i)}\right)$$

where (a) and (c) are due to the definition of conditional probability; and (b) comes from the independence assumption (3.87). Some of you might be wondering why $\{X_i\}_{i=1}^{m}$ are also independent. We can easily demonstrate this based on (3.87); check this in Problem 9.2. Here $\mathbb{P}_{X,Y}(x^{(i)}, y^{(i)})$ represents the probability distribution of the input-output pairs of the system:

$$\mathbb{P}_{X,Y}(x^{(i)}, y^{(i)}) := \mathbb{P}(X = x^{(i)}, Y = y^{(i)}). \qquad (3.89)$$

Similarly

$$\mathbb{P}_{X}(x^{(i)}) := \mathbb{P}(X = x^{(i)}). \qquad (3.90)$$

Recall the probability-interpretation-related assumption (3.82) made with regard to $\hat{y}$:

$$\hat{y} = \mathbb{P}(Y = 1 | X = x).$$

This implies that:

$$y = 1 : \quad \mathbb{P}_{Y|X}(y|x) = \hat{y};$$
$$y = 0 : \quad \mathbb{P}_{Y|X}(y|x) = 1 - \hat{y}.$$

Hence, a succinct representation for $\mathbb{P}_{Y|X}(y|x)$ reads:

$$\mathbb{P}_{Y|X}(y|x) = \hat{y}^y (1 - \hat{y})^{1-y}.$$

Using the notations of $(x^{(i)}, y^{(i)})$ and $\hat{y}^{(i)}$, we then get:

$$\mathbb{P}_{Y|X}\left(y^{(i)}|x^{(i)}\right) = (\hat{y}^{(i)})^{y^{(i)}} (1 - \hat{y}^{(i)})^{1-y^{(i)}}.$$

Plugging this into (3.88), we get:

$$\mathbb{P}\left(Y^{(1)} = y^{(1)}, \ldots, Y^{(m)} = y^{(m)} | X^{(1)} = x^{(1)}, \ldots, X^{(m)} = x^{(m)}\right)$$
$$= \prod_{i=1}^{m} (\hat{y}^{(i)})^{y^{(i)}} (1 - \hat{y}^{(i)})^{1-y^{(i)}}. \tag{3.91}$$

Applying this into (3.84) yields:

$$w^\star = \arg\max_w \prod_{i=1}^{m} (\hat{y}^{(i)})^{y^{(i)}} (1 - \hat{y}^{(i)})^{1-y^{(i)}}$$
$$\overset{(a)}{=} \arg\max_w \sum_{i=1}^{m} y^{(i)} \log \hat{y}^{(i)} + (1 - y^{(i)}) \log(1 - \hat{y}^{(i)}) \tag{3.92}$$
$$\overset{(b)}{=} \arg\min_w \sum_{i=1}^{m} -y^{(i)} \log \hat{y}^{(i)} - (1 - y^{(i)}) \log(1 - \hat{y}^{(i)})$$

where (a) is attributed to the non-decreasing nature of the $\log(\cdot)$ function and the positivity of $\prod_{i=1}^{m} (\hat{y}^{(i)})^{y^{(i)}} (1 - \hat{y}^{(i)})^{1-y^{(i)}}$; and (b) arises from the change in the objective function's sign.

In fact, the expression within the summation in the final equation of (3.92) adheres to the formula of an important notion in the field of information theory: *cross entropy* (Cover & Joy 2006). In particular, in the context of a loss function, it is referred to as the *cross entropy loss*:

$$\ell_{\mathsf{CE}}(y, \hat{y}) := -y \log \hat{y} - (1 - y) \log(1 - \hat{y}). \tag{3.93}$$

Therefore, the optimal loss function that maximizes likelihood is the cross entropy loss:

$$\ell^{\star}(\cdot, \cdot) = \ell_{\mathsf{CE}}(\cdot, \cdot).$$

Remarks on cross entropy loss (3.93) Let's briefly discuss the reasoning behind the naming of cross entropy loss (3.93). This naming is derived from the definition of *cross entropy*. Cross entropy is defined in relation to two random variables. To simplify, let's consider two binary random variables, say $X \sim \mathsf{Bern}(p)$ and $Y \sim \mathsf{Bern}(q)$. For these two random variables, cross entropy is defined as (Cover & Joy 2006):

$$H(p, q) := -p \log q - (1 - p) \log(1 - q). \tag{3.94}$$

Notice that the formula in (3.94) is identical to the term found within the summation in (3.92), except for the use of different notations. This is why it is referred to as the *cross entropy loss*. For those who are curious about why the formula in (3.94) is termed *cross entropy*, you will have an opportunity to explore the rationale in Problem 9.5.

How to solve (3.92)? From (3.92) and (3.81), we can write the optimization problem as:

$$\min_{w} \sum_{i=1}^{m} -y^{(i)} \log \frac{1}{1 + e^{-w^T x^{(i)}}} - (1 - y^{(i)}) \log \frac{e^{-w^T x^{(i)}}}{1 + e^{-w^T x^{(i)}}}. \tag{3.95}$$

Let $J(w)$ be the normalized version of the objective function:

$$J(w) := \frac{1}{m} \sum_{i=1}^{m} -y^{(i)} \log \hat{y}^{(i)} - (1 - y^{(i)}) \log(1 - \hat{y}^{(i)}). \tag{3.96}$$

The above optimization problem belongs to *convex optimization* (Boyd & Vandenberghe 2004; Suh 2022). Convex optimization, in simple terms, refers to a class of optimization problems that can be efficiently solved using computer algorithms. More formally, it pertains to problems in which the objective function is a *convex function* in the minimization problem. To provide a rough idea, a convex function can be visualized as a bowl-shaped function, as illustrated in Figure 3.30. This figure depicts a one-dimensional case for the variable w, for explanatory purposes. In a multidimensional scenario where $w := [w_1, \ldots, w_n]$, it implies that the function exhibits a bowl shape with respect to each individual component w_i where $i \in \{1, \ldots, n\}$. You can find the formal definition and the proof of the convexity of $J(w)$ later in this section; see Side Study #1 and #2, respectively.

One crucial aspect of convex functions (the bowl-shaped functions) is that the minimum point occurs at a unique stationary point, provided the minimum is finite. Although we will not provide a formal proof for this, it intuitively makes sense when considering the bowl-shaped curve in Figure 3.30. Leveraging this fact, we can assert that $w^\star$ represents the stationary point:

$$\nabla J(w^\star) = 0. \tag{3.97}$$

Hence, the primary concern is to determine this $w^\star$. However, there is a challenge in obtaining $w^\star$ analytically. The issue stems from the fact that there is no closed-form solution available (check!). But here's the good news. There are several algorithms that enable us to efficiently find this point without the need for a closed-form solution. One such prominent algorithm widely employed in the field is *gradient descent*.

Gradient descent (Lemaréechal 2012) Here's how the gradient descent algorithm operates. It's an iterative method. Let's assume that at the t-th iteration, we have an estimate for $w^\star$, say $w^{(t)}$. Typically, the initial estimate $w^{(0)}$ is randomly chosen. We then calculate the gradient of the function at this estimate: $\nabla J(w^{(t)})$. Subsequently, we update the estimate in a direction that is opposite to the gradient direction:

$$w^{(t+1)} \longleftarrow w^{(t)} - \alpha \nabla J(w^{(t)}) \tag{3.98}$$

where $\alpha > 0$ indicates the step size (also known as the learning rate). If you think about it, this update rule is quite intuitive. Suppose $w^{(t)}$ is positioned to the right of the optimal point $w^\star$, as illustrated in Figure 3.30.

In this scenario, it is evident that we should shift $w^{(t)}$ to the left to bring it closer to $w^\star$. The update rule precisely accomplishes this by subtracting $\alpha \nabla J(w^{(t)})$. Note that $\nabla J(w^{(t)})$ points in the rightward direction when $w^{(t)}$ is situated to the right of $w^\star$. We repeat this process until convergence is achieved. Interestingly, as $t \to \infty$, it does indeed converge:

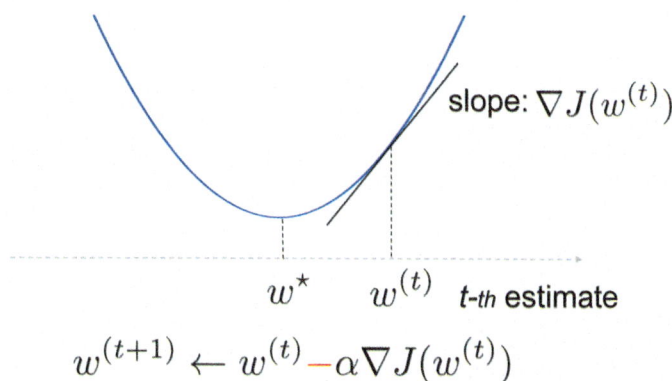

Fig. 3.30 Gradient descent

$$w^{(t)} \longrightarrow w^\star, \tag{3.99}$$

provided that the learning rate is appropriately chosen, such as one that decays exponentially with respect to t, for instance, e^{-t}. We will not delve into the proof of this convergence. In fact, the proof is not straightforward, and there is an extensive field in statistics dedicated to demonstrating the convergence of various algorithms.

Side study #1: Convex functions An informal yet intuitive definition of a convex function is as follows. We say that a function is convex if it is bowl-shaped, as illustrated in Figure 3.31.

What is the formal definition of convexity? The following insight can help us arrive at the definition. Consider two points, say x and y, as depicted in Figure 3.31. Now, think about a point that falls in between these two points, represented as $\lambda x + (1 - \lambda)y$ for $\lambda \in [0, 1]$. In the context of the bowl-shaped function, this suggests that the function evaluated at an λ-weighted linear combination of x and y is either less than or equal to the same λ-weighted linear combination of the function values evaluated at x and y:

$$f(\lambda x + (1 - \lambda)y) \le \lambda f(x) + (1 - \lambda)f(y). \tag{3.100}$$

This insight leads us to the following definition: A function f is said to be convex if (3.100) holds true for all $\lambda \in [0, 1]$ and for all values of x and y.

Side study #2: Proof of convexity Applying the definition of convex functions, it is possible to demonstrate that $J(w)$ is a convex function with respect to the optimization variable w. We can easily establish that convexity is maintained under addition. Why? Think about the definition of convexity. Thus, it suffices to prove the following two conditions:

$$\text{(i)} - \log \frac{1}{1 + e^{-w^T x}} \text{ is convex in } w;$$

Fig. 3.31 A geometric intuition behind a convex function

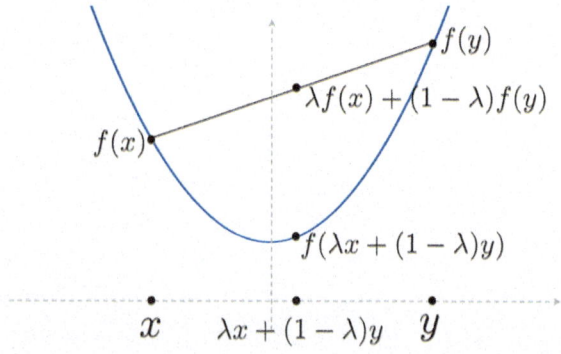

$$(ii) - \log \frac{e^{-w^T x}}{1 + e^{-w^T x}} \text{ is convex in } w.$$

Since the second function in the above can be represented as the sum of a linear function and the first function:

$$- \log \frac{e^{-w^T x}}{1 + e^{-w^T x}} = w^T x - \log \frac{1}{1 + e^{-w^T x}},$$

it suffices to prove the convexity of the first function.

Notice that the first function can be rewritten as:

$$- \log \frac{1}{1 + e^{-w^T x}} = \log(1 + e^{-w^T x}). \tag{3.101}$$

In fact, directly proving the convexity of (3.101) using the definition of convex functions can be somewhat intricate. However, there's an alternative approach based on the computation of the second derivative of a function, called the Hessian (Marsden & Tromba 2003). How do we calculate the Hessian? What are the dimensions of the Hessian? For a function $f : \mathbb{R}^d \to \mathbb{R}$, the gradient $\nabla f(x) \in \mathbb{R}^d$, and the Hessian $\nabla^2 f(x) \in \mathbb{R}^{d \times d}$. If you're not familiar with this topic, you can refer to a vector calculus book (Marsden & Tromba 2003) or consult resources like wikipedia.

A well-established fact states that if the Hessian of a function is positive semi-definite (PSD),[3] the function is convex. We will not delve into the proof here. Just remember this fact, as it is highly useful. Here we will apply this fact to prove the convexity of the function (3.101).

Taking a derivative of the right-hand-side in (3.101) w.r.t. w, we get:

$$\nabla_w \log(1 + e^{-w^T x}) = \frac{-x e^{-w^T x}}{1 + e^{-w^T x}}.$$

This is due to the chain rule of derivatives and the fact that $\frac{d}{dz} \log z = \frac{1}{z}$, $\frac{d}{dz} e^z = e^z$ and $\frac{d}{dw} w^T x = x$. Taking another derivative of the above, we obtain the Hessian as follows:

---

[3] We say that a *symmetric* matrix, say $Q = Q^T \in \mathbb{R}^{d \times d}$, is positive semi-definite if $v^T Q v \geq 0$, $\forall v \in \mathbb{R}^d$, i.e., all the eigenvalues of Q are non-negative. It is simply denoted by $Q \succeq 0$.

$$\nabla_w^2 \log(1 + e^{-w^T x}) = \nabla_w \left(\frac{-xe^{-w^T x}}{1 + e^{-w^T x}} \right)$$

$$\overset{(a)}{=} \frac{xx^T e^{-w^T x}(1 + e^{-w^T x}) - xx^T e^{-w^T x} e^{-w^T x}}{(1 + e^{-w^T x})^2} \qquad (3.102)$$

$$= \frac{xx^T e^{-w^T x}}{(1 + e^{-w^T x})^2}$$

$$\succeq 0$$

where (a) is due to the quotient role for derivative: $\frac{d}{dz}\frac{f(z)}{g(z)} = \frac{f'(z)g(z) - f(z)g'(z)}{g^2(z)}$. Here you might be wondering why, when taking the derivative of $-xe^{-w^T x}$ w.r.t w, we get $xx^T e^{-w^T x}$, instead of other combinations like xx, $x^T x^T$, or $x^T x$ in front of $e^{-w^T x}$. A useful rule-of-thumb is to test all the possibilities and choose the one that avoids syntax errors, specifically matrix dimension mismatches. For instance, the operation xx (or $x^T x^T$) is not a valid operation. Also, $x^T x$ is not appropriate because the Hessian matrix should be a square d-by-d matrix. The only candidate that remains without any syntax error is xx^T.

Look ahead

Thus far, we have established an optimization problem for a machine learning model, and found that the ML principle plays a crucial role in designing the optimal loss function. Furthermore, we've gained insights into how to tackle this problem using a well-known algorithm known as gradient descent. Moving forward, we will delve into the practical programming implementation of this algorithm. In the upcoming section, we'll explore these implementation details in the context of a simple classifier using TensorFlow.

3.10 Machine Learning: **TensorFlow** Implementation

Recap

In the preceding sections, we have formulated an optimization problem for machine learning:

$$\min_{w} \sum_{i=1}^{m} \ell_{\text{CE}} \left(y^{(i)}, \hat{y}^{(i)} \right) \tag{3.103}$$

where $\hat{y} := \frac{1}{1+e^{-w^T x}}$ indicates the prediction output with logistic activation and ℓ_{CE} denotes cross entropy loss:

$$\ell_{\text{CE}}(y, \hat{y}) = -y \log \hat{y} - (1 - y) \log(1 - \hat{y}). \tag{3.104}$$

We proved that cross entropy loss $\ell_{\text{CE}}(\cdot, \cdot)$ is the optimal loss function for maximizing likelihood. Additionally, we showed that the normalized version $J(w)$ of the above objective function exhibits convexity w.r.t. w. Consequently, we could leverage a well-known algorithm, gradient descent, to efficiently locate the unique stationary point.

$$J(w) := \frac{1}{m} \sum_{i=1}^{m} -y^{(i)} \log \hat{y}^{(i)} - (1 - y^{(i)}) \log(1 - \hat{y}^{(i)}). \tag{3.105}$$

Outline

In this section, we will delve into the implementation of the algorithm using a programming language in the context of a simple classifier. This section consists of three parts. In the first part, we will investigate what the simple classifier setting of our focus is.

In the second part, we will study four implementation details. The first pertains to the dataset used for training and testing. In machine learning, *testing* refers to the evaluation of a trained model's performance. For this purpose, we often employ an *unseen* dataset called *test dataset*. Here, *unseen* means that this dataset has not been utilized during the training phase. The second implementation detail addresses the construction of a deep neural network model, including the utilization of the ReLU activation function. The Perceptron, introduced in Sect. 3.8, served as the foundation for the first neural network. A *deep neural network* extends this concept by incorporating at least one hidden layer between the input and output layers (Ivakhnenko 1971). The

ReLU is a famous activation function which is often employed in hidden layers (Glorot et al. 2011). It stands for the Rectified Linear Unit, and its operation reads: $\text{ReLU}(x) = \max(0, x)$. The third implementation detail relates to the utilization of the softmax activation function at the output layer. This function is a natural extension of the logistic activation for multiple classes (more than two). The fourth implementation detail concerns the use of the Adam optimizer, an advanced variant of gradient descent that is widely adopted in practical applications (Kingma & Ba 2014).

In the last part of this section, we will delve into practical programming aspects of our classifier using a prominent deep learning framework: TensorFlow. To facilitate our programming tasks, we will leverage a higher-level programming language known as Keras, which is fully integrated with TensorFlow.

Handwritten digit classification The classifier we will concentrate on for our implementation exercise is the one that recognizes handwritten digits. In this task, the goal is to identify a digit from a handwritten image, as depicted in Figure 3.32. The figure illustrates an example in which an image of the digit 2 is correctly identified.

For model training, we make use of a widely recognized dataset, named MNIST (Modified National Institute of Standards and Technology) (LeCun et al. 1998). This dataset was created by reconfiguring examples from NIST's original dataset, hence suggesting the naming. It was prepared by one of the pioneers in the field of deep learning, Yann LeCun. The MNIST dataset comprises a total of $m = 60,000$ training images and $m_{\text{test}} = 10,000$ testing images. Each image $x^{(i)}$ is a grid of 28×28 pixels, with each pixel representing a gray-scale level ranging from 0 (white) to 1 (black). Additionally, each image is associated with a corresponding label $y^{(i)}$, which belongs to one of the 10 possible classes, i.e., $y^{(i)} \in \{0, 1, \ldots, 9\}$. You can refer to Figure 3.33 for a visual representation of the dataset.

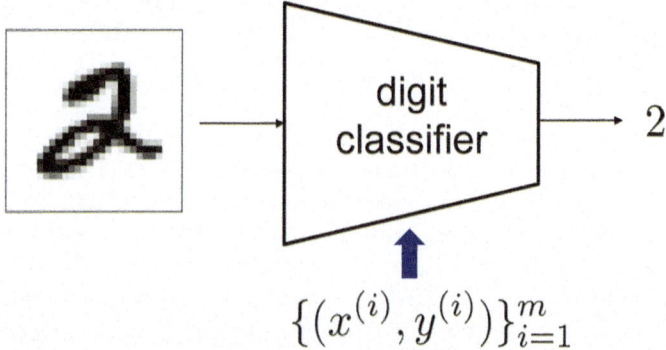

Fig. 3.32 Handwritten digit classification

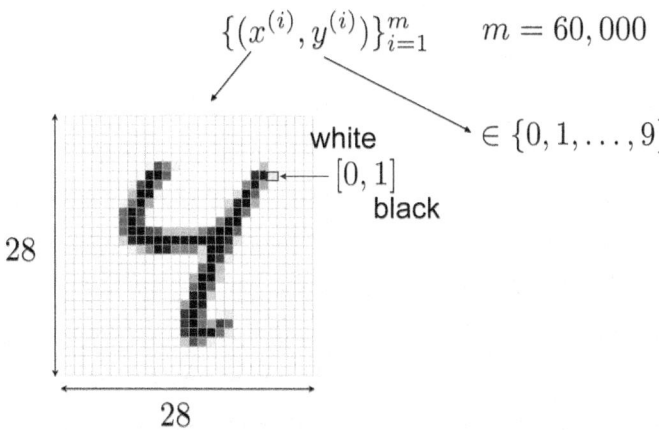

Fig. 3.33 The MNIST dataset: An input image is of 28-by-28 pixels, each indicating an intensity from 0 (white) to 1 (black); each label with size 1 takes one of the 10 classes from 0 to 9

A deep neural network model We utilize an advanced version of logistic regression as our model. There are two primary justifications for opting for this advanced version.

One reason is that logistic regression is a *linear* classifier, which means that the prediction function $f_w(\cdot)$ is confined to a linear function class. It is reasonable to assume that the performance of logistic regression, limited by this linear architecture, might not be satisfactory in numerous real-world applications. Indeed, this assumption holds true in many scenarios. Consequently, there has been a strong impetus to develop a more advanced neural network, akin to the Perceptron but with multiple layers. This extended architecture is referred to as a deep neural network (DNN). While the concept of DNNs was conceived in the 1960s (Ivakhnenko 1971), their potential benefits were not widely appreciated until a significant event in 2012. The big event was the winnning of Geoffrey Hinton (a prominent figure in the AI field, known as the Godfather of deep learning) and his two PhD students in the ImageNet recognition competition. What made this achievement remarkable was the fact that their DNN was capable of achieving *human-level* recognition performance, a feat never before realized (Krizhevsky et al. 2012). This milestone marked the inception of deep learning revolution. The task of digit classification of our interest here is one such application where a linear classifier falls short in terms of performance. Therefore, we will employ a DNN, yet a simple version with only two layers: one hidden layer and one output layer, as depicted in Figure 3.34. By convention, the input layer is not counted as a layer, so it is called a two-layer neural network instead of a three-layer one.

Each neuron in the hidden layer follows the same procedure as that in the Perceptron: it performs a linear operation followed by an activation function. In the early days, the activation function used was often the logistic function or its shifted

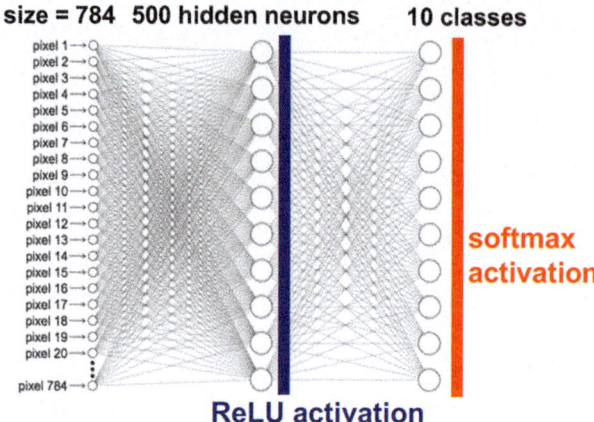

Fig. 3.34 A two-layer fully-connected neural network where input size is $28 \times 28 = 784$, the number of hidden neurons is 500 and the number of classes is 10. We employ ReLU activation at the hidden layer, and softmax activation at the output layer; see Figure 3.35 for details

version, known as the tanh function (which ranges between -1 and $+1$). However, in recent times, many practitioners have discovered that another activation function, named the *Rectified Linear Unit* (or ReLU for short), is more effective in terms of facilitating faster training while achieving superior or equivalent performance (Glorot et al. 2011). As mentioned earlier, its mathematical expression is $\text{ReLU}(x) = \max(0, x)$. Hence, a common rule of thumb in the deep learning field is to employ ReLU activation in all hidden layers. We adhere to this rule of thumb, as depicted in Figure 3.34.

Softmax activation at output layer The second rationale for utilizing an advanced version of logistic regression pertains to the number of classes in our targeted classifier. Logistic regression is primarily designed for binary classification tasks. However, in our digit classifier, we are dealing with a total of 10 classes. Consequently, the conventional logistic function is not directly applicable in this context. A natural extension of the logistic function for a general classifier with more than two classes is to employ a generalized version, called softmax. You can observe the operation of softmax in Figure 3.35.

Let z be the output of the last layer in a neural network prior to activation:

$$z := [z_1, z_2, \ldots, z_c]^T \in \mathbb{R}^c \qquad (3.106)$$

where c denotes the number of classes. The softmax function is then defined as:

$$\hat{y}_j := \left[\text{softmax}(z)\right]_j = \frac{e^{z_j}}{\sum_{k=1}^{c} e^{z_k}} \quad j \in \{1, 2, \ldots, c\}. \qquad (3.107)$$

Fig. 3.35 Softmax
activation employed at
output layer. This is a natural
extension of logistic
activation intended for the
two-class case

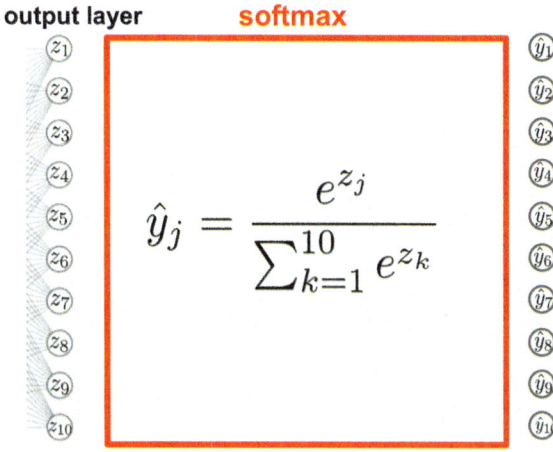

Note that this is a natural extension of the logistic function: for $c = 2$,

$$\hat{y}_1 := \big[\text{softmax}(z)\big]_1 = \frac{e^{z_1}}{e^{z_1} + e^{z_2}}$$

$$= \frac{1}{1 + e^{-(z_1 - z_2)}} \tag{3.108}$$

$$= \sigma(z_1 - z_2)$$

where $\sigma(\cdot)$ is the logistic function. Viewing $z_1 - z_2$ as the binary classifier output $\hat{y}$, this coincides exactly with the logistic function.

Here $\hat{y}_i$ can be interpreted as the probability that the ith example belongs to class i. Hence, like the binary classifier, one may want to assume:

$$\hat{y}_i = \mathbb{P}(y = [0, \dots, \underbrace{1}_{i\text{th position}}, \dots, 0]^T | x), \ i \in \{1, \dots, c\}. \tag{3.109}$$

As you might anticipate, based on this assumption, one can verify that the optimal loss function, in a sense of maximizing likelihood, is again the cross entropy loss:

$$\ell^*(y, \hat{y}) = \ell_{\text{CE}}(y, \hat{y}) = \sum_{j=1}^{c} -y_j \log \hat{y}_j$$

where y indicates a one-hot encoded label vector. For example, when the label is 2 in a scenario with $c = 10$ classes, the one-hot encoded vector y is represented as:

$$y = \begin{bmatrix} 0 \\ 0 \\ 1 \\ 0 \\ 0 \\ 0 \\ 0 \\ 0 \\ 0 \\ 0 \end{bmatrix} \begin{matrix} 0 \\ 1 \\ 2 \\ 3 \\ 4 \\ 5 \\ 6 \\ 7 \\ 8 \\ 9 \end{matrix}$$

The proof of this is almost the same as that in the binary classifier case. Therefore, we will omit the proof here. Instead you will have an opportunity to prove it in Problem 9.3.

Due to the above rationales, softmax activation has found extensive use in various classifiers in the field. Hence, we will also adopt this conventional activation function in our digit classifier.

Adam optimizer (Kingma & Ba 2014) Let us discuss a specific algorithm that we will employ in our setting. As mentioned earlier, we will use an advanced version of gradient descent, called the Adam optimizer. To see how the optimizer operates, let us first recall the vanilla gradient descent:

$$w^{(t+1)} \leftarrow w^{(t)} - \alpha \nabla J(w^{(t)})$$

where $w^{(t)}$ indicates the estimated weight in the t-th iteration, and α denotes the learning rate. Notice that the weight update relies only on the *current* gradient, reflected in $\nabla J(w^{(t)})$. Hence, in case $\nabla J(w^{(t)})$ fluctuates too much over iterations, the weight update oscillates significantly, thereby bringing about unstable training. To address this, people often use a variant algorithm that exploits *past* gradients for the purpose of stabilization. That is, the Adam optimizer.

Here is how Adam works. The weight update takes the following formula:

$$w^{(t+1)} = w^{(t)} + \alpha \frac{m^{(t)}}{\sqrt{s^{(t)}} + \epsilon} \tag{3.110}$$

where $m^{(t)}$ indicates a weighted average of the current and past gradients:

$$m^{(t)} = \frac{1}{1 - \beta_1^t} \left(\beta_1 m^{(t-1)} + (1 - \beta_1)(-\nabla J(w^{(t)})) \right). \tag{3.111}$$

Here $\beta_1 \in [0, 1]$ is a hyperparameter that captures the weight of past gradients, and hence it is called the *momentum*. So the notation m stands for momentum. The factor $\frac{1}{1-\beta_1^t}$ is applied in front, in an effort to stabilize training in initial iterations (small t). Check the detailed rationale behind this in Problem 9.8.

$s^{(t)}$ is a normalization factor that makes the effect of $\nabla J(w^{(t)})$ almost constant over t. In case $\nabla J(w^{(t)})$ is too big or too small, we may have significantly different scalings in magnitude. Similar to $m^{(t)}$, $s^{(t)}$ is defined as a weighted average of the current and past values:

$$s^{(t)} = \frac{1}{1 - \beta_2^t} \left(\beta_2 s^{(t-1)} + (1 - \beta_2)(\nabla J(w^{(t)}))^2 \right) \tag{3.112}$$

where $\beta_2 \in [0, 1]$ denotes another hyperparameter that captures the weight of past values, and s stands for *square*.

Notice that the dimensions of $w^{(t)}$, $m^{(t)}$ and $s^{(t)}$ are identical. So all the operations that appear in the above (including division in (3.110) and square in (3.112)) are component-wise. In (3.110), ϵ is a tiny value introduced to avoid division by 0 in practice (usually 10^{-8}).

TensorFlow: MNIST data loading Let us explore how to perform TensorFlow programming for implementing the simple digit classifier. To start, we need to load the MNIST dataset. MNIST is a well-known dataset and is conveniently available in the following package: `tensorflow.keras.datasets`. Furthermore, this package already provides both the training and testing datasets with an appropriate split ratio, so there is no need to be concerned about how to divide them. The only script required to import the MNIST dataset is as follows:

```
from tensorflow.keras.datasets import mnist
(X_train, y_train), (X_test, y_test) = mnist.load_data()
X_train = X_train/255.
X_test = X_test/255.
```

In this code, we divide the input (`X_train` or `X_test`) by its maximum value of 255, as part of a normalization process. This step is typically carried out as part of data preprocessing.

TensorFlow: A two-layer DNN In order to implement the simple DNN, illustrated in Figure 3.34, we rely upon two major packages:

> (i) `tensorflow.keras.models`;
>
> (ii) `tensorflow.keras.layers`.

The `models` package encompasses various functionalities related to the neural network architecture. One significant module within this package is the `Sequential` model, which can be thought of as a linear stack of neural network layers. The `layers` package contains a wide range of components that are essential for constructing a neural network. Examples include fully-connected dense layers and activation functions. With the help of these elements, we can easily assemble a model resembling the one depicted in Figure 3.34.

```
from tensorflow.keras.models import Sequential
from tensorflow.keras.layers import Dense, Flatten

model = Sequential()
model.add(Flatten(input_shape=(28,28)))
model.add(Dense(500, activation='relu'))
model.add(Dense(10,  activation='softmax'))
```

In this code snippet, the `Flatten` layer is used to transform a higher-dimensional entity, such as a 2D matrix, into a vector. In this example, a digit image with dimensions of 28-by-28 pixels is flattened into a vector of size $784(= 28 \times 28)$. The `add()` method is employed to append the desired layer to the end of the sequential model. The `Dense` layer signifies a fully-connected layer, and its input size is automatically inferred based on the preceding layer it is connected to in the sequential model. Therefore, the only specification required is the number of output neurons. In this instance, 500 denotes the number of hidden neurons, and you can also specify an activation function using an additional argument, for instance, `activation='relu'`. The output layer comprises 10 neurons, matching the number of classes in our classification task, and employs the **softmax** activation function.

TensorFlow: Training a model In the training phase, we first need to configure the optimization algorithm to be employed. In our case, we are using the **Adam** optimizer. As previously mentioned, **Adam** has three key hyperparameters: (i) the learning rate α; (ii) β_1 which controls the influence of past gradients; and (iii) β_2 which influences the impact of past gradient squares. The default values for these hyperparameters are: $(\alpha, \beta_1, \beta_2) = (0.001, 0.9, 0.999)$, and these values will be used if no specific values are provided.

We also need to define a loss function for our model, and we utilize the optimal choice, which is the cross entropy loss. Additionally, it is important to specify a performance metric that we'll monitor during both training and testing. A commonly used metric is accuracy. All of these settings, including the loss function and performance metric, can be configured using the `compile` method.

```
model.compile(optimizer='adam',
              loss='sparse_categorical_crossentropy',
              metrics=['acc'])
```

In this code, the option `optimizer='adam'` selects the default values for the learning rate and betas. If you wish to specify these hyperparameters manually, you can do so as follows:

```
opt=tensorflow.keras.optimizers.Adam(
        learning_rate=0.01,
        beta_1 = 0.92,
        beta_2 = 0.992)
```

We then replace the above option with `optimizer=opt` to use the manually defined optimizer. For the `loss` option in the `compile` method, we employ 'sparse_categorical_crossentropy', which is designed for cross entropy loss in multi-class classification scenarios, beyond the binary case.

Now we can bring this to train the model on MNIST data. During training, it is common to use a subset of the entire dataset to compute the gradient of the loss function. This subset is referred to as a *batch*. There are two additional terminologies to be aware of: (i) A *step* which denotes a procedure for computing the loss that covers only the examples within a single batch; (ii) An *epoch* which represents the entire process that encompasses all the examples. In our experiment, we utilize a batch size of 64 and conduct training over 20 epochs.

```
model.fit(X_train, y_train, batch_size=64, epochs=20)
```

TensorFlow: Testing the trained model In the testing phase, we first need to generate predictions from the model's output. This can be achieved using the predict() function, as demonstrated below:

```
model.predict(X_test).argmax(1)
```

Here argmax(1) returns the class associated with the highest softmax output among the 10 classes. To assess test accuracy, we make use of the evaluate() function:

```
model.evaluate(X_test, y_test)
```

Look ahead
Up to this point, we have explored the role of probabilistic modeling and the MAP/ML principles in the context of three applications: (i) communication; (ii) community detection in social networks; and (iii) machine learning. In the next upcoming sections, we will undertake a similar exploration for the final application: speech recognition. You will discover that speech recognition is a captivating application that encompasses nearly all of the key concepts and principles we have covered thus far.

Problem Set 9

Problem 9.1 (*Basic concepts regarding machine learning*)

(*a*) State the definition of an *algorithm*.

(*b*) State the definition of *machine learning*.

(*c*) State the definition of *artificial intelligence*.

(*d*) State the definition of *examples* (the terminology used in the machine learning field).

Problem 9.2 (*Independence*) In Sect. 3.9, we assume that

$$\text{data } \{(X_i, Y_i)\}_{i=1}^{n} \text{ are i.i.d.} \sim \mathbb{P}_{X,Y}(x, y). \tag{3.113}$$

Show that X_i's are also i.i.d.

Problem 9.3 (*Multiclass classifier* & softmax) In Sect. 3.8, we introduced a binary classifier. This problem delves into a more expansive context where the number of classes is not limited to two, but rather can be any arbitrary value, say $c \geq 2$. Let

$$z := [z_1, z_2, \ldots, z_c]^T \in \mathbb{R}^c \tag{3.114}$$

be the output of a neural network model prior to activation. To transform these real values into interpretable *probability* quantities within the range of 0 to 1, a commonly employed activation function is **softmax**:

$$\hat{y}_j := \left[\text{softmax}(z)\right]_j = \frac{e^{z_j}}{\sum_{k=1}^{c} e^{z_k}} \quad j \in \{1, 2, \ldots, c\}. \tag{3.115}$$

Note that this is a natural extension of the logistic function. See that for $c = 2$,

$$\begin{aligned}
\hat{y}_1 := \left[\text{softmax}(z)\right]_1 &= \frac{e^{z_1}}{e^{z_1} + e^{z_2}} \\
&= \frac{1}{1 + e^{-(z_1 - z_2)}} \\
&= \sigma(z_1 - z_2)
\end{aligned} \tag{3.116}$$

where $\sigma(\cdot)$ is the logistic function. If we regard $z_1 - z_2$ as the output $\hat{y}$ of a binary classifier, this coincides with logistic regression.

Consider a label Y of one-hot-vector type:

$$Y \in \{[1, 0, \ldots, 0]^T, [0, 1, 0, \ldots, 0]^T, \ldots, [0, \ldots, 0, 1]^T\}.$$

In this context, $\hat{y}_i$ can be understood as the probability that the ith example belongs to class i. Therefore, let us assume that

$$\hat{y}_i = \mathbb{P}(Y = [0, \ldots, \underbrace{1}_{i\text{th position}}, \ldots, 0]^T | X = x), \; i \in \{1, \ldots, c\}. \qquad (3.117)$$

Additionally, we assume that the random process $\{(X^{(i)}, Y^{(i)})\}_{i=1}^m$ for training examples is independent across i.

(a) Express the likelihood

$$\mathbb{P}\left(Y^{(1)} = y^{(1)}, \ldots, Y^{(m)} = y^{(m)} | X^{(1)} = x^{(1)}, \ldots, X^{(m)} = x^{(m)}\right) \qquad (3.118)$$

in terms of $y^{(i)}$'s and $\hat{y}^{(i)}$'s.

(b) Derive the *optimal loss* function that maximizes the likelihood (3.118).

Problem 9.4 (*Jensen's inequality*) A function $f(x)$ is said to be *convex* (bowl-shaped) if for any $\lambda \in [0, 1]$:

$$f(\lambda x + (1 - \lambda)y) \le \lambda f(x) + (1 - \lambda)f(y) \qquad \forall x, y. \qquad (3.119)$$

A function $g(x)$ is said to be *concave* if $-g(x)$ is convex. Suppose that a function f is convex and X is a discrete random variable.

(a) For a simple binary random variable case, show that

$$\mathbb{E}[f(X)] \ge f(\mathbb{E}[X]). \qquad (3.120)$$

(b) Now consider a general random variable case. Using the by-induction proof-technique, prove (3.120).

(c) For a general random variable case, identify conditions under which the equality in (3.120) holds.

Problem 9.5 (*Cross entropy*) In Sect. 3.9, we introduced a prominent notion in information theory, *cross entropy*, in the context of *binary* random variables. This problem extends the discussion to the general case beyond binary. Let X and Y be discrete random variables with pmfs $\mathbb{P}_X(x)$ and $\mathbb{P}_Y(y)$, respectively, where $x, y \in \mathcal{X}$. The cross entropy is defined as:

$$H(\mathbb{P}_X, \mathbb{P}_Y) := -\sum_{x \in \mathcal{X}} \mathbb{P}_X(x) \log_2 \mathbb{P}_Y(x) = \mathbb{E}\left[\log_2 \frac{1}{\mathbb{P}_Y(X)}\right] \qquad (3.121)$$

where the expectation is taken w.r.t. $\mathbb{P}_X$.

(*a*) Show that the function $-\log_2 x$ is convex in x.

   Hint: You may want to use the well-known fact related to the convex function mentioned in Side Study #2 in Sect. 3.9.

(*b*) Show that

$$H(\mathbb{P}_X, \mathbb{P}_Y) \geq H(X) \tag{3.122}$$

where $H(X)$ denotes the Shannon entropy (Cover & Joy 2006):

$$H(X) := -\sum_{x \in \mathcal{X}} \mathbb{P}_X(x) \log_2 \mathbb{P}_X(x) = \mathbb{E}\left[\log_2 \frac{1}{\mathbb{P}_X(X)}\right]. \tag{3.123}$$

Hint: Think about Jensen's inequality in Problem 9.4

(*c*) Identify conditions under which the equality in (3.122) holds.

Problem 9.6 (*KL divergence and mutual information*) In Sect. 3.9, we investigated an important notion, *cross entropy*. This problem now explores two additional prominent notions in information theory: *Kullback-Leibler (KL) divergence* and *mutual information*. Let X and Y be discrete random variables with $\mathbb{P}_X(x)$ and $\mathbb{P}_Y(y)$ where $x, y \in \mathcal{X}$.

(*a*) The KL divergence between $\mathbb{P}_X$ and $\mathbb{P}_Y$ is defined as (Cover & Joy 2006):

$$\mathsf{KL}(\mathbb{P}_X \| \mathbb{P}_Y) := \sum_{x \in \mathcal{X}} \mathbb{P}_X(x) \log \frac{\mathbb{P}_X(x)}{\mathbb{P}_Y(x)}. \tag{3.124}$$

Prove that $\mathsf{KL}(\mathbb{P}_X \| \mathbb{P}_Y) \geq 0$. Also identify conditions under which the equality holds.

(*b*) The mutual information between X and Y is defined as (Cover & Joy 2006):

$$I(X; Y) := \mathsf{KL}(\mathbb{P}_{X,Y} \| \mathbb{P}_X \mathbb{P}_Y) \tag{3.125}$$

where $\mathbb{P}_{X,Y}(x, y)$ denotes the joint distribution and $\mathbb{P}_X(x)\mathbb{P}_Y(y)$ indicates the product of individual probability distributions. Show that

$$I(X; Y) = \sum_{x \in \mathcal{X}} \mathbb{P}_X(x) \mathsf{KL}(\mathbb{P}_{Y|X=x} \| \mathbb{P}_Y) \tag{3.126}$$

where $\mathbb{P}_{Y|X=x}$ indicates the conditional distribution of Y given $X = x$.

Problem 9.7 (*Gradient descent*) Consider a function $J(w) = w^2 + 2w$ where $w \in \mathbb{R}$. Consider gradient descent with the learning rate $\alpha^{(t)} = \frac{1}{2^t}$ and an initial value of $w^{(0)} = 2$.

(a) Elaborate on the gradient descent algorithm, detailing the process by which the t-th estimate $w^{(t)}$ is updated from the previous estimate during each iteration.

(b) Utilize a Python code to execute gradient descent and generate a plot of $w^{(t)}$ as a function of t over a proper range, say $1 \leq t \leq 20$.

Problem 9.8 (*Optimizers*) Consider gradient descent:

$$w^{(t+1)} = w^{(t)} - \alpha \nabla J(w^{(t)})$$

where $w^{(t)}$ indicates the weights of an interested model at the t-th iteration; $J(w^{(t)})$ denotes the cost function evaluated at $w^{(t)}$; and α is the learning rate. Note that the weight update is influenced solely by the *current* gradient, reflected in $\nabla J(w^{(t)})$.

(a) (*Momentum optimizer* (Polyak 1964)) In the literature, there is a prominent variant of gradient descent that takes into account *past* gradients as well. Using such past information, one can damp an oscillating effect in the weight update that may incur instability in training. To capture past gradients and therefore address the oscillation problem, another quantity, often denoted by $m^{(t)}$, is usually introduced:

$$m^{(t)} = \beta m^{(t-1)} + (1 - \beta)\nabla(-J(w^{(t)})) \tag{3.127}$$

where β denotes another hyperparameter that captures the weight of the past gradients, simply called the *momentum*. Here m stands for the *momentum* vector. The variant of the algorithm (often called the *momentum optimization*) takes the following update for $w^{(t+1)}$:

$$w^{(t+1)} = w^{(t)} + \alpha m^{(t)}. \tag{3.128}$$

Show that

$$w^{(t+1)} = w^{(t)} - \alpha(1 - \beta)\sum_{k=0}^{t-1} \beta^k \nabla J(w^{(t-k)}) + \alpha \beta^t m^{(0)}.$$

(b) (*Bias correction*) Assuming that $\nabla J(w^{(t)})$ is the same for all t and $m^{(0)} = 0$, show that

$$w^{(t+1)} = w^{(t)} - \alpha(1 - \beta^t)\nabla J(w^{(t)}).$$

Note: For a large value of t, $1 - \beta^t \approx 1$, so it has almost the same scaling as that in the regular gradient descent. On the other hand, for a small value of t, $1 - \beta^t$ can be small, being far from 1. For instance, when $\beta = 0.9$ and $t = 2$, $1 - \beta^t = 0.19$. This motivates people to rescale the moment $m^{(t)}$ in (3.127) through division by $1 - \beta^t$. So in practice, we use:

$$\hat{m}^{(t)} = \frac{m^{(t)}}{1 - \beta^t};\tag{3.129}$$

$$w^{(t+1)} = w^{(t)} + \alpha \hat{m}^{(t)}.\tag{3.130}$$

This technique is so called the *bias correction*.

(c) (**Adam** optimizer (Kingma & Ba 2014)) Notice in (3.127) that a very large or very small value of $\nabla J(w^{(t)})$ affects the weight update in quite a different scaling. In an effort to avoid such a different scaling problem, people in practice often make *normalization* in the weight update (3.130) via a normalization factor, often denoted by $\hat{s}^{(t)}$ (Hinton et al. 2012):

$$w^{(t+1)} = w^{(t)} + \alpha \frac{\hat{m}^{(t)}}{\sqrt{\hat{s}^{(t)}} + \epsilon}\tag{3.131}$$

where the division is component-wise, and

$$\hat{m}^{(t)} = \frac{m^{(t)}}{1 - \beta_1^t},\tag{3.132}$$

$$m^{(t)} = \beta_1 m^{(t-1)} + (1 - \beta_1)(-\nabla J(w^{(t)})),\tag{3.133}$$

$$\hat{s}^{(t)} = \frac{s^{(t)}}{1 - \beta_2^t},\tag{3.134}$$

$$s^{(t)} = \beta_2 s^{(t-1)} + (1 - \beta_2)(\nabla J(w^{(t)}))^2.\tag{3.135}$$

Here $(\cdot)^2$ indicates a component-wise square; ϵ is a tiny value introduced to avoid division by 0 in practice (usually 10^{-8}); and s stands for *square*. This optimizer (3.131) is called the **Adam** optimizer. Explain the rationale behind the division by $1 - \beta_2^t$ in (3.135).

Problem 9.9 (*True or False?*)

(a) Consider an optimization problem for machine learning:

$$\min_w \sum_{i=1}^m \ell(y^{(i)}, f_w(x^{(i)}))\tag{3.136}$$

where $\{(x^{(i)}, y^{(i)})\}_{i=1}^m$ indicate input-output example pairs, and

$$f_w(x) = \frac{1}{1 + e^{-w^T x}}. \tag{3.137}$$

The optimal loss function (in a sense of maximizing the likelihood) is:

$$\ell^*(y, \hat{y}) = -y \log \hat{y} - (1 - y) \log(1 - \hat{y}). \tag{3.138}$$

(b) For two arbitrary distributions, say p and q, consider cross entropy $H(p, q)$. Then,

$$H(p, q) \geq H(q) \tag{3.139}$$

where $H(q)$ is the Shannon entropy w.r.t. q.

(c) For two arbitrary distributions, say p and q, consider cross entropy:

$$H(p, q) := -\sum_{x \in \mathcal{X}} p(x) \log q(x) = \mathbb{E}_p \left[\log \frac{1}{q(X)} \right] \tag{3.140}$$

where $X \in \mathcal{X}$ is a discrete random variable. Then,

$$H(p, q) = H(p) := -\sum_{x \in \mathcal{X}} p(x) \log p(x). \tag{3.141}$$

only when $q = p$.

(d) Consider a binary classifier in the machine learning setup where we are given input-output example pairs $\{(x^{(i)}, y^{(i)})\}_{i=1}^m$. Let $0 \leq \hat{y}^{(i)} \leq 1$ be the classifier output for the ith example. Let w be the parameters of the classifier. Define:

$$w_{\mathsf{CE}}^\star := \arg\min_w \frac{1}{m} \sum_{i=1}^m \ell_{\mathsf{CE}}\left(y^{(i)}, \hat{y}^{(i)}\right)$$

$$w_{\mathsf{KL}}^\star := \arg\min_w \frac{1}{m} \sum_{i=1}^m \mathsf{KL}\left(y^{(i)} \| \hat{y}^{(i)}\right)$$

where $\ell_{\mathsf{CE}}(\cdot, \cdot)$ denotes cross entropy loss and $\mathsf{KL}(y^{(i)} \| \hat{y}^{(i)})$ indicates the KL divergence between two binary random variables with parameters $y^{(i)}$ and $\hat{y}^{(i)}$, respectively. Then,

$$w_{\mathsf{CE}}^\star = w_{\mathsf{KL}}^\star.$$

(e) Consider a convex optimization problem:

$$\min_{w \in \mathbb{R}^d} f(w) \tag{3.142}$$

where $f(w)$ is convex in w. The gradient descent algorithm always ensures the convergence of the optimal point $w^\star$ regardless of an initial point $w^{(0)}$, as long as the learning rate $\alpha^{(t)}$ is properly set up.

(f) Suppose we execute the following code:

```python
import numpy as np
a = np.random.randn(4,3,3)
b = np.ones_like(a)
print(b[0].shape)
print(b.shape[0])
```

Then, the two prints yield the same results.

(g) Suppose that `image` is an MNIST image of `numpy array` type. Then, one can use the following commands to plot the image:

```python
import matplotlib.pyplot as plt
plt.imshow(image.squeeze(), cmap='gray_r')
```

3.11   Speech Recognition: Probabilistic Modeling

Recap

Over the past three sections, we have explored the link between machine learning and probability, established through the fundamental element employed in machine learning models: *data* $\{(x^{(i)}, y^{(i)})\}_{i=1}^{m}$. Since the nature of data can vary based on collection methods across various contexts, it can be viewed as a specific realization of a random process $\{(X^{(i)}, Y^{(i)})\}_{i=1}^{m}$, forming the bridge to probability. Furthermore, we illustrated the role of the ML principle in shaping a perceptron-based optimization for machine learning:

$$\min_{w} \sum_{i=1}^{m} \ell\left(y^{(i)}, \hat{y}^{(i)}\right) \tag{3.143}$$

where $\ell(\cdot, \cdot)$ represents a loss function, and $\hat{y}^{(i)} := f_w(x^{(i)})$ indicates the prediction function parameterized by the weights w. In terms of activation, we focused on the logistic function, which implements a smooth transition from the step function:

$$\hat{y} := f_w(x) = \frac{1}{1 + e^{-w^T x}}. \tag{3.144}$$

Subsequently, we demonstrated that the optimal loss function, in the sense of maximizing the likelihood function, is the *cross entropy loss*:

$$\ell_{CE}^{*}(y, \hat{y}) = -y \log \hat{y} - (1 - y) \log(1 - \hat{y}),$$

which results in a well-known classifier: *logistic regression*. Throughout the remainder of this book, we will do the same thing for the last application: *speech recognition*.

Outline

In this section, we will focus on the relationship between speech recognition and probability. The section consists of four parts. Initially, we will revisit the definition of speech recognition and investigate its associated system. Subsequently, we will delve into details of the system's input and output. Following that, we will study the system structure and formulate a *probabilistic* model accordingly. Finally, building upon this probabilistic model, we will demonstrate that speech recognition is an *inference* problem.

Speech recognition (Vaseghi 2008) An individual speaks into a microphone, capturing an analog sound waveform through sampling. The objective of speech recognition
is to decipher the meaning of the spoken words. In essence, it involves translating the
analog waveform, which encompasses the spoken words, into a written command,
ultimately represented in the form of *text*. Therefore, in speech recognition, our aim
is to decode spoken words into text.

Speech recognition system Later, we will demonstrate that speech recognition is an
inference problem. To introduce the interested entity X in the context of Figure 3.36,
we examine a speech recognition *system*, as depicted in Figure 3.37. Notice that the
input to the system is X that we aim to infer. Let us delve into the detailed structure
of the input X and the output Y in the system. To figure this out, we first ponder the
components that make up a text. Clearly, these components are language-dependent,
and for our discussion, we will focus on the English language.

The structure of an English text A text is essentially a sequence of words, in which
each word can be considered a natural unit for text. However, we can further break
down words into smaller units. For example, take the word "speech". One conceivable
smaller unit is an English alphabet. However, there is an issue in adopting this as a
basic unit. The issue arises from the fact that the actual input to the speech recognition
system is related to real sound, but the mapping between sound and the alphabet is
not one-to-one. For instance, the sound /i/ is represented by the letter "e" in the word
"nik_e", but the same sound may correspond to a different letter "i" in the word "b_it".
Here, the slash denotes a conventional notation for representing phonemes.

  On the other hand, from a phonetic perspective, a word can be broken down
into phonemes. For instance, the word "speech" comprises four phonemes: /s/, /p/,

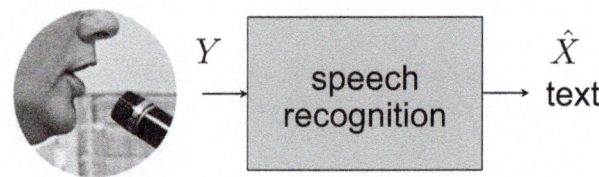

Fig. 3.36 Speech recognition: Transforming voice signals into a written text

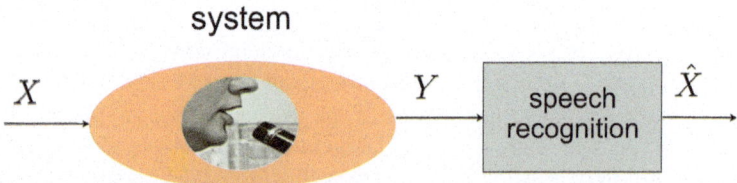

Fig. 3.37 Speech recognition system

24 consonants

Phoneme (sound)	Examples	Phoneme (sound)	Examples
/b/	banana, bubbles	/s/	sun, mouse
/c/	car, duck	/t/	turtle, little
/d/	dinosaur, puddle	/v/	volcano, halve
/f/	fish, giraffe	/w/	watch, queen
/g/	guitar, goggles	/x/	fox
/h/	helicopter	/y/	yo-yo
/j/	jellyfish, fridge	/z/	zip, please
/l/	leaf, bell	/sh/	shoes, television
/m/	monkey, hammer	/ch/	children, stitch
/n/	nail, knot	/th/	mother
/p/	pumpkin, puppets	/th/	thong
/r/	rain, write	/ng/	sing, ankle

20 vowels

Phoneme (sound)	Examples	Phoneme (sound)	Examples
Short Vowel Sounds			
/a/	apple	/oo/	moon, screw
/e/	elephant, bread	**Other Vowel Sounds** 'oo'	book, could
/i/	igloo, gym	/ou/	house, cow
/o/	octopus, wash	/oi/	coin, boy
/u/	umbrella, won	**'r' Controlled Vowel Sounds** /ar/	star, glass
Long Vowel Sounds...			
/ae/	rain, tray	/or/	fork, board
/ee/	tree, me	/er/	herb, nurse
/ie/	light, kite	/air/	chair, pear
/oa/	boat, bow	/ear/	spear, deer
/ue/	tube, emu	"schwa" (close to /u/)	teacher, picture

Fig. 3.38 44 phonemes in English: (1) 24 consonants; and (2) 20 vowels

system

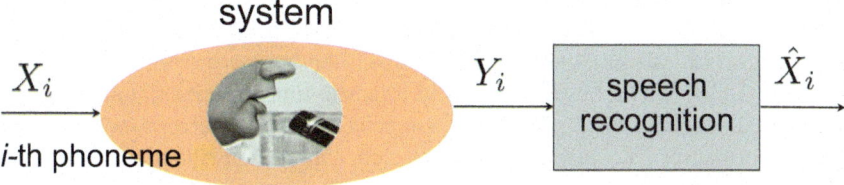

X_i *i*-th phoneme Y_i speech recognition $\hat{X}_i$

Fig. 3.39 Speech recognition system: Each component of the input X_i indicates the ith phoneme that takes one of 44 phonemes (24 consonants and 20 vowels)

/i/, and /ch/. Phonemes can effectively serve as the smallest phonetic units in a language. There are two categories of phonemes: (1) consonants and (2) vowels. English encompasses 44 phonemes (24 consonants and 20 vowels). Refer to Figure 3.38.

In light of this, speech recognition can be viewed as the task of figuring out the sequence of phonemes that constitutes a text. See Figure 3.39 for an illustration. The sequence $\{X_i\}_{i=1}^{n}$ of phonemes is the one we aim to decode, treating the phonemes as random variables. Let $X_i \in \mathcal{X}$ denote the ith phoneme of $\{X_i\}_{i=1}^{n}$, where $\mathcal{X}$ represents the set of phonemes with an alphabet size of $|\mathcal{X}| = 44$.

Inside the system Now, let's figure out what the output Y_i is in the speech recognition system. The output $\{Y_i\}_{i=1}^{n}$ is the information that the speech recognition block will process, so it should be indicative of something related to the actual sound. To figure this out, we first need to establish the connection between $\{X_i\}_{i=1}^{n}$ and the actual sound signal captured by the microphone in the system. Refer to Figure 3.40. A user wishing to convey the meaning of the text $\{X_i\}_{i=1}^{n}$ verbally speaks into the

system

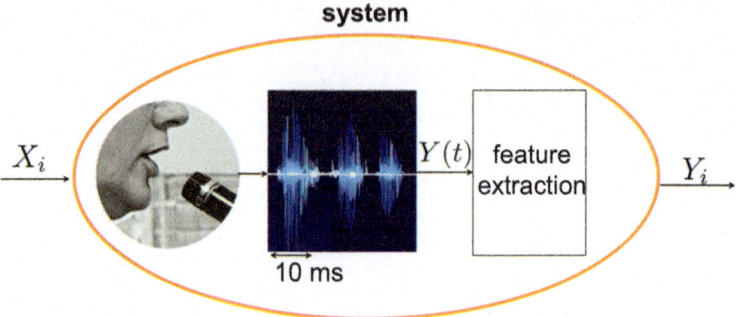

Fig. 3.40 Inside the speech recognition system

microphone, producing an analog waveform, say $Y(t)$. However, rather than dealing with the continuous-time analog signals $Y(t)$, we seek *discrete*-time values that can be represented as a sequence $\{Y_i\}_{i=1}^{n}$. Fortunately, there is a method to convert $Y(t)$ into discrete-time signals.

This translation is based on the following observation: Each phoneme typically spans around 10 ms, as illustrated in Figure 3.40. While this duration may vary based on individual speaking pace, the 10 ms serves as an average across many phonemes from different speakers. We segment the analog waveform into 10 ms intervals, aiming to extract key components (discrete-time quantities) from the signal within each interval. For simplicity, we assume no pauses between phonemes. In the presence of pauses, they can be readily detected, and the corresponding signals can be segmented accordingly.

Notice that the crucial information in speech is often prominent in the *frequency* domain. Therefore, a natural approach is to employ a well-known transformation technique that highlights frequency components. That is, the *Fourier transform*:

$$Y(f) := \int_{-\infty}^{+\infty} Y(t)e^{-j2\pi ft}dt. \tag{3.145}$$

However, we encounter two issues here. First, the input to the Fourier transform is an *infinite-time-horizon* signal, whereas we aim to extract a specific component corresponding only to the signal within each 10 ms time interval. To address this, one approach is to extract only a portion of $Y(f)$ by employing a time-windowed Fourier transform:

$$Y_i(f) := \int_{10\,ms\cdot(i-1)}^{10\,ms\cdot i} Y(t)e^{-j2\pi ft}dt. \tag{3.146}$$

The second challenge is that $Y_i(f)$ remains a *continuous* quantity in f. One workaround is to extract the corresponding Fourier coefficients from it, achieved by selecting $Y_i(f)$ at specific frequencies f:

$$Y_i := \begin{bmatrix} Y_i(f_1) \\ Y_i(f_2) \\ \vdots \\ Y_i(f_k) \end{bmatrix} \tag{3.147}$$

where f_j's represent specific frequencies associated with significant spectral components, and k denotes the number of such frequencies. As illustrated above, there are typically multiple Fourier coefficients for the signal within each 10 ms interval. However, for simplicity, we assume that there is only one significant spectral component (i.e., $k = 1$). Here we refer to that single component as a feature. We denote Y_i as the ith feature corresponding to the ith phoneme. Refer to Figure 3.40 for the entire procedure inside the system.

Relation between $\{X_i\}_{i=1}^n$ and $\{Y_i\}_{i=1}^n$ Figure 3.41 provides an overview of the complete process involving X_i (the ith phoneme) and Y_i (the ith feature). Now, how do $\{X_i\}_{i=1}^n$ and $\{Y_i\}_{i=1}^n$ interrelate? The system introduces a considerable amount of randomness, primarily stemming from two major sources: (1) distinct voice characteristics (e.g., accent) and (2) noise (e.g., thermal noise due to random movements of electrons in the electrical circuit). This inherent randomness introduces uncertainty into $\{Y_i\}_{i=1}^n$, establishing a *probabilistic* relationship between the input and the output. Consequently, we can perceive speech recognition as an *inference* problem.

Look ahead
Given that speech recognition is an inference problem, the optimal inference approach is once again the MAP estimation. In the upcoming section, we will delve into the exploration of the optimal MAP estimator for speech recognition.

system

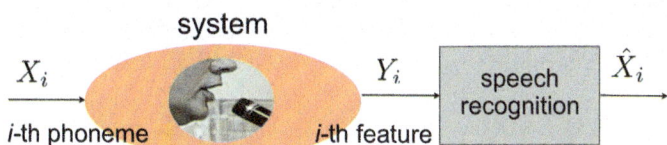

X_i Y_i speech recognition $\hat{X}_i$

i-th phoneme i-th feature

Fig. 3.41 Speech recognition system: X_i indicates the ith phoneme and Y_i denotes the corresponding feature (spectral information)

3.12 Speech Recognition: The MAP Principle

Recap
In the preceding section, we established that speech recognition is an inference problem. The objective of speech recognition is to deduce a sequence $\{X_i\}_{i=1}^n$ of phonemes (enabling us to recognize what a speaker is saying) from a sequence $\{Y_i\}_{i=1}^n$ of features (pertaining to the actual spoken words). See Figure 3.42 for illustration.

Additionally, we identified a significant amount of randomness in the system arising from distinct voice characteristics of a speaker and system noise (e.g., thermal noise). This randomness establishes a probabilistic relationship between the system's output (observation) and the input (the entity of interest for inference), affirming that speech recognition is indeed an inference problem.

Outline
In this section, we will explore the optimal algorithm based on the MAP principle. The section comprises three parts. First, we will derive the optimal MAP estimator. To achieve an explicit MAP solution, it is necessary to concisely represent the "a priori probability" and the "likelihood". Fortunately, there exists a favorable statistical structure within $\{(X_i, Y_i)\}_{i=1}^n$ that facilitates an efficient representation of these two quantities. Hence, in the second part, we will delve into this structure. Finally, we will leverage this structure to simplify the MAP estimator.

The optimal inference for speech recognition Given that speech recognition is an inference problem, the optimal inference is the one that maximizes conditional correct decision probability:

$$\mathbb{P}(X_1 = \hat{X}_1, \ldots, X_n = \hat{X}_n | Y_1 = y_1, \ldots, Y_n = y_n). \qquad (3.148)$$

As we figured out several times, it coincides with the MAP rule which finds the one that maximizes the a posteriori probability:

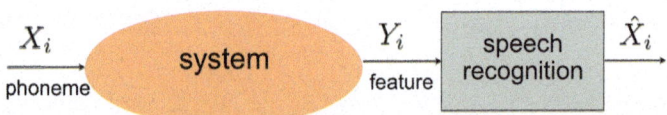

Fig. 3.42 A block diagram of the speech recognition system and recovery block

$$\hat{\mathbf{x}}_{\mathsf{MAP}} = \arg \max_{\hat{x}_1,\ldots,\hat{x}_n \in \mathcal{X}} \mathbb{P}(X_1 = \hat{x}_1, \ldots, X_n = \hat{x}_n | Y_1 = y_1, \ldots, Y_n = y_n)$$

$$= \arg \max_{\hat{x}_1,\ldots,\hat{x}_n \in \mathcal{X}} \frac{\mathbb{P}_{\mathbf{X}}(\hat{x}_1, \ldots, \hat{x}_n) f(y_1, \ldots, y_n | \hat{x}_1, \ldots, \hat{x}_n)}{f(y_1, \ldots, y_n)} \quad (3.149)$$

$$= \arg \max_{\hat{x}_1,\ldots,\hat{x}_n \in \mathcal{X}} \mathbb{P}_{\mathbf{X}}(\hat{x}_1, \ldots, \hat{x}_n) f(y_1, \ldots, y_n | \hat{x}_1, \ldots, \hat{x}_n)$$

when the second equality follows from the definition of conditional probability and the continuous nature of y_i's. Here $\mathcal{X}$ represents the set of all the phonemes that each X_i can take on, usually called the alphabet.

Two quantities that we need to know about To compute the MAP solution (3.149), we need to determine two crucial quantities: (1) A priori probability $\mathbb{P}_{\mathbf{X}}(\hat{x}_1, \ldots, \hat{x}_n)$; (2) conditional pdf (likelihood) $f(y_1, \ldots, y_n | \hat{x}_1, \ldots, \hat{x}_n)$, which characterizes the statistical relationship between the input and the output. In light of communication systems, the speech recognition system can be interpreted as a *channel*.

A straightforward method to determine $\mathbb{P}_{\mathbf{X}}(\hat{x}_1, \ldots, \hat{x}_n)$ involves examining the quantity for each sequence. However, this straightforward approach presents a complexity challenge, given that the number of possible patterns for the sequence $\{X_i\}_{i=1}^n$ grows exponentially with n:

$$|\mathcal{X}|^n = 44^n. \quad (3.150)$$

Remember that there are 44 phonemes in English: 24 consonants and 20 vowels. The total number of probability values needed to fully specify $\mathbb{P}_{\mathbf{X}}(\hat{x}_1, \ldots, \hat{x}_n)$ becomes overwhelmingly large, especially for a significant n, making it challenging to obtain a priori knowledge. Moreover, there are numerous distinct values for the likelihood $f(y_1, \ldots, y_n | \hat{x}_1, \ldots, \hat{x}_n)$. Even worse, $(y_1, \ldots, y_n)$ are continuous values.

However, there exists an advantageous statistical structure in $\{(X_i, Y_i)\}_{i=1}^n$, enabling an efficient representation of both the a priori probability and the likelihood. By leveraging this statistical structure, we can model $\{(X_i, Y_i)\}_{i=1}^n$, allowing us to specify $\mathbb{P}_{\mathbf{X}}(\hat{x}_1, \ldots, \hat{x}_n)$ and $f(y_1, \ldots, y_n | \hat{x}_1, \ldots, \hat{x}_n)$ with a significantly smaller number of parameters.

A random process $\{X_i\}_{i=1}^n$ A simple way to model the sequence of phonemes is to assume that these random variables are independent. We have encountered an independent process several times before; for example, the additive white Gaussian noise discussed in the communication application is an i.i.d. random process. However, the independence assumption is not suitable for speech recognition. Certain phonemes are more likely to follow others; for instance, the phoneme /th/ is more likely to be followed by /e/ rather than /s/. Therefore, assuming independence among the random variables appears to be a less than ideal choice.

A generalized Markov model Then, how are the components in $\{X_i\}_{i=1}^n$ related each other? It turns out that a random process introduced in Sect. 2.1, the *generalized Markov model*, is well-suited to capture dependencies across phonemes. Recall its definition: $\{X_i\}_{i=1}^n$ is a generalized Markov process with ℓ memories if

$$\mathbb{P}(x_{i+1}|x_i, \ldots, x_{i-\ell+1}, x_{i-\ell}, \ldots, x_1) = \mathbb{P}(x_{i+1}|x_i, \ldots, x_{i-\ell+1}).$$

We also verified that another properly defined random process, specifically $S_i := (X_i, \ldots, X_{i-\ell+1})$, is a single-memory Markov process. Therefore, it is sufficient to focus on the case $\ell = 1$ with an appropriate rearrangement. For illustrative purposes, we assume that the sequence of phonemes is a single-memory Markov process.

Joint distribution $\mathbb{P}_{\mathbf{X}}(x_1, \ldots, x_n)$ Utilizing the graphical model introduced in Sect. 2.1, we can depict the single-memory Markov process as:

$$X_1 - X_2 - X_3 - \cdots - X_{n-1} - X_n. \tag{3.151}$$

It is named a Markov chain due to its chain-like appearance. Leveraging this statistical structure, one can now formulate the joint distribution as:

$$\begin{aligned}
\mathbb{P}_{\mathbf{X}}(x_1, &\ldots, x_n) \\
&\overset{(a)}{=} \mathbb{P}(x_1)\mathbb{P}(x_2|x_1)\mathbb{P}(x_3|x_2, x_1) \cdots \mathbb{P}(x_n|x_{n-1}, \ldots, x_1) \\
&\overset{(b)}{=} \mathbb{P}(x_1)\mathbb{P}(x_2|x_1)\mathbb{P}(x_3|x_2) \cdots \mathbb{P}(x_n|x_{n-1}) \\
&= \mathbb{P}(x_1) \prod_{i=2}^{n} \mathbb{P}(x_i|x_{i-1})
\end{aligned} \tag{3.152}$$

where (a) stems from the definition of conditional probability; and (b) comes from the Markov property. Hence, it suffices to be aware of only $\mathbb{P}(x_1)$ and $\mathbb{P}(x_i|x_{i-1})$ to compute $\mathbb{P}_{\mathbf{X}}(x_1, \ldots, x_n)$.

How to figure out $\mathbb{P}(x_1)$ **and** $\mathbb{P}(x_i|x_{i-1})$**?** We only need to specify 44 values for $\mathbb{P}(x_1)$ and 44^2 values for $\mathbb{P}(x_i|x_{i-1})$. The sum $44 + 44^2$ is much smaller than the enormous number 44^n required to specify the joint distribution when the statistical structure is not exploited. In practice, one can estimate individual pmf $\mathbb{P}(x_1)$ and transition probability $\mathbb{P}(x_i|x_{i-1})$ from any large text by computing the following sample means:

$$\mathbb{P}(/s/) = \mathbb{P}(X_1 = /s/) \approx \frac{\# \text{ of occurences of ``}s\text{''}}{\# \text{ of phonemes in the interested text}}; \tag{3.153}$$

$$\mathbb{P}(/t/|/s/) = \mathbb{P}(X_i = /t/|X_{i-1} = /s/) \approx \frac{\# \text{ of ``}t\text{'' that follows ``}s\text{''}}{\# \text{ of occurrences of ``}s\text{''}}. \tag{3.154}$$

It turns out the Law of Large Numbers (LLN), which we learned about w.r.t. the i.i.d. random process, can be extended to the Markov process as well. We will not provide the proof, as it diverges from the current narrative (also, the proof is not straightforward). If you desire a more in-depth understanding, you might consider taking a course on random processes. With the aid of the extended LLN, the estimates mentioned above converge around the ground-truth distributions as the number of phonemes in the text increases.

Likelihood function $f(y_1, \ldots, y_n|x_1, \ldots, x_n)$ Now, let us determine how to acquire knowledge about the likelihood. We discover that a crucial observation regarding the system allows us to identify the statistical structure of $\{Y_i\}_{i=1}^n$, thus offering a concrete method for computing $f(y_1, \ldots, y_n|x_1, \ldots, x_n)$. Recall the internal workings of the system; refer to Figure 3.43. The key observation here is that Y_i can be regarded as a noisy version of X_i, and the noise is independent of any other random variables involved in the system. The mathematical representation of this observation is that given X_i, Y_i is mutually independent of all the other random variables. For instance,

$$Y_1 \perp (X_2, \ldots, X_n, Y_2, \ldots, Y_n)|X_1$$

where the symbol $\perp$ means "mutual independence". This property leads to the graphical model for $\{Y_i\}_{i=1}^n$ as in Figure 3.44.

Note that if we remove node X_i, node Y_i will be disconnected from the rest of the graph. This reflects the fact that Y_i depends on other random variables only through X_i. An interesting question about the model arises: Is the observation sequence $\{Y_i\}_{i=1}^n$ a Markov model? No. Why? But the *underlying sequence* $\{X_i\}_{i=1}^n$ that we want to figure out is a Markov model. This is why it is referred to as the *Hidden Markov Model*, HMM for short (Rabiner & Juang 1986).

Exploiting this statistical property, one can express the conditional pdf as:

$$\begin{aligned}
&f(y_1, \ldots, y_n|x_1, \ldots, x_n) \\
&= f(y_1|x_1, \ldots, x_n)f(y_2, \ldots, y_n|x_1, \ldots, x_n, y_1) \\
&\overset{(a)}{=} f(y_1|x_1)f(y_2, \ldots, y_n|x_1, \ldots, x_n, y_1) \\
&\quad \vdots \\
&\overset{(b)}{=} f(y_1|x_1)f(y_2|x_2)\cdots f(y_n|x_n) \\
&= \prod_{i=1}^n f(y_i|x_i)
\end{aligned} \tag{3.155}$$

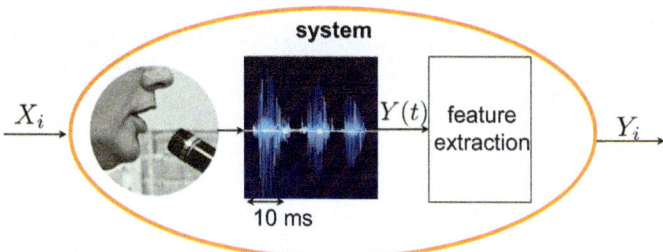

Fig. 3.43 Inside the speech recognition system

$$Y_1 \quad Y_2 \quad Y_2 \qquad\qquad Y_{n-1} \qquad Y_n$$
$$| \qquad | \qquad | \qquad\qquad\quad | \qquad\qquad |$$
$$X_1 - X_2 - X_3 - \cdots - X_{n-1} - X_n$$

Fig. 3.44 A Hidden Markov Model (HMM) for the output $\{Y_i\}_{i=1}^n$ of the speech recognition system

where (a) and (b) holds because, given x_i, y_i is independent of everything else for any i.

Note that computing the likelihood $f(y_1, \ldots, y_n | x_1, \ldots, x_n)$ requires only knowledge of the individual likelihood $f(y_i | x_i)$. Here, a question arises: How can we obtain knowledge of the individual likelihood $f(y_i | x_i)$? If the ith feature y_i is a *discrete* value, then we need to specify only the number $|\mathcal{X}| \times |\mathcal{Y}|$ of possible values for $f(y_i | x_i)$. However, the value of the feature is generally a *continuous* value, although it can be quantized so that it can be represented by a discrete random variable. Therefore, we need to figure out the *functional relationship* between y_i and x_i to specify the likelihood function.

How to determine $f(y_i | x_i)$? Recall that the likelihood function depends on the *randomness* that occurs in the system. The randomness originates from the following two sources: (i) different voice characteristics, and (ii) noise introduced in an electrical circuit used in the system. If the system has no noise and a speaker is given, we can consider Y_i simply as a deterministic function of X_i. For example, given $X_i = /a/$, Y_i is a deterministic function of $/a/$, say μ_a. See Figure 3.45.

However, due to the noise in the system, the feature Y_i would be randomly distributed around the true feature μ_a corresponding to the phoneme $/a/$. Recall what we learned about noise in the communication application. The primary source of the noise is the random movement of electrons due to heat, known as thermal noise. We understand that thermal noise can be modeled as additive white Gaussian noise (AWGN). Hence, given $X_i = /a/$, we can model Y_i as:

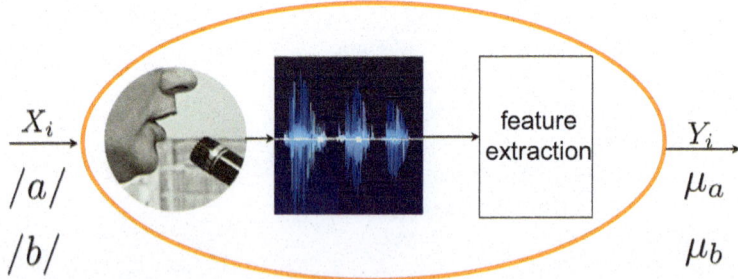

Fig. 3.45 Relationship between X_i and Y_i when a speaker is given in the noiseless case

$$Y_i = \mu_a + Z_i \tag{3.156}$$

where Z_i's are i.i.d. $\sim \mathcal{N}(0, \sigma^2)$. Similarly, given $X_i = /b/$, Y_i can be modeled as the true feature concerning $/b/$, say μ_b, plus additive Gaussian noise. Here, for simplicity, we will assume that we know the variance of the additive noise σ^2. One way to estimate the variance is via the Maximum Likelihood Estimation (MLE) for Gaussian distribution, as we learned in Sect. 2.6. That is, to infer the variance from multiple measurements Y_i's fed by null signals. More concretely, suppose $X_i = 0$ for $i \in \{1, \ldots, n\}$. Then, under the statistical modeling of Y_i as described above, we obtain:

$$Y_i = Z_i, \quad i \in \{1, \ldots, n\}. \tag{3.157}$$

Then, the sample mean of Y_i^2's serves as the MLE for σ^2:

$$\hat{\sigma}_{\mathsf{ML}}^2 = \frac{Y_1^2 + \cdots + Y_n^2}{n}. \tag{3.158}$$

Even if we know the noise variance, the likelihood function $f(y_i|x_i)$ is not yet specified. This is because the true features such as μ_a and μ_b are *user-specific* parameters. Hence, these parameters need to be estimated to fully specify the likelihood function. There is one very popular approach that enables the estimation, and we will explore this approach next.

Machine learning for estimating (μ_a, μ_b) Assume for a moment that we knew the sequence of phonemes. This may seem like an unusual assumption, given that the sequence of phonemes is precisely what we aim to infer. However, we could request the speaker to articulate some predetermined phonemes for us at the outset. In fact, certain speech recognition software adopts this approach. Then, using this known sequence, we could estimate μ_a and μ_b based on the provided phonemes. In this method, we learn the parameters by obtaining input-output example pairs from the speaker initially. What does this procedure remind you of? Yes, it is *machine learning*.

Here is a detailed explanation of how machine learning works in this context. Suppose we ask the user to pronounce "*aaaaaaaa*" for eight time slots. This would result in:

$$Y_i = \mu_a + Z_i, \quad i \in \{1, 2, \ldots, 8\}. \tag{3.159}$$

Given that the user is given, μ_a is constant, implying $Y_i \sim \mathcal{N}(\mu_a, \sigma^2)$. The objective here is to estimate μ_a based on eight i.i.d. samples $(Y_1, \ldots, Y_8)$. Once again, what does this remind you of? Yes, the optimal approach to estimate μ_a is the Maximum Likelihood Estimation (MLE):

$$\hat{\mu}_a^{\mathsf{ML}} = \frac{Y_1 + \cdots + Y_8}{8}. \tag{3.160}$$

Simplified optimal algorithm Substituting (3.152) and (3.155) into the objective function in (3.149), we obtain:

$$\mathbb{P}_{\mathbf{X}}(x_1, \ldots, x_n) f(y_1, \ldots, y_n | x_1, \ldots, x_n) = \mathbb{P}(x_1) \prod_{i=2}^{n} \mathbb{P}(x_i | x_{i-1}) \prod_{i=1}^{n} f(y_i | x_i).$$

Plugging this into (3.149) yields:

$$\hat{\mathbf{x}}_{\mathsf{MAP}} = \arg \max_{\hat{x}_1, \ldots, \hat{x}_n \in \mathcal{X}} \mathbb{P}(x_1) \prod_{i=2}^{n} \mathbb{P}(x_i | x_{i-1}) \prod_{i=1}^{n} f(y_i | x_i). \qquad (3.161)$$

Look ahead
It turns out that there is an efficient method to compute the simplified MAP solution (3.161). In the next section, we will delve into this approach.

3.13 Speech Recognition: The Viterbi Algorithm I

Recap

In the preceding two sections, we have established that speech recognition can be interpreted as an inference problem and subsequently derived the optimal MAP estimator. Further, we simplified the MAP estimator by leveraging the favorable statistical structure inherent in $\{(X_i, Y_i)\}_{i=1}^n$, where X_i represents the ith phoneme and Y_i denotes the ith feature (spectral information) (Figure 3.46).

Recall that the MAP estimator is given by:

$$\hat{\mathbf{x}}_{\mathsf{MAP}} = \arg \max_{x_1,\ldots,x_n \in \mathcal{X}} \mathbb{P}_{\mathbf{X}}(x_1,\ldots,x_n) f(y_1,\ldots,y_n|x_1,\ldots,x_n). \quad (3.162)$$

Assuming a single-memory Markov model for $\{X_i\}_{i=1}^n$, we simplified the a priori probability as:

$$\mathbb{P}_{\mathbf{X}}(x_1,\ldots,x_n) = \mathbb{P}(x_1) \prod_{i=2}^n \mathbb{P}(x_i|x_{i-1}).$$

Exploiting the Hidden Markov Model (HMM) structure inherent in $\{Y_i\}_{i=1}^n$, we showed that the likelihood is the product of individual components:

$$f(y_1,\ldots,y_n|x_1,\ldots,x_n) = \prod_{i=1}^n f(y_i|x_i).$$

Incorporating the above two into (3.162), we simplified the MAP solution to:

$$\hat{\mathbf{x}}_{\mathsf{MAP}} = \arg \max_{x_1,\ldots,x_n \in \mathcal{X}} \mathbb{P}(x_1) \prod_{i=2}^n \mathbb{P}(x_i|x_{i-1}) \prod_{i=1}^n f(y_i|x_i). \quad (3.163)$$

One naive way to derive $\hat{\mathbf{x}}_{\mathsf{MAP}}$ involves conducting an exhaustive search across all possible sequence patterns for $\{x_i\}_{i=1}^n$. However, this method, known as *exhaustive search*, poses a significant challenge. The challenge arises from the vast number of possible sequence patterns, totaling $|\mathcal{X}|^n = 44^n$. The computational complexity increases *exponentially* with n.

Outline

In the next two sections, including this one, we will study another widely adopted method designed to tackle the computational challenge. That is, the *Viterbi algorithm*. This section will explore several concepts that form the basis

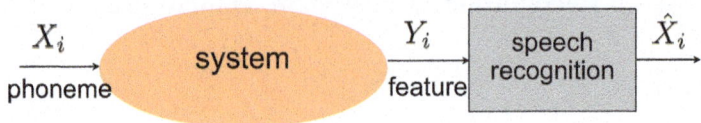

X_i phoneme → system → Y_i feature → speech recognition → $\hat{X}_i$

Fig. 3.46 A block diagram of the speech recognition system and recovery block

of the algorithm, and the next section will delve into a detailed understanding of how the algorithm operates. This section consists of three parts. In fact, "Viterbi" is the last name of the algorithm's inventor. First of all, we will introduce the inventor and emphasize the key feature of the algorithm concerning complexity. The Viterbi algorithm proves to be a highly *generic* algorithm applicable across various domains, exhibiting a specific form of optimization. In the second part, we will transform the original optimization problem (3.163) into another formulation that adheres to the specific structure. Lastly, we will explore three crucial concepts that serve to explain how the algorithm works.

The key feature of the Viterbi algorithm The inventor of the algorithm is Andrew Viterbi. See Figure 3.47 for his portrait. He stands as a luminary in the fields of communication and information theory, also known as a co-founder of Qualcomm. Interestingly, he developed the algorithm while addressing a distinct yet captivating problem that arises in communication (Forney 1973). Subsequently, it was realized by many people that the algorithm holds applicability across a broad spectrum of

Fig. 3.47 Andrew Viterbi is a giant in the communication and information theory fields

problems, extending beyond communication to encompass various applications such as speech recognition we are currently exploring.

The key aspect of the algorithm that merits attention is its capacity for *linear* growth in complexity with respect to n. It is in stark contrast to the exponential complexity of 44^n encountered in the exhaustive search. It is worth noting that, as you will discover later, the algorithm is not directly tied to the probability concepts emphasized in this book. Nevertheless, its potency lies in its remarkable effectiveness coupled with significantly reduced complexity. This is the reason that we introduce the Viterbi algorithm in this book. Familiarity with this algorithm is invaluable particularly to those who engage in endeavors related to optimization. The algorithm's utility will consistently prove beneficial in various applications.

Translation into the canonical form Let us translate the original optimization (3.163) into another with the canonical structure that the algorithm relies upon. To this end, we first massage the optimization (3.163):

$$\hat{\mathbf{x}}_{\text{MAP}} = \arg \max_{x_1,\dots,x_n \in \mathcal{X} } \mathbb{P}(x_1) \prod_{i=2}^{n} \mathbb{P}(x_i|x_{i-1}) \prod_{i=1}^{n} f(y_i|x_i).$$

Setting $x_0 = 0$ (nothing), we can simplify the optimization as:

$$\hat{\mathbf{x}}_{\text{MAP}} = \arg \max_{x_1,\dots,x_n \in \mathcal{X}} \prod_{i=1}^{n} \mathbb{P}(x_i|x_{i-1}) f(y_i|x_i)$$

$$= \arg \max_{x_1,\dots,x_n \in \mathcal{X}} \sum_{i=1}^{n} \log \{\mathbb{P}(x_i|x_{i-1}) f(y_i|x_i)\}$$

$$= \arg \min_{x_1,\dots,x_n \in \mathcal{X}} \sum_{i=1}^{n} \log \underbrace{\left\{ \frac{1}{\mathbb{P}(x_i|x_{i-1}) f(y_i|x_i)} \right\}}_{=:c(s_i,y_i)}$$

where the second equality comes from the fact that applying an increasing function $\log(\cdot)$ does not alter the maximizer; and the last equality is a consequence of reversing the sign of the objective function. Notice that the ith component in the summation in the last line is a function of (x_{i-1}, x_i, y_i). So one can denote the ith component by $c(s_i, y_i)$ where

$$s_i := \begin{bmatrix} x_{i-1} \\ x_i \end{bmatrix}. \tag{3.164}$$

Here we call s_i the *state*. The rationale behind this naming will become clear later. For notational simplicity, we define:

$$c_i(s_i) := c(s_i, y_i). \tag{3.165}$$

The retention of the subscript i in c is due to the fact that the ith component is also a function of y_i. Notice that decoding $(x_1, x_2, \ldots, x_n)$ is interchangeable with decoding $(s_1, s_2, \ldots, s_n)$. Additionally, keep in mind that y_i values are given, while x_i values serve as the optimization variables. Thus, the original optimization problem (3.163) is equivalent to:

$$\hat{s}_{MAP} = \arg \min_{s_1, \ldots, s_n} \sum_{i=1}^{n} c_i(s_i) \tag{3.166}$$

where

$$c_i(s_i) = \log \left\{ \frac{1}{\mathbb{P}(x_i|x_{i-1}) f(y_i|x_i)} \right\}. \tag{3.167}$$

The above optimization problem (3.166), which we call the *canonical optimization*, forms the foundational basis for the Viterbi algorithm.

Concept #1: Cost We are now ready to study three concepts that form the basis of the algorithm. The first concept pertains to the interested quantity $c_i(s_i)$ in the optimization (3.166). Here, $c_i(s_i)$ can be viewed as something *negative*, as a smaller value indicates a more favorable situation. Accordingly, we refer to it as the *cost*. The choice of notation using c becomes apparent in light of this interpretation.

Concept #2: State s_i The second concept is the state s_i defined in (3.164), which possesses certain properties. To facilitate the explanation of these properties, let us assume that x_i can only be one of two possible phonemes, say $/a/$ and $/b/$. For notational simplicity, we will omit the slash symbol: $x_i \in \{a, b\}$. In this case, we observe that s_i can assume one of the following six candidates:

$$\begin{bmatrix} 0 \\ b \end{bmatrix}, \begin{bmatrix} 0 \\ a \end{bmatrix}, \underbrace{\begin{bmatrix} a \\ a \end{bmatrix}}_{s1}, \underbrace{\begin{bmatrix} b \\ a \end{bmatrix}}_{s2}, \underbrace{\begin{bmatrix} b \\ b \end{bmatrix}}_{s3}, \underbrace{\begin{bmatrix} a \\ b \end{bmatrix}}_{s4}. \tag{3.168}$$

Note that the first two states only occur at the beginning ($i = 1$), assuming that $x_0 = 0$. Consequently, these two states are negligible compared to the others for a large value of n. To simplify, we disregard these two states. To this end, we intentionally set $x_0 = a$, reducing the possible states to only four. The number "four" is a clearly *finite* number. There exists a terminology that designates an entity with respect to such states, capable of taking only a *finite* number of possible candidates. That is, a *Finite State Machine* (FSM).

The FSM possesses an intriguing property that can be visualized. To understand this, observe that each state s_i can transition to another based on the value of x_{i+1}. Consequently, one can envision the state transitions as illustrated in Figure 3.48. In this representation, the labels a or b above a transition arrow denote the value of x_{i+1} given the current state s_i. For example, if $x_{i+1} = b$ in the current state $s1$, we

Fig. 3.48 A finite state machine with four states. A transition occurs depending on the value of x_{i+1}

$$s_i := \begin{bmatrix} x_{i-1} \\ x_i \end{bmatrix}$$

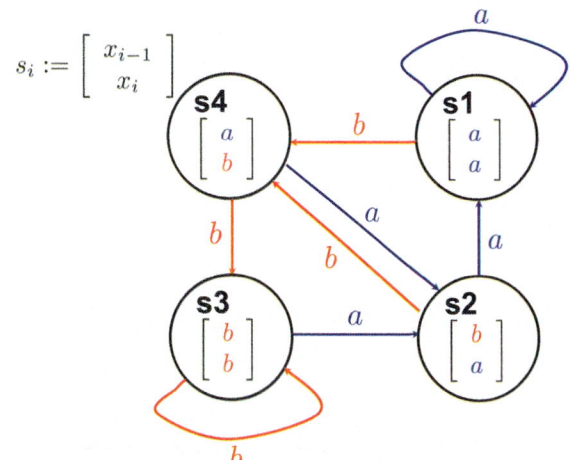

follow the red transition arrow to reach the state $s4$ labeled as $[a; b]$. This visual representation is termed a *state transition diagram*.

While the state transition diagram effectively illustrates how each state transitions to another, it falls short in capturing *time evolution*. This is precisely where another concept related to FSM comes into play. That is, the *trellis diagram*. In the subsequent discussion, we will describe how the trellis diagram looks like, and subsequently elaborate on how it aids in solving the optimization problem (3.166).

Concept #3: Trellis diagram The trellis diagram exhibits both state transitions and time evolution. To clearly understand how it works, let us walk you through details with the help of Figs. 3.49 and 3.50.

Consider the state at time 1:

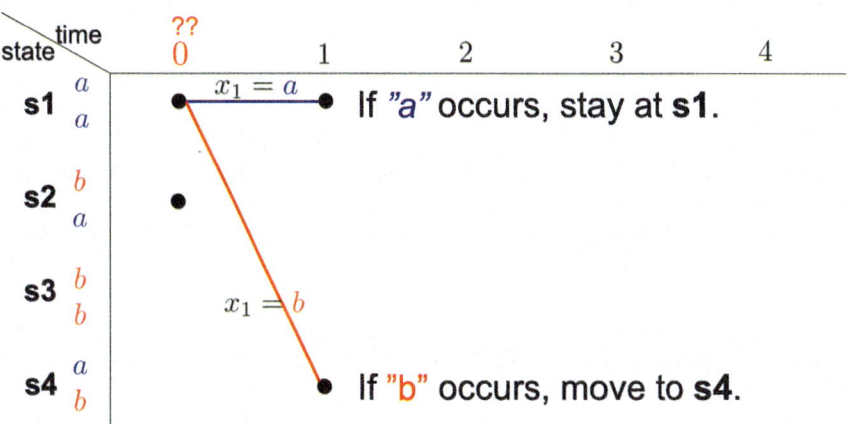

Fig. 3.49 A trellis diagram at $i = 1$

Fig. 3.50 A trellis diagram at $i = 2$

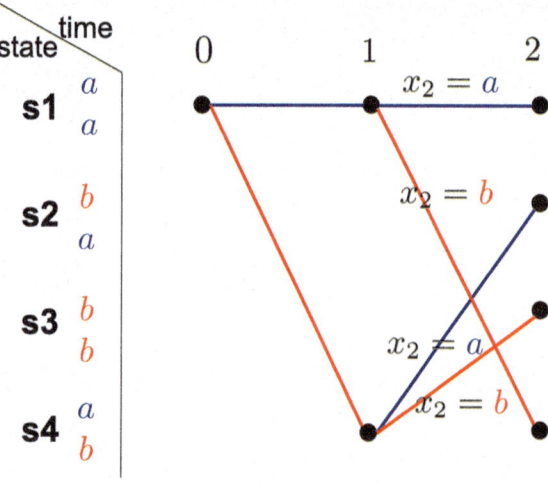

$$S_1 = \begin{bmatrix} x_0 \\ x_1 \end{bmatrix} = \begin{bmatrix} a \\ x_1 \end{bmatrix} = \begin{cases} \begin{bmatrix} a \\ a \\ a \\ a \end{bmatrix} = s1, \text{ if } x_1 = a; \\ \begin{bmatrix} a \\ a \\ a \\ b \end{bmatrix} = s4, \text{ if } x_1 = b, \end{cases}$$

where the second equality is due to our assumption $x_0 = a$. In Figure 3.49, at time 1, s_1 can take one of the two possible states, $s1$ and $s4$, represented by the two dots. At time 0, s_0 has two potential options depending on the value of x_{-1}: $s1$ and $s2$. To eliminate ambiguity at time 0, we further assume $x_{-1} = a$, fixing the initial state as $s1$. This ensures a consistent starting point from $s1$. As mentioned earlier, if $x_1 = a$, then $s_1 = s1$; otherwise, $s_1 = s4$.

For a more comprehensive understanding of how it works, we examine an additional time slot, as depicted in Figure 3.50. Assume $s_1 = s1$. If $x_2 = a$, it remains in the same state $s1$ (indicated by the blue transition arrow); otherwise, it transitions to $s4$ (indicated by the red arrow). Conversely, with $s_1 = s4$, if $x_2 = a$, it transitions to $s2$; otherwise, it moves to $s3$.

Cost calculation In the optimization (3.166), our focus is on the cost $c_i(s_i)$. A natural question arises: How can we compute the cost in (3.166) from the trellis diagram? To address this, examine a specific example for $n = 4$; refer to Figure 3.51.

In this example, where $(x_1, x_2, x_3, x_4) = (a, b, a, a)$, the trellis path follows the blue-red-blue-blue transition arrows, resulting in the sequence of state changes: $s1 \rightarrow s1 \rightarrow s4 \rightarrow s2 \rightarrow s1$. To simplify cost calculation, we associate a cost with each corresponding state node. For instance, we place $c_1([a; a])$ (indicated in green in Figure 3.51) near the black dot at $s1$ in time 1. Similarly, we assign $c_2([a; b])$, $c_3([b; a])$, and $c_4([a; a])$ to their respective black dots. By aggregating the costs associated with all the black dots, we can easily compute the cost for this particu-

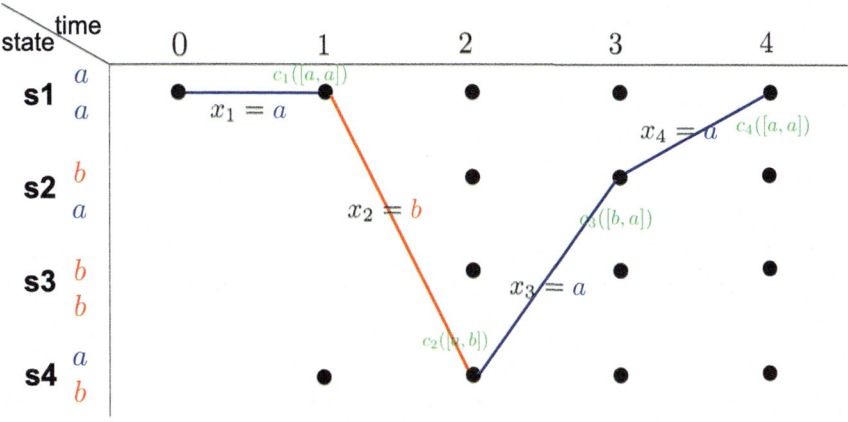

Fig. 3.51 A trellis diagram for the sequence $(x_1, x_2, x_3, x_4) = (a, b, a, a)$

lar sequence. Extending this process to all possible sequence patterns allows us to calculate the optimal path as:

$$\hat{s}_{\text{MAP}} = \arg \min_{s_1, s_2, s_3, s_4} \{c_1(s_1) + c_2(s_2) + c_3(s_3) + c_4(s_4)\}.$$

As previously noted, a naive exhaustive search demands 2^4 cost calculations, making the complexity prohibitively expensive, especially for large values of n.

Look ahead

In the subsequent section, we will delve into the Viterbi algorithm, an efficient method adept at leveraging the structure of the trellis diagram to find $\hat{s}_{\text{MAP}}$.

3.14 Speech Recognition: The Viterbi Algorithm II

Recap

In the preceding section, we explored three concepts that we claimed elucidate the mechanism of the Viterbi algorithm designed for computing the optimal MAP estimator:

$$\hat{s}_{MAP} = \arg \min_{s_1,\ldots,s_n} \sum_{i=1}^{n} c_i(s_i) \qquad (3.169)$$

where

$$c_i(s_i) := \log\left\{\frac{1}{\mathbb{P}(x_i|x_{i-1})f(y_i|x_i)}\right\} \quad \text{and} \quad s_i := \begin{bmatrix} x_{i-1} \\ x_i \end{bmatrix}. \qquad (3.170)$$

The first concept is the cost $c_i(s_i)$, which denotes an interested quantity in the optimization problem (3.169). The second concept pertains to the state s_i concerning a finite state machine characterized by the following states:

$$\underbrace{\begin{bmatrix} a \\ a \end{bmatrix}}_{s1}, \underbrace{\begin{bmatrix} b \\ a \end{bmatrix}}_{s2}, \underbrace{\begin{bmatrix} b \\ b \end{bmatrix}}_{s3}, \underbrace{\begin{bmatrix} a \\ b \end{bmatrix}}_{s4}.$$

The third concept involves the trellis diagram, which provides a visual representation of how each state evolves over time. Figure 3.52 presents an example of the trellis diagram for the sequence $(x_1, x_2, x_3, x_4) = (a, b, a, a)$, under the assumption that $x_0 = x_{-1} = a$.

Outline

In this section, we will study how the Viterbi algorithm works in detail. The second comprises three parts. Initially, we will emphasize a key observation that provides crucial insight into the algorithm. Subsequently, we will explore how the algorithm operates. Lastly, we will demonstrate that computational complexity exhibits a linear increase with n, as previously claimed.

A key observation The inspiration for the Viterbi algorithm stems from a key observation. To see this clearly, consider two possible sequence patterns presented in Figure 3.53:

(i) $(x_1, x_2, x_3, x_4) = (b, a, b, a)$ (marked in purple);

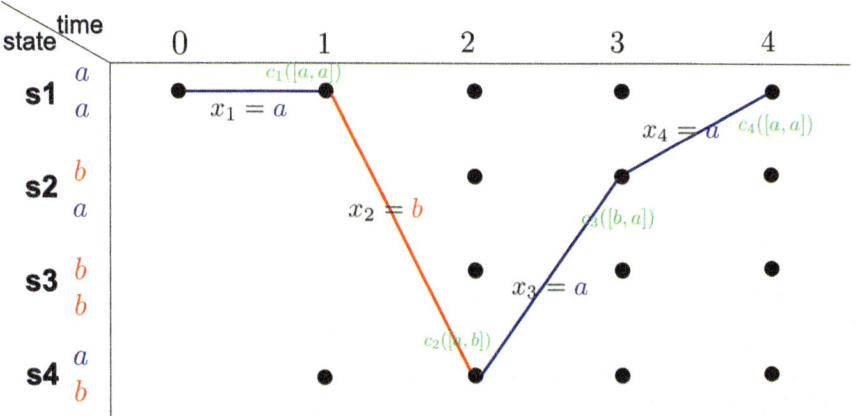

Fig. 3.52 A trellis diagram for the sequence $(x_1, x_2, x_3, x_4) = (a, b, a, a)$

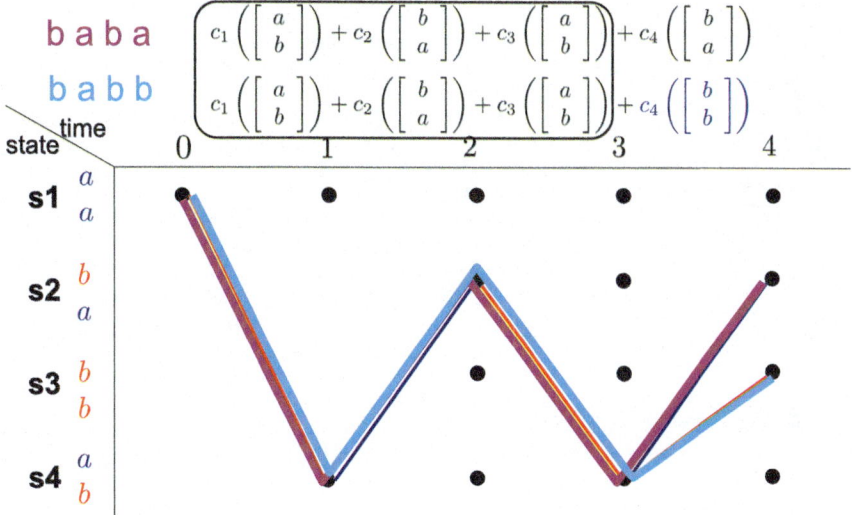

Fig. 3.53 Key observation: Two trellis paths are significantly overlapped; hence, the two corresponding costs are identical except for the cost w.r.t. the last state vector

(ii) $(x_1, x_2, x_3, x_4) = (b, a, b, b)$ (marked in blue).

The key observation here is the substantial overlap between the two trellis paths. Consequently, the corresponding costs are identical, except for the cost associated with the last state vector. This observation motivated Viterbi to come up with the following idea.

Idea of the Viterbi algorithm The idea is to store an *aggregated cost* up to time t successively, which is then utilized to compute a subsequent aggregated cost for the

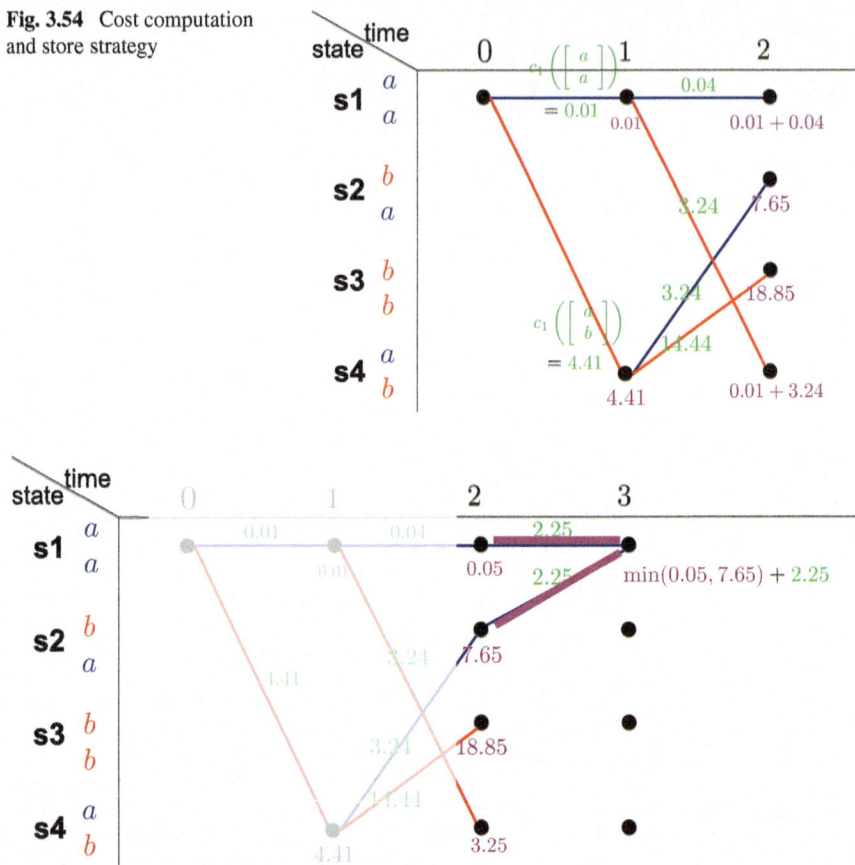

Fig. 3.54 Cost computation and store strategy

Fig. 3.55 The code idea of the Viterbi algorithm

next time slot. To grasp the details of this process, consider a simple example. Refer to Figs. 3.54 and 3.55.

For illustrative purposes, we examine a simple scenario with $n = 4$, where cost computations have been completed using (3.170) for all possible states and $i \in \{1, \ldots, n\}$. Here is an example for such cost computations.

$$
\begin{bmatrix}
c_1([a; a]) & c_2([a; a]) & c_3([a; a]) & c_4([a; a]) \\
c_1([b; a]) & c_2([b; a]) & c_3([b; a]) & c_4([b; a]) \\
c_1([b; b]) & c_2([b; b]) & c_3([b; b]) & c_4([b; b]) \\
c_1([a; b]) & c_2([a; b]) & c_3([a; b]) & c_4([a; b])
\end{bmatrix}
=
\begin{bmatrix}
0.01 & 0.04 & 2.25 & 17.64 \\
4.41 & 3.24 & 0.25 & 4.84 \\
16.81 & 14.44 & 6.25 & 0.04 \\
4.41 & 3.24 & 0.25 & 4.84
\end{bmatrix}.
$$

The cost $c_1([a; a]) = 0.01$ is positioned along the blue transition arrow, as depicted in Figure 3.54. Subsequently, we store this cost at the state $s1$ node at time 1 (marked by a black dot). The storage is denoted by a purple-colored number in close proximity to the black dot. The same procedure is repeated concerning the cost when $x_1 = b$. The cost $c_1([a; b]) = 4.41$ is positioned along the corresponding red transition arrow. Subsequently, we store 4.41 in close proximity to the black dot associated with the state $s4$. Moving on, examine the cost $c_2([a; a]) = 0.04$. Consequently, the aggregated cost up to time 2 for the state $s1$ would be $0.01 + 0.04$, as shown in Figure 3.54. Similarly, the aggregated costs for the other states ($s2, s3, s4$) would be $(4.41 + 3.24, 4.41 + 14.44, 0.01 + 3.24)$, respectively.

The core idea of the Viterbi algorithm becomes evident from time 3 onward. Refer to Figure 3.55 for a visual representation of this idea. Consider the state $s1$ at time 3. This state occurs when $x_3 = a$ and originates from two possible prior states: $s1$ and $s2$. Examine the cost $c_3([a; a]) = 2.25$. The aggregated cost, assuming it comes from the prior state $s1$ (storing 0.05 for the aggregated cost up to time 2), would be: $0.05 + 2.25 = 2.3$. On the other hand, the aggregated cost concerning the prior state $s2$ would be: $7.65 + 2.25 = 9.9$. To determine the optimal path, considering that we aim to find the path with the *minimum* aggregated cost, we discard the path with the larger cost 9.9 in the competition. Consequently, we only need to store the *minimum* value between the two costs at the state $s1$ in time 3, ignoring the less favorable path. Therefore, we store 2.3 at the node, as illustrated in Figure 3.56.

We repeat the same process for the other states. In the case of state $s2$, the *lower* path emerges as the winner, leading us to store the corresponding aggregated cost of 3.5 at the state $s2$ node, while eliminating the upper losing path. The same procedure is applied to states $s3$ and $s4$. This iterative process continues until the last time slot, and you can observe all the associated computations in Figure 3.56.

Determining the path that results in the minimum aggregated cost from the illustration is straightforward. Examine the four aggregated costs at the last time slot: 19.94, 5.14, 0.34, 7.14. Choose the minimum cost, which is 0.34. To identify the corresponding sequence pattern, utilize *backtracking* since we retain only the survivor paths in the picture while eliminating the losing paths. The survivor

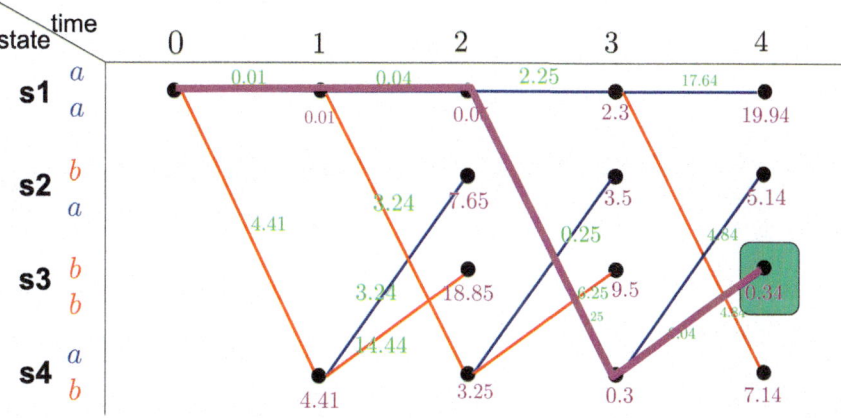

Fig. 3.56 Choose s* that minimizes the aggregated cost

paths form the thick purple trajectory in Figure 3.56, corresponding to the sequence $(x_1, x_2, x_3, x_4) = (a, a, b, b)$.

Complexity Recall our claim that the complexity of the Viterbi algorithm grows linearly with n. To better understand this, consider the basic operation that takes place at each node. Focus on the operation related to the state $s1$ node at time 3, as depicted in Figure 3.57. Given the precomputed values of the a priori probability and the likelihood function (which can be stored in a table), the cost computation involves: 1 logarithmic operation, 1 multiplication, 1 addition, and 1 comparison. The comparison operation is needed for selecting the minimum between the two. This operation is repeated for the other states across all time slots, reflected in the black dots. Since the total number of black dots is $4n$, it is evident that the complexity of the Viterbi algorithm grows linearly with n.

> **Look ahead**
> We have explored the Viterbi algorithm, which provides an efficient and optimal MAP solution. This solution is notable for its low computational complexity, scaling linearly with the total number n of time slots. In the forthcoming section, we will study how to implement the Viterbi algorithm via Python.

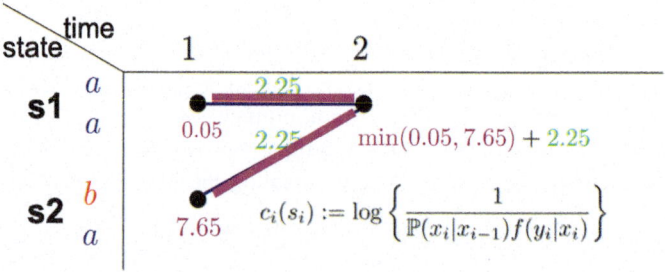

Fig. 3.57 Operations per state: 1 logarithmic operation, 1 multiplication, 1 addition, and 1 comparison

3.15   Speech Recognition: **Python** Implementation of the Viterbi Algorithm

Recap

Over the preceding two sections, we have delved into the Viterbi algorithm. The optimal solution offered by the Viterbi algorithm boasts a low computational complexity that scales linearly with the total number n of time slots. The algorithm's foundation lies in a finite state machine, leveraging the structure of the associated trellis diagram. This strategic approach facilitates the effective identification of the optimal sequence in the following optimization:

$$\mathbf{s}^{\star} = \arg\min_{s_1,\dots,s_n} \sum_{i=1}^{n} c_i(s_i) \tag{3.171}$$

where

$$c_i(s_i) := \log\left\{\frac{1}{\mathbb{P}(x_i|x_{i-1})f(y_i|x_i)}\right\} \quad \text{and} \quad s_i := \begin{bmatrix} x_{i-1} \\ x_i \end{bmatrix}. \tag{3.172}$$

Outline

In this section, we will implement the Viterbi algorithm using **Python**. This section consists of three parts. First, we will establish basic coding components in the context of a simple scenario where $n = 4$; $\mathcal{X} = \{+1, -1\}$ (corresponding to two phonemes, say a and b); $x_0 = x_{-1} = +1$; and the observed sequence is $(y_1, y_2, y_3, y_4) = (2.1, 1.8, 0.5, -2.2)$. The transition probabilities $\mathbb{P}(x_i|x_{i-1})$ and observation likelihoods $f(y_i|x_i)$ adhere to a specific setting as follows:

$$\mathbb{P}(x_i|x_{i-1}) = \frac{1}{3}\left(\mathbf{1}\{x_i = x_{i-1}\} + 1\right);$$
$$f(y_i|x_i) = \frac{1}{\sqrt{2\pi\sigma^2}}e^{-\frac{(y_i-x_i)^2}{2\sigma^2}}. \tag{3.173}$$

Following that, we will utilize the coding components to construct a function that finds the optimal sequence within the specified scenario. Finally, employing this developed function, we will conduct a simple experiment to demonstrate that the Viterbi algorithm well reconstructs the input sequence.

Simplification of the optimization problem (3.171)   Under the specified setting (3.173), the cost $c_i(s_i)$ can be computed as:

$$c_i(s_i) = \log 3 - \log\left(1 + \mathbf{1}\{x_i = x_{i-1}\}\right) + \frac{1}{2}\log(2\pi\sigma^2) + \frac{1}{2\sigma^2}(y_i - x_i)^2$$

$$= \log 3 + \frac{1}{2}\log(2\pi\sigma^2) - \log\left(1 + \mathbf{1}\{s_i[1] = s_i[0]\}\right) + \frac{1}{2\sigma^2}(y_i - s_i[1])^2.$$

Since the first two terms in the last line above are independent of the state s_i, the optimization problem (3.171) can be simplified to:

$$s^\star = \arg\min_{s_1,\ldots,s_n} \sum_{i=1}^{n} \underbrace{\left\{ -\log\left(1 + \mathbf{1}\{s_i[1] = s_i[0]\}\right) + \frac{1}{2\sigma^2}(y_i - s_i[1])^2 \right\}}_{=:c_i'(s_i)}. \tag{3.174}$$

In the above expression, we represent the simplified cost as $c_i'(s_i)$.

State matrix S and the observation sequence vector y We define two quantities to be used as inputs for a forthcoming function. The first is the state matrix $\mathbf{S}$ that consolidates all available states. Recall the earlier definition of the state: $s_i := [x_{i-1}, x_i]^T$. Depending on the binary values of x_{i-1} and x_i, there are four states: $s1 = [+1, +1]$, $s2 = [-1, +1]$, $s3 = [-1, -1]$, and $s4 = [+1, -1]$. The state matrix $\mathbf{S}$ is defined as the result of stacking these four states row-wise:

$$\mathbf{S} = \begin{bmatrix} +1 & +1 \\ -1 & +1 \\ -1 & -1 \\ +1 & -1 \end{bmatrix}. \tag{3.175}$$

The second quantity is the observation squence vector $\mathbf{y} = [y_1, \ldots, y_n]^T$. In the considered example,

$$\mathbf{y} = \begin{bmatrix} 2.1 \\ 1.8 \\ 0.5 \\ -2.2 \end{bmatrix}. \tag{3.176}$$

Below is a code snippet for constructing these quantities:

```python
import numpy as np
S = np.array([[+1,+1],[-1,+1],[-1,-1],[+1,-1]])
y = np.array([2.1, 1.8, 0.5, -2.2])
print(S.shape)
print(y.shape)
(4, 2)
(4,)
```

The state_to_idx function One basic coding component frequently utilized is the state_to_idx function, which provides the row index in the state matrix $\mathbf{S}$

corresponding to a given state vector (choosing from $s1, s2, s3$, and $s4$). For example, when given $s1 = [+1, +1]$, it should yield 0 according to (3.175). Here is the code implementation for this function:

```python
def state_to_idx(s_current, S):
    # Check if each row in S (axis=1) matches s_current
    # For s_current=[+1,+1], it reads [True,False,False,False]
    idx_check = np.all(S == s_current, axis=1)
    # Return the index of the matched row
    # For s_current=[+1,+1], it returns 0
    return np.where(idx_check)[0][0]

s_current = np.array([[1,1]])
print(s_current)
print(state_to_idx(s_current, S))
```

```
[[1 1]]
0
```

The prev_state function: Return the candidates of the previous state Another basic coding component is the prev_state function, which returns all the candidates for the previous state concerning a given current state. A crucial operation in the Viterbi algorithm involves storing an aggregated cost at each node in the trellis diagram. For this purpose, we first need to retrieve the aggregated cost stored in the previous state. Thus, it necessitates the identification of the previous state. To understand how to implement the prev_state function, consider an instance where the current state is $[x_{i-1}, x_i] = [+1, +1] = s1$. In this scenario, the previous state is $[x_{i-2}, x_{i-1}] = [+1, +1] = s1$ if $x_{i-2} = +1$, and it is $[x_{i-2}, x_{i-1}] = [-1, +1] = s2$ when $x_{i-2} = -1$. Below is the code implementation for the prev_state function:

```python
def prev_state(s_current,S):
    # s_current = [x_{i-1},x_i]
    # Output: Possible candidates for the previous state
    #           [+1, x_{i-1}]] and [-1,x_{i-1}]]

    # Find the row index matching s_current
    current_idx = state_to_idx(s_current,S)
    # Retrieve x_{i-1} using the current_idx
    x_i1 = S[current_idx,0].squeeze()
    # Return the two condidates for [x_{i-2},x_{i-1}]
    # depending on x_{i-2}=+1 or x_{i-2}=-1
    cand_plus = np.array([+1,x_i1])
    cand_minus = np.array([-1,x_i1])
    return [cand_plus, cand_minus]

s_current = np.array([[+1,+1]])
print(prev_state(s_current, S))
```

```
[array([1, 1]), array([-1,  1])]
```

Compute the current cost at every node in the trellis diagram Next, consider the simplified cost in (3.174) for every $i \in \{1, 2, 3, 4\}$ and for all possible patterns of s_i:

$$c'_i(s_i) = -\log(1 + \mathbf{1}\{s_i[1] = s_i[0]\}) + \frac{1}{2\sigma^2}(y_i - s_i[1])^2. \tag{3.177}$$

Let us evaluate these for $\sigma^2 = 0.1$. For instance, at $i = 1$,

$$
\begin{aligned}
c'_1([+1; +1]) &= \{-\log(1 + 1) + 5(2.1 - 1)^2\} = 5.36; \\
c'_1([-1; +1]) &= \{-\log(1) + 5(2.1 - 1)^2\} = 6.05; \\
c'_1([-1; -1]) &= \{-\log(1 + 1) + 5(2.1 + 1)^2\} = 47.36; \\
c'_1([+1; -1]) &= \{-\log(1) + 5(2.1 + 1)^2\} = 48.05.
\end{aligned}
\tag{3.178}
$$

We build a matrix containing all of the simplified costs as elements:

$$
\begin{aligned}
\texttt{trellis} &= \begin{bmatrix}
c'_1([+1; +1]) & c'_2([+1; +1]) & c'_3([+1; +1]) & c'_4([+1; +1]) \\
c'_1([-1; +1]) & c'_2([-1; +1]) & c'_3([-1; +1]) & c'_4([-1; +1]) \\
c'_1([-1; -1]) & c'_2([-1; -1]) & c'_3([-1; -1]) & c'_4([-1; -1]) \\
c'_1([+1; -1]) & c'_2([+1; -1]) & c'_3([+1; -1]) & c'_4([+1; -1])
\end{bmatrix} \\[2mm]
&= \begin{bmatrix}
5.36 & 2.51 & 0.56 & 50.51 \\
6.05 & 3.2 & 1.25 & 51.2 \\
47.36 & 38.51 & 10.56 & 6.51 \\
48.05 & 39.2 & 11.25 & 7.2
\end{bmatrix}.
\end{aligned}
$$

One way to construct this matrix is as follows.

```
sigma2= 0.1
# state 1: s1
cs1=-np.log(1+int(S[0][0]==S[0][1])) \
    +1/(2*sigma2)*(y-S[0][1])**2
# state 2: s2
cs2=-np.log(1+int(S[1][0]==S[1][1])) \
    +1/(2*sigma2)*(y-S[1][1])**2
# state 3: s3
cs3=-np.log(1+int(S[2][0]==S[2][1])) \
    +1/(2*sigma2)*(y-S[2][1])**2
# state 4: s3
cs4=-np.log(1+int(S[3][0]==S[3][1])) \
    +1/(2*sigma2)*(y-S[3][1])**2
trellis=np.array([cs1,cs2,cs3,cs4])
print(trellis)
```

```
[[ 5.35685282  2.50685282  0.55685282 50.50685282]
 [ 6.05        3.2         1.25        51.2       ]
 [47.35685282 38.50685282 10.55685282  6.50685282]
 [48.05       39.2        11.25        7.2        ]]
```

Initialize `survived_costs` at time 1   We are now ready to compute the costs
that are survived from competition among the candidates for the previous state
costs. These are the values that we indicated in purple in Figure 3.58.

We first compute the costs at time 1.

```
# Initialize "costs" to be stored
survived_costs = np.zeros_like(trellis)
# Initialize "paths" (of a dictionary type) that would store
# the index of the previous winner state
# key: (i,t) --> state i at time t
# value: the index of the previous winner state
paths = {}

num_states = S.shape[0]
num_times = y.shape[0]

# Initialize costs at time 0
for i in range(num_states):
    survived_costs[i,0] = trellis[i,0]
    paths[(i,0)] = None # Specify the end of path as ''None''

print(survived_costs)
```
```
[[ 5.35685282  0.          0.          0.          ]
 [ 6.05        0.          0.          0.          ]
 [47.35685282  0.          0.          0.          ]
 [48.05        0.          0.          0.          ]]
```

Notice that the values at $(1, 1)$ and $(4, 1)$ entries in the `survived_costs` coincide
with 5.36 and 48.05 (marked in purple) that we manually calculated in Figure 3.58.
The only distinction relative to Figure 3.58 is that we do have costs at $(2, 1)$ and

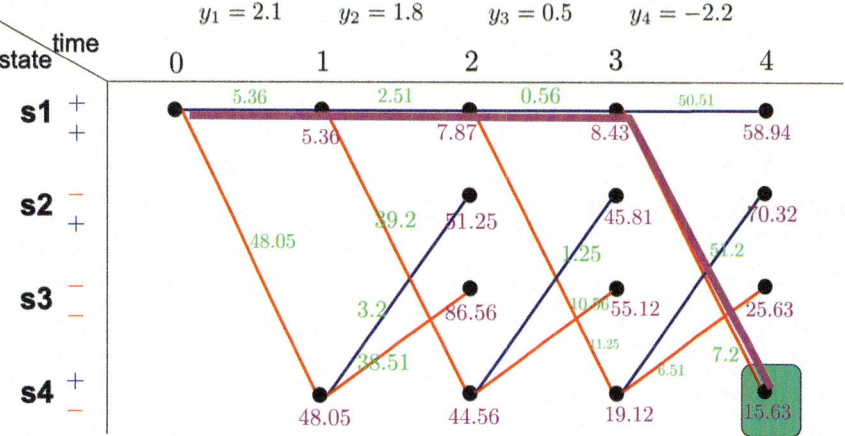

Fig. 3.58 Illustration of the Viterbi algorithm in action. The values marked in purple represent the
costs that are survived from competition among the candidates from the previous states

(3, 1) entries. This discrepancy arises because we assume $x_0 = 0$ here, whereas we set $x_0 = +1$ with the state at $t = 0$ being s1 in Figure 3.58.

Compute `survived_costs` **at other time slots**   Next, we calculate the survived costs at other time slots $i \in \{2, 3, \ldots, n\}$.

```python
# Iterate over other time slots i=2,...,n
for t in range(1, num_times):
    # Iterate over states
    for i in range(num_states):
        # Find two candidates for the previous state
        prev_state_cand = prev_state(S[i,:],S)
        # Find the minimum cost among the two candidates
        min_cost, prev = min(
            (survived_costs[state_to_idx(s,S),t-1] + trellis[i,t],
                                 state_to_idx(s,S))
                                for s in prev_state_cand)
        # Store the minimum cost as a survival
        survived_costs[i,t] = min_cost
        # Store the index of the previous winner state
        paths[(i,t)] = prev

print(survived_costs)
print(paths)
```

```
[[ 5.35685282   7.86370564   8.42055846 58.92741128]
 [ 6.05        50.55685282 45.80685282 70.31370564]
 [47.35685282 85.86370564 55.11370564 25.62055846]
 [48.05        44.55685282 19.11370564 15.62055846]]
{(0, 0): None, (1, 0): None, (2, 0): None, (3, 0): None,
(0, 1): 0, (1, 1): 2, (2, 1): 2, (3, 1): 0,
(0, 2): 0, (1, 2): 3, (2, 2): 3, (3, 2): 0,
(0, 3): 0, (1,3): 3, (2, 3): 3, (3, 3): 0}
```

We observe that the components in `survived_costs` closely resemble the purple-colored values in Figure 3.58. The slight differences arise because we rounded the values in the figure to two decimal places.

Find the minimum cost and the corresponding input sequence   After obtaining `paths`, we can identify the optimal sequence $\mathbf{s}^\star$ by backtracking the winning states across all time instants.

```python
optimal_path = []
t = num_times -1
# The index of the winner state in the last time slot
state_idx = np.argmin(survived_costs[:,t])

# Backtrack the winner states across all time instants
while state_idx is not None:
    # Print the winner state at each time
    print(f"winner state: {state_idx} at time {t+1}")
    # Insert the winner state at the 0th position
    # This way, we enable ``last input first ouput''
```

```
        optimal_path.insert(0,S[state_idx,:])
        # Move to the previous winner state
        state_idx = paths[(state_idx,t)]
        t -= 1

print(optimal_path)
```

```
winner state: 3 at time 4
winner state: 0 at time 3
winner state: 0 at time 2
winner state: 0 at time 1
[array([1, 1]), array([1, 1]), array([1, 1]), array([1, -1])]
```

Here, `optimal_path` represents a list of (x_0, x_1), (x_1, x_2), (x_2, x_3), (x_3, x_4), as per the last-input-first-output rule. Let us convert this into the input sequence: (x_1, x_2, x_3, x_4).

```
seq_estimate = np.array([pair[1] for pair in optimal_path])
print(seq_estimate)
```

```
[ 1  1  1 -1]
```

A final function for the Viterbi algorithm   Utilizing all the previously discussed coding components, we construct a function that executes the Viterbi algorithm for arbitrary-length observation sequence vector **y**.

```
def state_to_idx(s_current, S):
    # Check if each row of S (axis=1) matches with s_current
    # For s_current=[+1,+1], it reads [True,False,False,False]
    idx_check = np.all(S == s_current, axis=1)
    # Return the index of the matched row
    # For s_current=[+1,+1], it returns 0
    return np.where(idx_check)[0][0]

def prev_state(s_current,S):
    # s_current = [x_{i-1},x_i]
    # Output: Possible candidates for the previous state
    #          [+1, x_{i-1}] and [-1,x_{i-1}]

    # Find the row index matching s_current
    current_idx = state_to_idx(s_current,S)
    # Retrieve x_{i-1} using the current_idx
    x_i1 = S[current_idx,0].squeeze()
    # Return the two condidates for [x_{i-2},x_{i-1}]
    # depending on x_{i-2}=+1 or x_{i-2}=-1
    cand_plus = np.array([+1,x_i1])
    cand_minus = np.array([-1,x_i1])
    return [cand_plus, cand_minus]

def viterbi_algorithm(S,y,sigma2):
    # Compute the current cost at every node
    # state 1: s1
    cs1=-np.log(1+int(S[0][0]==S[0][1])) \
```

```
        +1/(2*sigma2)*(y-S[0][1])**2
# state 2: s2
cs2=-np.log(1+int(S[1][0]==S[1][1])) \
        +1/(2*sigma2)*(y-S[1][1])**2
# state 3: s3
cs3=-np.log(1+int(S[2][0]==S[2][1])) \
        +1/(2*sigma2)*(y-S[2][1])**2
# state 4: s3
cs4=-np.log(1+int(S[3][0]==S[3][1])) \
        +1/(2*sigma2)*(y-S[3][1])**2
trellis=np.array([cs1,cs2,cs3,cs4])

# Initialize "costs" to be stored
survived_costs = np.zeros_like(trellis)
# Initialize "paths" (of a dictionary type) that would
# store the index of the previous winner state
# key: (i,t) --> state i at time t
# value: the index of the previous winner state
paths = {}

num_states = S.shape[0]
num_times = y.shape[0]

# Initialize costs at time 1
for i in range(num_states):
    survived_costs[i,0] = trellis[i,0]
    paths[(i,0)] = None # Specify the end of path

# Iterate over other time slots i=2,...,n
for t in range(1, num_times):
    # Iterate over states:
    for i in range(num_states):
        # Find the two candidates for the previous state
        prev_state_cand = prev_state(S[i,:],S)
        # Find the minimum cost among the two candiates
        min_cost, prev = min(
    (survived_costs[state_to_idx(s,S),t-1] + trellis[i,t],
                            state_to_idx(s,S))
                            for s in prev_state_cand)
        # Store the minimum cost as a survival
        survived_costs[i,t] = min_cost
        # Store the index of the winner state
        paths[(i,t)] = prev

optimal_path = []
t = num_times -1
# The index of the winner path in the last time slot
state_idx = np.argmin(survived_costs[:,t])

# Backtrack the winner states over all time instants
while state_idx is not None:
    # Insert the winner row at the 0th position
    # This way, we enable ``last input first ouput''
```

```
        optimal_path.insert(0,S[state_idx,:])
        # Move to the previous winner state
        state_idx = paths[(state_idx,t)]
        t -= 1

    # Convert 'optimal_path' into the input sequence
    seq_estimate = np.array([pair[1] for pair in optimal_path])
    return seq_estimate
```

Testing the Viterbi algorithm We conduct an experiment to validate the functionality of the Viterbi algorithm. Consider a scenario with $n = 10$ and $\sigma^2 = 0.1$. To generate $\mathbb{P}(x_i|x_{i-1}) = \frac{1}{3}(\mathbf{1}\{x_i = x_{i-1}\} + 1)$, we introduce the Bernoulli process $\{W_i\}_{i=1}^n$ with a parameter of $\frac{2}{3}$. Utilizing W_i, we define X_i as follows:

$$X_i = (2W_i - 1)X_{i-1}, \quad i \in \{1, 2, \ldots, n\} \tag{3.179}$$

where we set $X_0 = +1$. To generate Y_i according to $f(y_i|x_i) = \frac{1}{\sqrt{2\pi\sigma^2}}e^{-\frac{(y_i-x_i)^2}{2\sigma^2}}$, we introduce an additive white Gaussian noise $Z_i \sim \mathcal{N}(0, \sigma^2)$:

$$Y_i = X_i + Z_i, \quad i \in \{1, 2, \ldots, n\}. \tag{3.180}$$

Below is a code for simulating this setup.

```python
from scipy.stats import bernoulli
import numpy as np

n = 10 # input sequence length
W = bernoulli(2/3)

sigma2 = 0.1 # noise variance

# Generate the Bernoulli process W
W_Bern = W.rvs(n)

# Generate the input sequence
X = np.zeros(n) # initialization
X[0]=2*W_Bern[0]-1
for i in range(1,n): X[i] = (2*W_Bern[i]-1)*X[i-1]

# Pass through an AWGN with N(0,sigma2)
sigma2=0.1
Y = X + np.sqrt(sigma2)*np.random.randn(n)

print(W_Bern)
print(X)
print(Y)
```

```
[1 1 1 0 1 1 1 1 1 1]
[1 1 1 -1 -1 -1 -1 -1 -1 -1]
```

```
[ 1.04661279   0.77728314   1.68026144  -1.16066161  -0.92059665
 -0.92873195  -1.73723733  -0.83182916  -0.89149197  -0.95462836]
```

We feed these values into the implemented `viterbi_algorithm` function to estimate the input sequence.

```
S = np.array([[+1,+1],[-1,+1],[-1,-1],[+1,-1]])

print(X)
print(viterbi_algorithm(S,Y,sigma2))
```

```
[ 1   1   1  -1  -1  -1  -1  -1  -1  -1]
[ 1   1   1  -1  -1  -1  -1  -1  -1  -1]
```

We observe that the estimates match the original signals.

Closing

Finally, we wish to share an insight that might be beneficial for student readers in shaping their future careers. A key takeaway from this book underscores the pivotal role of fundamental concepts and principles, particularly in the realms of probability and the MAP/ML principles. Hence, we strongly encourage you to focus on building a solid foundation. The fundamental concepts we emphasize are particularly pertinent to modern AI technologies that might pique your interest. Becoming an expert in the AI field demands a diverse skill set, and one critical component is a good understanding of mathematics, with special emphasis on the following four fundamental branches.

The first branch is *probability*. Recall that data in machine learning can be interpreted as a specific realization of a random process. The second is *optimization*, recognizing that the objectives of machine learning are often attained through optimization. The third is *linear algebra*, which provides essential tools for translating objective functions and/or constraints into manageable formulas. Many seemingly complex mathematical expressions can be represented as simple terms involving matrix multiplications and additions. The last is *information theory*, a field that offers valuable insights into the optimal architecture of machine learning models. Consider, for example, the role of cross entropy, an information-theoretic notion, in the design of optimal loss functions.

These fundamentals are crucial for engaging in activities relevant to AI. Our advice is to cultivate strength in these foundational areas. A caveat is that these fundamentals are often best developed during your time in school. It is a bit exaggerated, but it seems indeed the case according to the experiences of our own and many others. If you are students, you may be able to understand what this means after you graduate. The time and mental capacity to deeply understand certain principles might become more limited, and your stamina might not be as robust as it is now.

Are these fundamentals enough? Unfortunately the answer is no. There is another crucial skill you should possess. Consider the ultimate outcome of machine learning: "algorithms". In essence, these algorithms are a set of instructions, often requiring extensive computations. Manual computation is nearly impossible, necessitating the use of a computer. This is where another vital skill comes into play. That is, the proficiency in *programming tools* like Python and TensorFlow, which we have utilized throughout this book. We highly recommend developing strength in programming as well. Swift implementations through excellent programming skills will empower you to bring your ideas to life and advance them. A caveat is that programming tools evolve over time. This evolution is driven by advancements in computation resources and power. Therefore, it is essential to stay informed and adapt to these changes.

Problem Set 10

Problem 10.1 (*Speech recognition is an inference problem*) Consider the speech recognition system discussed in Sect. 3.11.

(*a*) Draw the speech recognition system. Also explain what are the components of the input and the output in the system.

(*b*) State the definition of an *inference problem*.

(*c*) Show that speech recognition is an inference problem.

Problem 10.2 (*The Markov process*)

(*a*) State the definition of a *Markov process*.

(*b*) In Sect. 3.12, we learned that the sequence of phonemes $\{X_i\}_{i=1}^n$ in a text can be modeled using the concept of a *Markov process*. Elaborate on why this is the case.

Problem 10.3 (*A graphical model: Exercise #1*) Describe the statistical structure of the joint distribution associated with a graphical model in Figure 3.59.

Problem 10.4 (*A graphical model: Exercise #2*) Consider $(X_1, \ldots, X_7)$ with the following joint distribution:

$$\mathbb{P}(X_1, \ldots, , X_7)$$
$$= \mathbb{P}(X_1)\mathbb{P}(X_2|X_1)\mathbb{P}(X_3|X_1)\mathbb{P}(X_4|X_2)\mathbb{P}(X_5|X_2)\mathbb{P}(X_6|X_3)\mathbb{P}(X_7|X_3).$$

Draw the associated graphical model.

Problem 10.5 (*A Hidden Markov Model*)

(*a*) State the definition of a *Hidden Markov Model*.

(*b*) In Sect. 3.12, we learned that the sequence of features $\{Y_i\}_{i=1}^n$ in the speech recognition system can be effectively modeled using a *Hidden Markov Model*. Elaborate on the reasoning behind this modeling choice.

Fig. 3.59 Illustration of a graphical model

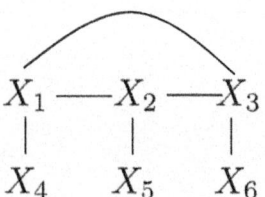

Problem 10.6 (*The MAP/ML principles*) Consider $Y_i = X + Z_i$ where X is a continuous random variable with the pdf $f_X(x)$ and $x \in \mathbb{R}$. Here, Z_i's are i.i.d. with $\sim \mathcal{N}(0, \sigma^2)$ for $i \in \{1, \ldots, n\}$, and they are independent of X. Given the observations $(Y_1, Y_2, \ldots, Y_n)$, our goal is to estimate $\hat{X}$.

(*a*) Show that the optimal estimator is:

$$\hat{X}_{\text{opt}} = \arg\max_{x \in \mathcal{X}} f_X(x) f_{Y_1, \ldots, Y_n}(y_1, \ldots, y_n | X = x). \tag{3.181}$$

(*b*) Suppose that $X \sim \mathcal{N}(0, \sigma_x^2)$. Derive the MAP estimator $\hat{X}_{\text{MAP}}$.

(*c*) Derive the ML estimator $\hat{X}_{\text{ML}}$.

Problem 10.7 (*The Viterbi algorithm*) In Sect. 3.12, we derived the optimal MAP estimator for speech recognition:

$$\hat{\mathbf{x}}_{\text{MAP}} = \arg\min_{x_1, \ldots, x_n \in \mathcal{X}} \sum_{i=1}^{n} \log\left\{ \frac{1}{\mathbb{P}(x_i | x_{i-1}) f(y_i | x_i)} \right\} \tag{3.182}$$

where $\{x_i\}_{i=1}^{n}$ indicates the sequence of phonemes and $\{y_i\}_{i=1}^{n}$ denotes the sequence of corresponding features. Let

$$s_i := \begin{bmatrix} x_{i-1} \\ x_i \end{bmatrix}, \quad c_i(s_i) := \log\left\{ \frac{1}{\mathbb{P}(x_i | x_{i-1}) f(y_i | x_i)} \right\}.$$

Assume that $\mathcal{X} = \{a, b\}$, $x_0 = x_{-1} = a$ and $n = 5$.

(*a*) Show that the MAP estimator (3.182) is equivalent to:

$$\hat{s}_{\text{MAP}} = \arg\min_{s_1, \ldots, s_n} \sum_{i=1}^{n} c_i(s_i). \tag{3.183}$$

(*b*) Identify the Finite State Machine by enumerating all possible states. Additionally, draw the state transition diagram and trellis diagram.

(*c*) Suppose we are given the costs:

$$\begin{bmatrix} c_1([a; a]) & c_2([a; a]) & c_3([a; a]) & c_4([a; a]) & c_5([a; a]) \\ c_1([b; a]) & c_2([b; a]) & c_3([b; a]) & c_4([b; a]) & c_5([b; a]) \\ c_1([b; b]) & c_2([b; b]) & c_3([b; b]) & c_4([b; b]) & c_5([b; b]) \\ c_1([a; b]) & c_2([a; b]) & c_3([a; b]) & c_4([a; b]) & c_5([a; b]) \end{bmatrix}$$

$$= \begin{bmatrix} 0.64 & 2.56 & 0.01 & 2.25 & 2.89 \\ 1.44 & 0.16 & 3.61 & 0.25 & 0.09 \\ 10.24 & 5.76 & 15.21 & 6.25 & 5.29 \\ 1.44 & 0.16 & 3.61 & 0.25 & 0.09 \end{bmatrix}.$$

Run the Viterbi algorithm by hand or computer to draw the trellis diagram having cost on each edge. Also find the shortest path at each stage.

(d) Using part (c), derive the MAP solution $\hat{s}_{MAP}$ and the corresponding sequence of phonemes $\{x_i\}_{i=1}^n$.

Problem 10.8 (*The Viterbi algorithm in communication*) Consider a particular yet practically-relevant communication channel, known as *inter-symbol interference (ISI) channel*:

$$Y_i = X_i + X_{i-1} + Z_i, \quad i \in \{1, \ldots, 5\} \tag{3.184}$$

where Z_i's are i.i.d. $\sim \mathcal{N}(0, \sigma^2)$. At each time, we send $X_i = +1$ or $X_i = -1$ equally likely. Assume that X_i's are i.i.d, being independent of Z_i's. Suppose that the received signals are given as:

$$(y_1, y_2, y_3, y_4, y_5) = (1.2, 0.4, -1.9, -0.5, 0.3). \tag{3.185}$$

Define the state s_i as:

$$s_i := \begin{bmatrix} x_{i-1} \\ x_i \end{bmatrix}. \tag{3.186}$$

Say that $[+1; +1]$ is $s1$; $[-1; +1]$ is $s2$; $[-1; -1]$ is $s3$; and $[+1; -1]$ is $s4$. Assume that $x_0 = x_{-1} = +1$, so the initial state s_0 is $s1$.

(a) Show that given $(y_1, y_2, \ldots, y_5)$, the MAP estimate for $(s_1, s_2, \ldots, s_5)$ is:

$$s_{MAP}^* = \arg \min_{s_1, \ldots, s_5} \sum_{i=1}^{5} (y_i - [1 \ 1]s_i)^2. \tag{3.187}$$

(b) Compute

$$c_i(s_i) := (y_i - [1 \ 1]s_i)^2 \tag{3.188}$$

for $s_i \in \{[+1; +1], [-1; +1]; [-1; -1]; [+1; -1]\}$ and $i \in \{1, \ldots, 5\}$.

(c) Run the Viterbi algorithm to draw the trellis diagram and then find the shortest path at each stage.

(d) Using part (c), derive the MAP solution s_{MAP}^* and the corresponding sequence of transmitted signals x_i's.

Problem 10.9 (*True or False?*)

(*a*) Let $\{W\}_{i=1}^{n}$ be the Bernoulli process with parameter p. Suppose:

$$X_i = (2W_i - 1)X_{i-1}, \quad i \in \{1, 2, \ldots, n\} \qquad (3.189)$$

where $X_0 = +1$. Then, $\{X\}_{i=1}^{n}$ is a Markov process.

(*b*) Suppose that $X_1 \sim \text{Bern}(p)$, $X_2 \sim \text{Bern}(\frac{1}{2})$ and $S = X_1 \oplus X_2$. Then, S follows $\text{Bern}(\frac{1}{2})$ no matter what p is.

Appendix A
Python Basics

Python is friendly and easy to learn.

A.1 Jupyter Notebook

Outline
Python requires another software platform for its execution, and that platform is Jupyter Notebook. This section provides an introduction to basic stuffs regarding Jupyter Notebook. The section consists of four parts. First, we will explain the role of Jupyter Notebook in light of Python. Subsequently, we will delve into the installation process and the steps to launch a file for scripting code. Following that, we will explore various user-friendly interfaces that facilitate the scripting of Python code. Lastly, we will acquaint ourselves with several frequently used shortcuts for writing and executing code.

What is Jupyter notebook? Jupyter Notebook serves[1] as a powerful tool for writing and executing Python code. Its primary advantage lies in the ability to execute individual lines of code rather than the entire script. This feature proves invaluable for easy debugging, particularly when dealing with lengthy code segments.

---

[1] This tutorial is adapted from Appendix A in (Suh, 2022, 2023a, b).

© The Editor(s) (if applicable) and The Author(s), under exclusive license to Springer Nature Singapore Pte Ltd. 2025
C. Suh, *Probability for Information Technology*,
https://doi.org/10.1007/978-981-97-4032-1

```
a=1
b=2
a+b
```

```
a=1
```

```
b=2
```

```
a+b
```
3

There are two common approaches to using Jupyter Notebook. The first involves running the code on a server (or cloud) machine, while the second method utilizes a local machine. Here, we will focus on the latter.

Installation & launch Utilizing a local machine necessitates the installation of a software tool called Anaconda. You can download and install the latest version by visiting the following: https://www.anaconda.com/products/individual.

You can choose the appropriate Anaconda installer based on your machine's operating system. Refer to the three versions displayed in Figure A.1. During the installation process, you might encounter certain errors. One common issue is related to "non-ASCII characters". To address this, ensure that the destination folder path for Anaconda does not include any non-ASCII characters, such as Korean characters. Another frequent error message pertains to permission issues for accessing the specified path. To avoid this, run the Anaconda installer in "run as administrator" mode. You can access this mode by right-clicking on the downloaded executable file.

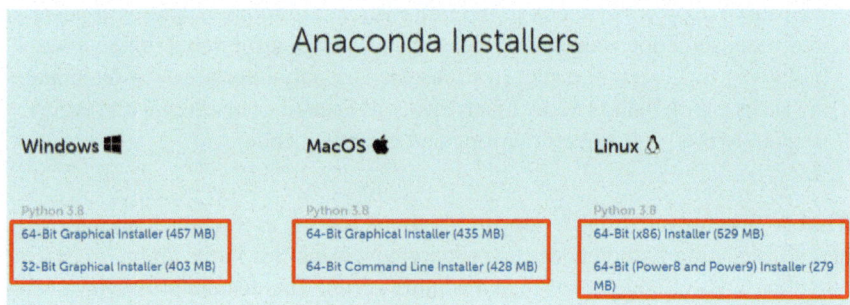

Fig. A.1 Three versions of Anaconda installers

Fig. A.2 How to launch **Jupyter notebook** in the **Anaconda** prompt

		Name ↓	Last Modified	File size
□ □ 3D Objects			4 months ago	
□ □ Contacts			4 months ago	
□ □ Desktop			a month ago	
□ □ Documents			3 hours ago	
□ □ Downloads			4 months ago	
□ □ Dropbox			a day ago	
□ □ Favorites			3 months ago	
□ □ Links			4 months ago	
□ □ Music			4 months ago	
□ □ OneDrive			3 hours ago	
□ □ Saved Games			4 months ago	
□ □ Searches			4 months ago	
□ □ Videos			4 months ago	
□ □ gsview64.ini			4 months ago	43 B
□ □ Sti_Trace.log			12 days ago	0 B

Fig. A.3 Web browser of a successfully launched **Jupyter notebook**

To launch **Jupyter Notebook**, you can utilize the **Anaconda** prompt (for Windows) or the terminal (for macOS and Linux). The process is straightforward-simply type `jupyter notebook` in the prompt and press `Enter`. Subsequently, a **Jupyter Notebook** window will appear. If the window does not open automatically, you can alternatively copy and paste the URL (indicated by the arrow in Figure A.2) into your web browser to open it manually. If it works properly, you should be able to see the window as in Figure A.3.

Creating a new notebook file is also simple. First, navigate to the folder where you intend to save the notebook file. Next, click on the `New` tab located at the top right (highlighted in a blue box), and then select the `Python 3` tab (highlighted in a red box). Refer to Figure A.4 for the location of these tabs.

Interface Jupyter Notebook comprises two essential components necessary for code execution. The first is a computational engine called the `Kernel`, which can be

Fig. A.4 How to create a **Jupyter notebook** file on the web browser

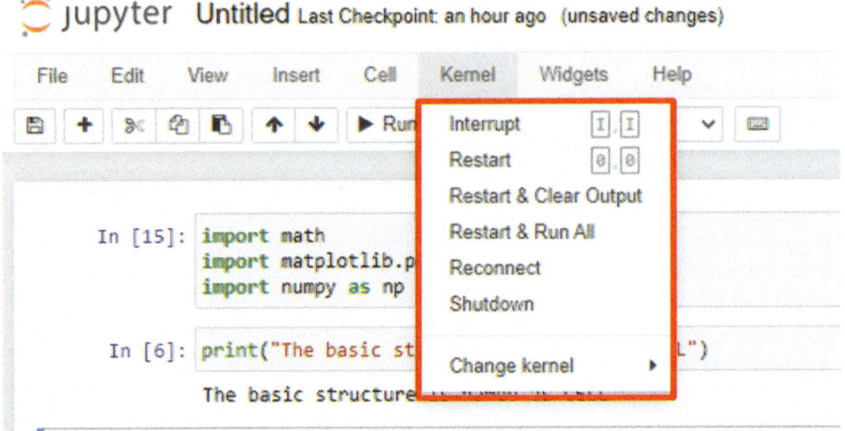

Fig. A.5 `Kernel` is a computational engine which serves to run the code. There are several relevant functions under the `Kernel` tap

controlled through various functions in the `Kernel` tab. Refer to Figure A.5 for details.

The second component is a unit referred to as a `cell`, where you can compose a script. The cell operates in two modes. The first is the edit mode, which allows you to type either: (i) a code script for running a program, or (ii) any text, similar to a normal text editor. Code scripts can be written under the `Code` tab (highlighted in a red box) as illustrated in Figure A.6. Text editing can be performed under the `Markdown` tab, marked in a blue box. The other mode is the command mode. In this mode, we can edit the entire notebook. For instance, tasks such as copying or deleting cells, as well as rearranging cells, can be performed.

Shortcuts There are numerous shortcuts that are highly useful for editing and navigating a notebook. We highlight three types of frequently used shortcuts. Firstly, there's a set of shortcuts for transitioning between the edit and command modes. Typing `Esc` switches from edit to command mode, while `Enter` performs the reverse. Secondly, for inserting or deleting a cell, the shortcut `a` inserts a new cell *above* the current cell, `b` does the same but inserts it *below* the current cell, and `d+d` deletes a cell. These shortcuts should be executed under the command mode to serve their intended purposes. Lastly, a set of shortcuts is dedicated to executing a cell. Arrow keys facilitate movement between different cells. Shift + Enter runs the current cell and moves to the next one. To remain in the current cell after execution, Ctrl + Enter is used instead.

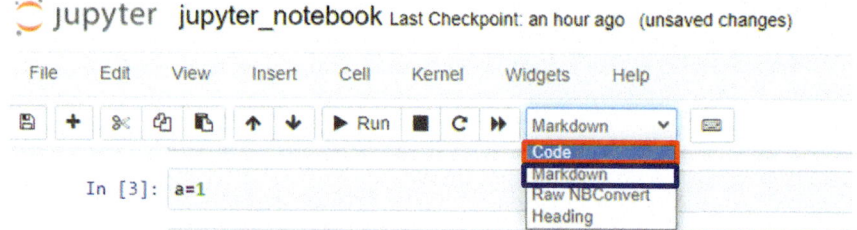

Fig. A.6 How to choose the Code or Markdown option in the edit mode

A.2   Basic Syntaxes of Python

> **Outline**
> In this section, we will learn some basic Python syntaxes required to write a script that incorporates probabilistic concepts, principles, and algorithmic implementation. In particular, three concepts are emphasized: (i) class; (ii) package; and (iii) function. Additionally, we will acquaint you with a set of Python packages that are both valuable and pertinent to the content covered in this book.

A.2.1   Data Structure

There are two prominent data-structure components in Python: (i) list; and (ii) set.

(i) List List is a built-in data type which allows us to store multiple elements in a single variable. The elements are listed with an order and the list allows for duplication. Please see below for examples of the use.

```python
x = [1, 2, 3, 4]    # construct a simple list
print(x)
```
```
[1, 2, 3, 4]
```

```python
x.append(5)    # add an item at the end
print(x)
```
```
[1, 2, 3, 4, 5]
```

```python
x.pop()    # delete an item located in the last
print(x)
```
```
[1, 2, 3, 4]
```

```python
# checking if a particular element exists in the list
if 3 in x:
    print(True)
if 5 in x:
    print(True)
else:
    print(False)
```
```
True
False
```

```
# A single-line construction of a list
y = [x for x in range(1,10)]
print(y)
```

```
[1, 2, 3, 4, 5, 6, 7, 8, 9]
```

```
# Retrieving all the elements through a "for" loop
for i in x:
    print(i)
```

```
1
2
3
4
```

(ii) Set Set is another built-in data type which plays a similar role as List. Two key distinctions are: (i) it is unordered; and (ii) it does not allow for duplication. See below for some examples of how to use.

```
x = set({1, 2, 3})    # construct a set
print(f"x: {x}, type of x: {type(x)}")
```

```
x: {1, 2, 3}, type of x: <class 'set'>
```

The f in front of strings in the print command tells Python to look at the values inside {·}.

```
x.add(1)    # add an existing item
print(x)
```

```
{1, 2, 3}
```

```
x.add(4)    # add a new item
print(x)
```

```
{1, 2, 3, 4}
```

```
x.remove(2)  # remove an item "2"
print(x)
```

```
{1, 3, 4}
```

```
# checking if a particular element exists in the list
if 1 in x:
    print(True)
if 5 in x:
    print(True)
else:
    print(False)
```

```
True
False
```

```
# Retrieving all the elements through a "for" loop
for i in x:
    print(i)
```

```
1
2
3
4
```

A.2.2 Package

We will introduce five packages which are particularly instrumental in scripting codes for the problems that appear in this book: (i) math; (ii) random; (iii) itertools; (iv) numpy; and (v) scipy.

(i) math This module provides a collection of useful math expressions such as exponent, log, square root and power. See some relevant examples below.

```python
import math
```

```python
math.exp(1)    # exp(x)
```

```
2.718281828459045
```

```python
print(math.log(1, 10))    # log(x, base)
print(math.log(math.exp(20)))   # natural logarithm
print(math.log2(4))        # base-2 logarithm
print(math.log10(1000))     # base-10 lograithm
```

```
0.0
20.0
2.0
3.0
```

```python
print(math.sqrt(16))    # square root
print(math.pow(2,4))  # x raised to y (same as x**y)
print(2**4)
```

```
4.0
16.0
16
```

```python
print(math.cos(math.pi)) # cosine of x radians
print(math.dist([1,2],[3,4]))     # Euclidean distance
```

```
-1.0
2.8284271247461903
```

```
# The erf() function can be used to compute traditional
# statistical functions such as the CDF of
# the standard Gaussian distribution
def phi(x):
    # CDF of the standard Gaussian distribution
    return (1.0 + math.erf(x/math.sqrt(2.0)))/2.0

phi(1)
```

0.8413447460685428

(ii) random This module yields random number generation. See below for some examples.

```
import random
```

```
random.randrange(start=1, stop=10, step=1)
# a random number in range(start, stop, step)
random.randrange(10) # integer from 0 to 9 inclusive
```

5

```
# returns random integer n such that a<=n<=b
random.randint(1, 10)
```

7

(ii) itertools This package offers a concise method for exploring all possible cases in various combinatorics-related scenarios.

```
from itertools import permutations, combinations
```

```
# generating all permutations of [1, 2, 3]
p = permutations([1, 2, 3])

for i in p:
    print(i)
```

```
(1, 2, 3)
(1, 3, 2)
(2, 1, 3)
(2, 3, 1)
(3, 1, 2)
(3, 2, 1)
```

```
# generating all length-2 combinations of [1, 2, 3]
c = combinations([1, 2, 3], 2)
```

```
for i in c:
    print (i)
```

```
(1, 2)
(1, 3)
(2, 3)
```

```
# generating all length-3 combinations of [1, 2, 3, 4, 5]
c = combinations([1, 2, 3, 4, 5], 3)

for i in c:
    print (i)
```

```
(1, 2, 3)
(1, 2, 4)
(1, 2, 5)
(1, 3, 4)
(1, 3, 5)
(1, 4, 5)
(2, 3, 4)
(2, 3, 5)
(2, 4, 5)
(3, 4, 5)
```

(iv) numpy NumPy stands out as the most widely used package for managing matrices and vectors, providing a plethora of useful functions. Here, we enumerate several frequently employed functions.

(a) numpy.array() numpy.array() is a specialized array data structure in numpy. This differs from Python data type array().

```
import numpy as np
```

```
np.array([1, 2, 3])    # construct an array
```

```
array([1, 2, 3])
```

```
np.array([[1, 2], [3, 4]])    # construct a 2D array
```

```
array([[1, 2],
       [3, 4]])
```

```
x = np.ones((2,2))
# construct an all-one matrix with size of 2-by-2
x = np.zeros((2,2))
# construct an all-zero matrix with size of 2-by-2
print(np.ones_like(x))
# all-one matrix with the same shape and type of input
print(np.zeros_like(x))
# all-zero matrix with the same shape and type of input
```

```
[[1. 1.]
 [1. 1.]]
[[0. 0.]
 [0. 0.]]
```

```
# range of x
x_grid=np.arange(0,1,0.0001)
# or one can use:
x_grid2=np.linspace(0,1,0.0001)
```

```
# concatenation of two numpy arrays
x1 = np.array([1,2])
x2 = np.array([3,4])
xc = np.concatenate((x1,x2)) # column-wise
xr = np.vstack((x1,x2)) # row-wise
print(xc)
print(xr)
```

```
[1 2 3 4]
[[1 2]
 [3 4]]
```

```
# sign function
x = np.array([1.2,-3,2,-4.2])
s = np.sign(x)
print(s)
```

```
[ 1. -1.  1. -1.]
```

(b) numpy.random() This module is designed for conducting random sampling from diverse probability distributions. Below, we outline a few widely recognized examples. For further insights, you may wish to refer to: https://numpy.org/doc/1.16/reference/routines.random.html

```
# sampling a number from standard Gaussian distribution
np.random.normal(loc = 0, scale = 1)
# loc: mean, scale: standard deviation
np.random.randn() # plays the same role
```

```
-2.5459976698222495
```

```
# sampling multiple numbers as per the standard Gaussian
np.random.normal(0, 1, size = (2, 2))
# Here the size determines the output shape
np.random.randn(2,2) # plays the same role
```

```
array([[-1.8133258 , -1.01151295],
       [-0.37375747,  0.36005748]])
```

```
np.random.rand(2,2) # Uniform over [0,1]
```

```
array([[0.06535694, 0.2507505 ],
       [0.17559137, 0.60967901]])
```

```
# Uniform over [0.8,1]
np.random.uniform(0.8,1,(2,2))
```

```
array([[0.89902277, 0.85310313],
       [0.96578371, 0.85695091]])
```

```
# integer uniformly chosen from {0,1,2} 10 times independently
np.random.randint(3,size=10)
```

```
array([1, 2, 0, 2, 1, 1, 0, 1, 2, 0])
```

```
# Binomial distribution
np.random.binomial(10000,0.5) # 10000 trials of Bern(0.5)
```

```
5042
```

(c) numpy.linalg This package provides numerous beneficial functions related to linear algebra. Here are a select few.

```
from numpy import linalg
```

```
x = np.random.randn(2,2)
print(linalg.det(x))       # Determinant of a matrix x
print(linalg.inv(x))       # Inverse of a matrix x
print(linalg.norm(x))      # Matrix or vector norm
print(linalg.svd(x))       # Singular value decomposition
print(linalg.eig(x))       # Eigenvalue decomposition
```

```
0.7125655927348966
[[ 0.77007826 -0.38835738]
 [ 2.33455331  0.64504946]]
1.832010151997132
(array([[-0.2060815,   0.97853483],
        [ 0.97853483,  0.2060815 ]]),
array([1.78814528, 0.39849424]),
array([[-0.96330981,  0.2683919 ],
       [ 0.2683919 ,  0.96330981]]))
(array([0.50418566+0.67702467j, 0.50418566-0.67702467j]),
array([[0.02479485-0.37684352j, 0.02479485+0.37684352j],
       [0.92594502+0.j         , 0.92594502-0.j          ]]))
```

(d) numpy.fft In communication, speech recognition, and signal processing, one of the essential operations is the Discrete Fourier Transform (DFT). Consider time-domain discrete signals x_i where $i \in \{0, 1, \ldots, N - 1\}$. The corresponding frequency-domain signals are expressed as follows:

$$X[k] = \frac{1}{\sqrt{N}} \sum_{m=0}^{N-1} x[m]e^{-j\frac{2\pi}{N}mk} \qquad k \in \{0, 1, \ldots, N-1\}.$$

This can be easily implemented using the built-in function $\mathtt{fft}$ located in $\mathtt{numpy.fft}$.

$$\mathtt{fft}(x) = \sum_{m=0}^{N-1} x[m]e^{-j\frac{2\pi}{N}mk}.$$

To align with the defined Discrete Fourier Transform (DFT), it's necessary to divide the result obtained from $\mathtt{fft}$ by $\sqrt{N}$. Likewise, the inverse function $\mathtt{ifft}$ performs the reverse operation.

$$\mathtt{ifft}(X) = \frac{1}{N} \sum_{k=0}^{N-1} X[k]e^{j\frac{2\pi}{N}mk}.$$

```
from numpy.fft import fft
from numpy.fft import ifft

x_time = np.random.randn(8)
X_freq = fft(x_time)/np.sqrt(8)
x_time_rec = np.sqrt(8)*ifft(X_freq)
print(x_time)
print(x_time_rec)
```

```
[ 0.19398987  0.92755053  1.14652418  1.05737049 -0.66500356
  0.43650243 1.04576987 -0.95167376]
[ 0.19398987+0.j 0.92755053+0.j 1.14652418+0.j 1.05737049+0.j
 -0.66500356+0.j 0.43650243+0.j 1.04576987+0.j -0.95167376+0.j]
```

(e) resizing Resizing is frequently employed to transform the dimensions from one scale to another.

```
x = np.random.randn(4,4,1)
y = x.view(dtype=np.float_).reshape(-1,2)
# '-1' can be inferred from the context: Shape of (8,2)
print(y)
z = x.squeeze()
print(z.shape)
```

```
[[-0.85719316  2.99692221]
 [ 1.16327996 -0.11955541]
 [-0.76229609  0.79871494]
 [ 0.99757568  0.69329723]
 [-1.52198295 -0.74430996]
 [ 0.17174063  0.25343301]
 [ 0.07151011 -2.90945412]]
```

```
[ 1.1874155  -0.64209109]]
(4, 4)
```

(v) scipy This module offers an extensive array of probability distributions and associated statistics. Here, we highlight several of them. For further details, please consult: https://docs.scipy.org/doc/scipy/reference/stats.html

```python
from scipy import stats
```

```python
# A random variable with the standard Gaussian
X = stats.norm(loc = 0, scale = 1)
# loc:mean, scale:standard deviation
print(X.cdf(np.array([-1, 0, 1])))
# computes the CDF at each numpy array
print(X.rvs(size = 3))
# generating a sequence of random variables
```

```
[0.15865525 0.5         0.84134475]
[ 0.39460402 -0.8042592   -0.71404882]
```

```python
# Another random variable with the uniform distribution
Y = stats.uniform(loc = 0, scale = 1)
# uniform distribution in [loc, loc + scale]
print(Y.cdf(np.array([-1, 0, 0.5, 1])))
print(Y.rvs(size = 3))
```

```
[0.   0.   0.5 1. ]
[0.72953474 0.67879248 0.47947748]
```

For binary random variables, we employ a built-in function, `bernoulli` in `scipy.stats`.

```python
from scipy.stats import bernoulli
X = bernoulli(0.5)
X_samples = X.rvs(10)
print(X_samples)
```

```
[0 1 0 0 1 0 0 1 0 0]
```

To model a binomial distribution, we utilize the pre-existing `binom` function within the `scipy.stats` module.

```python
from scipy.stats import binom
n,p=10,0.1

i_values=list(range(n))
dist = [binom.pmf(i,n,p) for i in i_values]
print(dist)
```

```
[0.3486784401000001, 0.38742048899999965, 0.19371024450000007,
0.05739562799999998, 0.011160261000000001, 0.0014880347999999995,
0.00013778100000000007, 8.748000000000003e-06,
3.6449999999999996e-07, 8.999999999999995e-09]
```

The `scipy.stats` module also includes built-in functions for entropy and the Kullback-Leibler (KL) divergence.

```python
from scipy.stats import entropy

pX1 = np.array([1/2, 1/2]) # numpy.array
pX2 = [1/2, 1/2]   # list
print(entropy(pX1, base=2))
print(entropy(pX2, base=2))
```

```
1.0
1.0
```

Here, the input distribution can accept either a `numpy.array` or a `list`.

```python
from scipy.special import rel_entr

# Compute p(x,y)
pXY = np.array([1/4, 1/4, 1/3, 1/6])
# Compute p(x)p(y)
pXpY = np.array([pX[0]*pY[0],pX[0]*pY[1],
                 pX[1]*pY[0],pX[1]*pY[1]])
kl_builtin = rel_entr(pXY,pXpY)
print(sum(kl_builtin))
# To convert into log base 2
print(sum(kl_builtin)/np.log(2))
```

```
0.014362591564146779
0.020720839623908218
```

For computing the KL divergence, the `rel_entr` function in `scipy.special` can be utilized. Note that `rel_entr` employs the natural logarithm instead of log base 2. It returns a list of values representing $p(x) \log \frac{p(x)}{q(x)}$. Therefore, proper conversion may be necessary.

In communication problems, it is common to calculate the Q-function:

$$Q(a) := \int_a^\infty \frac{1}{\sqrt{2\pi}} e^{-\frac{z^2}{2}} dz.$$

This is useful for analyzing the probability of error in communication. The integration can be numerically computed using the `erfc` function defined in `scipy.special`.

$$\texttt{erfc}(x) := \int_x^\infty \frac{2}{\sqrt{\pi}} e^{-t^2} dt.$$

The relationship between $Q(a)$ and $\texttt{erfc}(x)$ is:

$$Q(a) := \int_a^\infty \frac{1}{\sqrt{2\pi}} e^{-\frac{z^2}{2}} dz$$

$$\overset{(a)}{=} \int_{\frac{a}{\sqrt{2}}}^\infty \frac{1}{\sqrt{\pi}} e^{-t^2} dt$$

$$= \frac{1}{2} \cdot \texttt{erfc}\left(\frac{a}{\sqrt{2}}\right)$$

where (a) comes from the change of variable $t := \frac{z}{\sqrt{2}}$ $(dz = \sqrt{2}dt)$.

```python
from scipy.special import erfc

a = 10
Qfunc = 1/2*erfc(a/np.sqrt(a/2))
print(Qfunc)
```

```
1.2698142947354283e-10
```

A.2.3 Visualization

The widely used library for drawing graphs is $\texttt{matplotlib.pyplot}$. Here is how to utilize it:

```python
import matplotlib.pyplot as plt
```

```python
x_value = [x for x in range(10)]
y_value = [y for y in range(10, 20)]

plt.figure(figsize=(4,4),dpi=150) # figure size and resolution
plt.plot(x_value, y_value, color='blue', label='line')
plt.xlabel('x')                    # labeling x-axis
plt.ylabel('y')                    # labeling y-axis
plt.title('sample curve')
plt.legend()
plt.show() # No need to use show() in jupyter notebook.
```

Fig. A.7 Plotting a simple
function via
`matplotlib.pyplot`

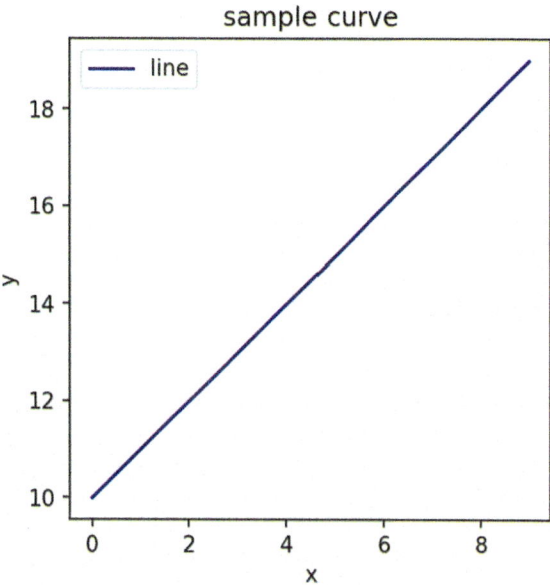

We can also plot multiple curves on a single graph.

```
# we can plot multiple curves at once

x = [x for x in range(10)]
y_1 = [3*y for y in range(10)]
y_2 = [2*y for y in range(10)]

plt.figure(figsize=(4,4), dpi=150)
plt.plot(x, y_1, color='blue', label='y=3x')    # plot_1
plt.plot(x, y_2, color='red', label='y=2x')     # plot_2
plt.xlabel('x')      # labeling x-axis
plt.ylabel('y')      # labeling y-axis
plt.title('two sample curves')
plt.legend()
plt.show()
```

Fig. A.8 Plotting multiple functions and adding a legend

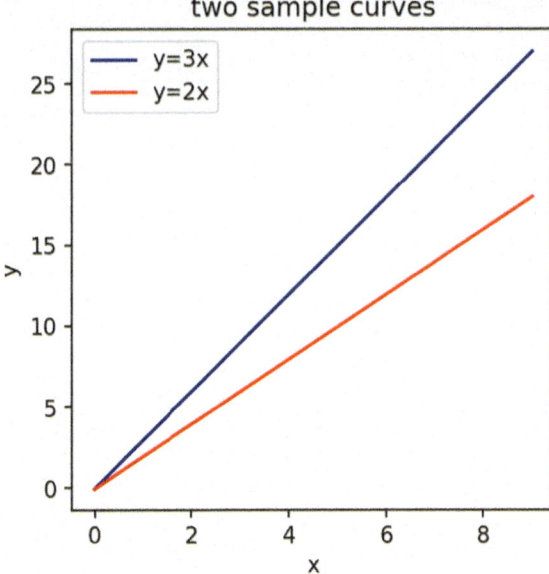

To visualize the probability distribution of a random variable, the `seaborn` statistical visualization package is commonly employed.

```python
import seaborn as sns
import matplotlib.pyplot as plt
from scipy.stats import bernoulli

X = bernoulli(0.5)
X_samples = X.rvs(1000)

plt.figure(figsize=(4,4), dpi=150)
sns.histplot(X_samples)
plt.xlabel('Values of a random variable')
plt.ylabel('Histogram')
plt.show()
```

Fig. A.9 Plotting a histogram of independent realizations of a random variable

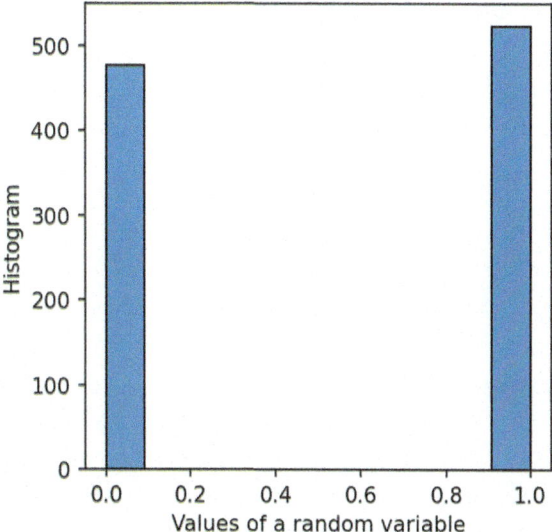

In communication problems, we often depict the probability of error, which is typically very small, e.g., 10^{-5}. To emphasize differences among these small probability values, we utilize a logarithmic scale for the error probability. To achieve this, we employ the function `plt.yscale('log')`.

```python
import numpy as np
from scipy.special import erfc
import matplotlib.pyplot as plt

SNRdB = np.arange(0,21,1)
SNR = 10**(SNRdB/10)

# Q-function
Qfunc = 1/2*erfc(np.sqrt(SNR/2))

plt.figure(figsize=(4,4), dpi=150)
plt.plot(SNRdB, Qfunc, label='Q(sqrt(SNR))')
plt.yscale('log')
plt.xlabel('SNR (dB)')
plt.grid(linestyle=':', linewidth=0.5)
plt.title('Q function')
plt.legend()
plt.show()
```

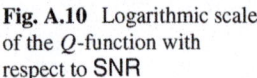

Fig. A.10 Logarithmic scale of the *Q*-function with respect to SNR

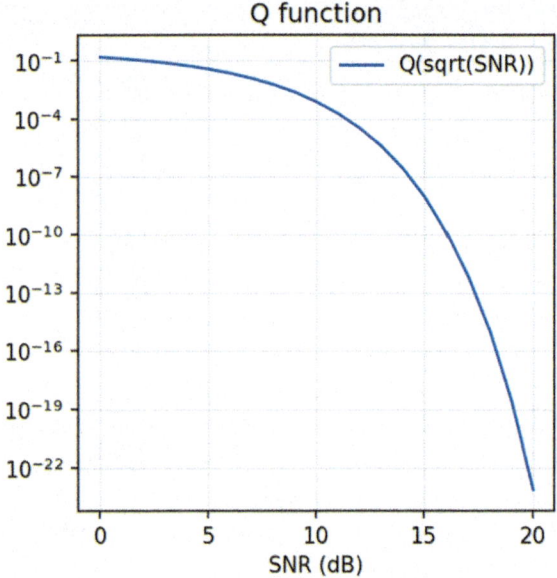

To adjust the font size of axis tick values, we employ the `matplotlib.rc` command. Here is an example code.

```python
# To adjust the font size of axis tick values
import matplotlib
import matplotlib.pyplot as plt
import numpy as np

x= np.linspace(0,1,100)
y=np.sqrt(x)
plt.figure(figsize=(10,5),dpi=200)
matplotlib.rc('xtick',labelsize=10)
matplotlib.rc('ytick',labelsize=10)
plt.subplot(1,2,1)
plt.plot(x,y)
plt.title('tick size = 10')
matplotlib.rc('xtick',labelsize=20)
matplotlib.rc('ytick',labelsize=20)
plt.subplot(1,2,2)
plt.plot(x,y)
plt.title('tick size = 20')
plt.show()
```

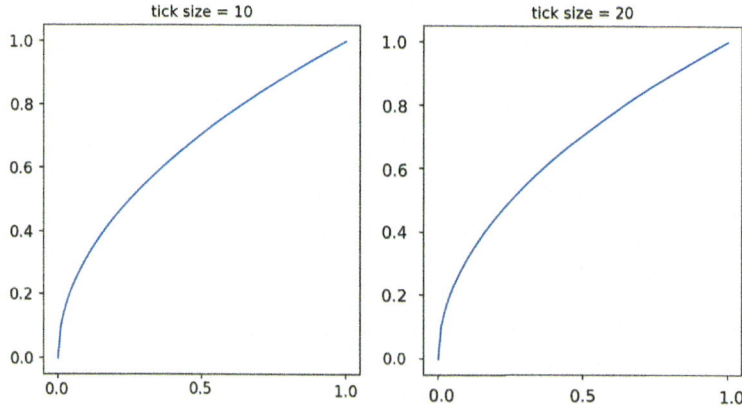

Fig. A.11 How to adjust the font size of axis tick values

Appendix B
TensorFlow and **Keras** Basics

Keras enables a seamless transition from conceptualization to implementation with minimal steps.

Outline

In Part III, we explored various IT applications, including machine learning and deep learning. Deep learning involves leveraging a deep neural network (DNN) as a fundamental model for predictive modules. Numerous software tools, such as machine learning frameworks and application programming interfaces (APIs), facilitate the implementation of deep learning. Examples include TensorFlow, Keras, PyTorch, DL4J, Caffe, and MXNet. Each framework has its own advantages and disadvantages, depending on specific design goals such as usability, training speed, functionality, and scalability in distributed training.

This book focuses on usability, placing emphasis on user-friendly experimentation. Therefore, we delve into the most high-level API, which is Keras. The Keras API streamlines the journey from concept to implementation with minimal steps. In this appendix, we cover four fundamental aspects of Keras. Keras is seamlessly integrated with TensorFlow, meaning that Keras is included in TensorFlow installation.

In the first part, we will guide you through the process of installing TensorFlow. Deep learning implementation typically involves three key steps: (i) data preparation and processing; (ii) building a neural network model; and (iii) training and testing the model. In the second part, we will explore an easy approach to handle data provided by Keras. Subsequently, we will demonstrate how to construct a neural network model using popular packages like keras.models and keras.layers. Finally, we will delve into the training and testing procedures. For clarity and ease of understanding, we will demonstrate the entire procedures via a simple example.

© The Editor(s) (if applicable) and The Author(s), under exclusive license to Springer
Nature Singapore Pte Ltd. 2025
C. Suh, *Probability for Information Technology*,
https://doi.org/10.1007/978-981-97-4032-1

Fig. B.1 Handwritten digit
classification

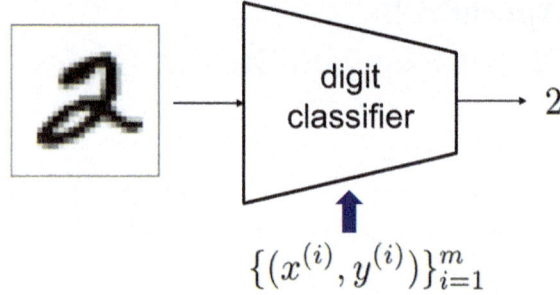

$$\{(x^{(i)}, y^{(i)})\}_{i=1}^{m}$$

Installation Utilizing Keras necessitates the installation of TensorFlow.[2] The installation process is straightforward:

```
pip install tensorflow
```

TensorFlow 2 packages seamlessly integrate with Keras. To ensure a proper installation, a `pip` version higher than 19.0 (or higher than 20.3 for macOS) is required. You can upgrade `pip` to the latest version using the command: `pip install -upgrade pip`. To verify a successful installation, try importing `keras` with the following command.

```
from tensorflow import keras
```

If there are no errors, you are ready to proceed. In case of any issues, refer to the installation guidelines at:

https : //www.tensorflow.org/install

A simple task We will focus on a simple task: handwritten digit classification. In this task, the objective is to identify the digit represented in a handwritten image. Refer to Figure B.1 for an example image. The figure showcases a scenario where an image of the digit 2 is correctly identified.

Preparing and processing data A widely used dataset for the digit classification task is the MNIST (Modified National Institute of Standards and Technology) dataset. This dataset comprises $m = 60,000$ training images and $m_{\text{test}} = 10,000$ testing images. Each image, say $x^{(i)}$, is a 28×28 pixel array, with each pixel representing a grayscale level ranging from 0 (white) to 1 (black). Additionally, each image is associated with a label, say $y^{(i)}$, belonging to one of the 10 classes $y^{(i)} \in \{0, 1, \ldots, 9\}$. Refer to Figure B.2 for a visual representation.

One advantage of using Keras is that popular datasets, including MNIST, are conveniently available in `keras.datasets` sub-package. Moreover, both training

---

[2] This tutorial is adapted from Appendix B in (Suh, 2022, 2023a,b).

Fig. B.2 The MNIST dataset: An input image is of 28-by-28 pixels, each indicating an intensity from 0 (white) to 1 (black); each label with size 1 takes one of the 10 classes from 0 to 9

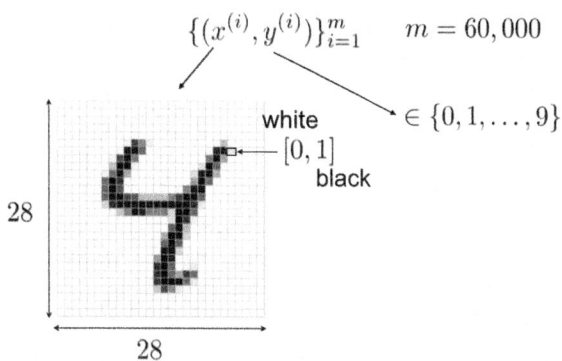

and testing datasets are preloaded with an appropriate split ratio, eliminating the need for manual splitting. The only script we need to write is:

```
from tensorflow.keras.datasets import mnist
(X_train, y_train), (X_test, y_test) = mnist.load_data()
X_train = X_train/255.
X_test = X_test/255.
```

```
Downloading data from https://storage.googleapis.com/
tensorflow/tf-keras-datasets/mnist.npz
11493376/11490434 [==========================] - 1s 0us/step
11501568/11490434 [==========================] - 1s 0us/step
```

We normalize the input (`X_train` or `X_test`) by dividing it by its maximum value of 255. This normalization step is part of data preprocessing. In situations where `keras.datasets` does not provide the dataset of interest, it is crucial to be familiar with data preprocessing techniques. This often involves the use of another prominent library, namely **pandas**, which is particularly adept at handling .csv files. While we will not delve into the details of using **pandas** here, for further information, you may refer to:

https : //pandas.pydata.org/

For data visualization, we use `matplotlib.pyplot`. Here is a simple code for plotting one sample image.

```
import matplotlib.pyplot as plt

plt.imshow(X_train[0], cmap = 'gray_r')
plt.colorbar()
plt.title('{}'.format(y_train[0], fontsize=30))
```

Refer to Figure B.3 for the output. The option `cmap = 'gray_r'` ensures a white background with a black letter, while `cmap = 'gray'` produces the inverted

Fig. B.3 A sample image in the MNIST dataset

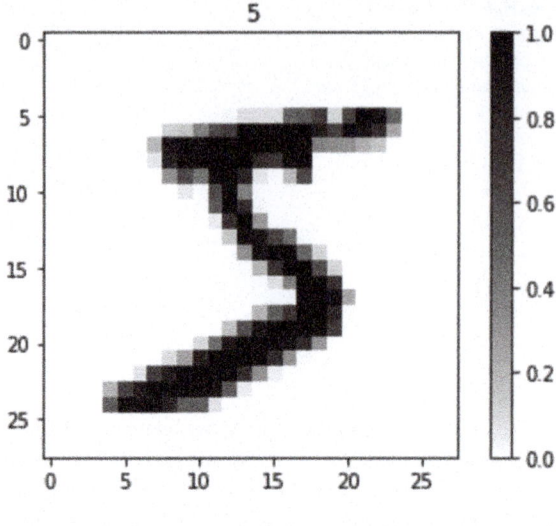

Fig. B.4 Plotting many image samples in a single figure

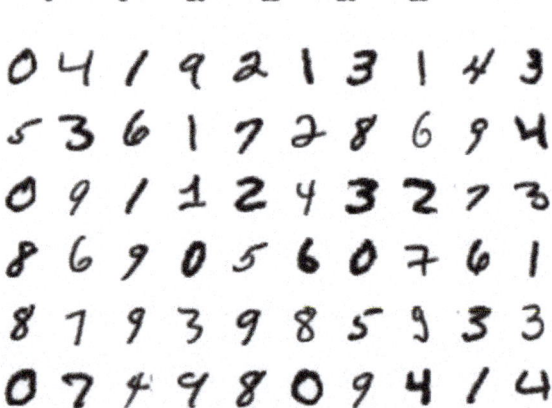

version with a white letter on a black background. The `colorbar()` function displays the color bar on the right, as shown in Figure B.3.

We can display multiple images in a single figure. Here is an example showcasing the visualization of 60 images.

```
num_of_images = 60
for index in range(1,num_of_images+1):
    plt.subplot(6,10, index)
    plt.axis('off')
    plt.imshow(X_train[index], cmap = 'gray_r')
```

See Figure B.4 for the output.

Building a neural network model We employ a two-layer neural network discussed in Sect. 3.10. We introduce a hidden layer with 500 neurons. See Figure B.5 for an illustration. We employ the ReLU at the hidden layer, and softmax at the output layer.

input size = 784 500 hidden neurons   10 classes

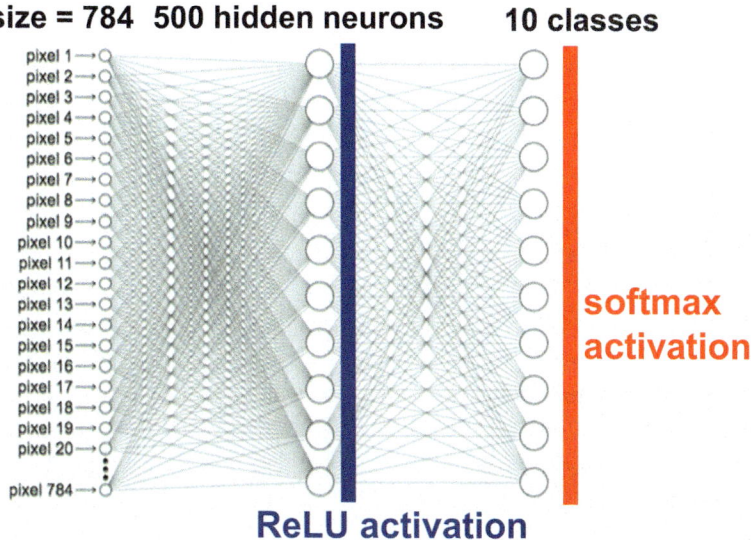

ReLU activation

Fig. B.5 A two-layer fully-connected neural network where input size is $28 \times 28 = 784$, the number of hidden neurons is 500 and the number of classes is 10. We employ ReLU activation at the hidden layer, and softmax activation at the output layer

Keras includes two major packages:

(i) tensorflow.keras.models;

(ii) tensorflow.keras.layers.

The models package encompasses various functionalities related to a neural network. A key module is Sequential, representing a neural network entity and described as a linear stack of layers. Within the layers package, numerous components of a neural network are available, including fully-connected dense layers and activation functions. Utilizing these elements, we can easily construct a model as illustrated in Figure B.5.

```python
from tensorflow.keras.models import Sequential
from tensorflow.keras.layers import Dense, Flatten

model = Sequential()
model.add(Flatten(input_shape=(28,28)))
model.add(Dense(500, activation='relu'))
model.add(Dense(10,  activation='softmax'))
model.summary()
```

```
Model: "sequential_1"
```

```
Layer (type)                    Output Shape                Param #
=================================================================
 flatten (Flatten)              (None, 784)                 0

 dense (Dense)                  (None, 500)                 392500

 dense_1 (Dense)                (None, 10)                  5010
=================================================================
Total params: 397,510
Trainable params: 397,510
Non-trainable params: 0
```

The `Flatten` layer transforms a higher-dimensional entity, such as a 2D matrix, into a flattened vector. In this instance, a digit image sized 28-by-28 is flattened into a vector of size $784 (= 28 \times 28)$. The `add()` method is utilized to append a particular layer to the end of the sequential model. The `Dense` layer signifies a fully-connected layer, with the input size automatically determined by the preceding layer it is attached to. The only parameter to specify is the number of output neurons. In this case, 500 denotes the number of hidden neurons. Additionally, an activation function can be set using another argument, for example, `activation='relu'`. The output layer is equipped with 10 neurons (matching the number of classes) and utilizes the softmax activation function to represent the probability of an output belonging to a specific class. The `summary()` method provides a comprehensive list of all the layers, specifying their sizes and the number of associated parameters.

Training a model Firstly, we need to establish an algorithm for use. A widely employed algorithm is gradient descent, and here, we ill utilize its enhanced version introduced in Sect. 3.10, namely, the Adam optimizer. The Adam optimizer can be regarded as an intelligent modification of gradient descent, offering more stable training. As mentioned earlier, Adam involves three crucial hyperparameters: (i) the learning rate α; (ii) β_1 (representing the weight of past gradients); and (iii) β_2 (indicating the weight of the square of past gradients). The default values are $(\alpha, \beta_1, \beta_2) = (0.001, 0.9, 0.999)$, and these are used if no specific values are provided.

Additionally, we must specify a loss function. As discussed in Sect. 3.9 (also in Problem 9.3 for the more-than-two class case), the optimal choice, in a sense of maximizing likelihood, is cross entropy. For evaluating performance during training and testing, a performance metric can be defined. Accuracy is a commonly used metric. All of these can be set using another method, `compile`.

```
model.compile(optimizer='adam',
              loss='sparse_categorical_crossentropy',
              metrics=['acc'])
```

The option `optimizer='adam'` selects the default values for the learning rate and β-parameters. For a manual selection, we define:

```
opt=tensorflow.keras.optimizers.Adam(
        learning_rate=0.01,
        beta_1 = 0.92,
        beta_2 = 0.992)
```

We then replace the previous option with `optimizer=opt`. Regarding the `loss` option in `compile`, we utilize `'sparse_categorical_crossentropy'`, indicating cross entropy loss for more than two classes.

Now, we can proceed to train the model on the MNIST data. During training, a portion of the entire examples is used to compute the gradient of the loss function. This subset is known as a *batch*. Two additional terminologies are relevant here. One is the *step*, denoting a loss computation procedure across examples within a single batch. The other is the *epoch*, encompassing the entire process involving all examples. In our experiment, we utilize a batch size of 64 and run for 20 epochs.

```
history=model.fit(X_train, y_train, batch_size=64, epochs=20)
```
```
Epoch 1/20
938/938 [===] - 2s 2ms/step - loss: 0.0025 - acc: 0.9992
Epoch 2/20
938/938 [===] - 2s 2ms/step - loss: 0.0059 - acc: 0.9981
Epoch 3/20
938/938 [===] - 2s 2ms/step - loss: 0.0031 - acc: 0.9990
Epoch 4/20
938/938 [===] - 2s 2ms/step - loss: 0.0074 - acc: 0.9976
Epoch 5/20
938/938 [===] - 2s 2ms/step - loss: 0.0025 - acc: 0.9993
Epoch 6/20
938/938 [===] - 2s 2ms/step - loss: 0.0043 - acc: 0.9984
Epoch 7/20
938/938 [===] - 2s 2ms/step - loss: 0.0044 - acc: 0.9984
Epoch 8/20
938/938 [===] - 2s 2ms/step - loss: 0.0010 - acc: 0.9998
Epoch 9/20
938/938 [===] - 2s 2ms/step - loss: 1.2813e-04 - acc: 1.0
Epoch 10/20
938/938 [===] - 2s 2ms/step - loss: 3.5169e-05 - acc: 1.0
Epoch 11/20
938/938 [===] - 2s 2ms/step - loss: 2.1899e-05 - acc: 1.0
Epoch 12/20
938/938 [===] - 2s 2ms/step - loss: 1.6756e-05 - acc: 1.0
Epoch 13/20
938/938 [===] - 2s 2ms/step - loss: 1.2778e-05 - acc: 1.0
Epoch 14/20
938/938 [===] - 2s 2ms/step - loss: 9.8947e-06 - acc: 1.0
Epoch 15/20
938/938 [===] - 2s 2ms/step - loss: 0.0082 - acc: 0.9981
Epoch 16/20
938/938 [===] - 2s 2ms/step - loss: 0.0090 - acc: 0.9971
Epoch 17/20
938/938 [===] - 2s 2ms/step - loss: 0.0016 - acc: 0.9995
Epoch 18/20
938/938 [===] - 2s 2ms/step - loss: 3.9583e-04 - acc: 0.9999
Epoch 19/20
938/938 [===] - 2s 2ms/step - loss: 7.6672e-05 - acc: 1.0
Epoch 20/20
938/938 [===] - 2s 2ms/step - loss: 2.4958e-05 - acc: 1.0
```

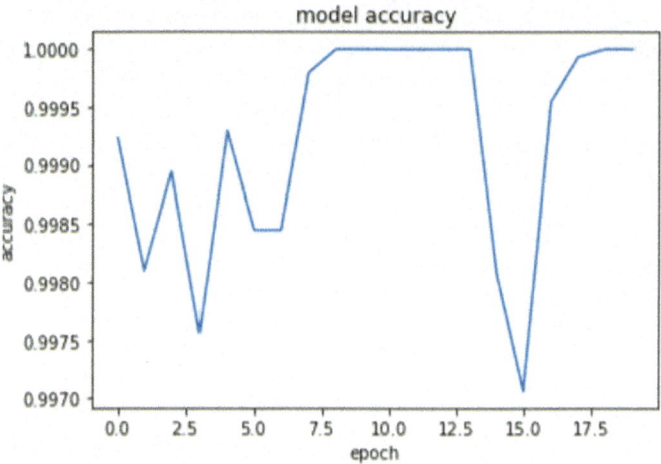

Fig. B.6 Accuracy as a function of epochs

An advantageous aspect of the `fit()` function is that it returns a dictionary containing the metrics collected during training. We can inspect these collected metrics by:

```
# list all data in history object
print(history.history.keys())
```

```
dict_keys(['loss', 'acc'])
```

With this data, we can create a plot depicting the accuracy curve as a function of epochs (Figure B.6).

```
plt.plot(history.history['acc'])
plt.title('model accuracy')
plt.xlabel('epoch')
plt.ylabel('accuracy')
```

Testing the trained model For testing, we first need to generate predictions from the model output. This can be accomplished using the `predict()` function, as demonstrated below:

```
model.predict(X_test).argmax(1)
```

```
array([7, 2, 1, …, 4, 5, 6], dtype=int64)
```

Here, `argmax(1)` identifies the class corresponding to the highest softmax output among the 10 classes. To assess the test accuracy, we employ the `evaluate()` function:

```
model.evaluate(X_test, y_test)
```

```
313/313 [===] - 0s 751us/step - loss: 0.1001 - acc: 0.9847
```

```
[0.10007859766483307, 0.9847000241279602]
```

Saving and loading Saving the trained model and later loading the saved model is a straightforward process. See below:

```
model.save('saved_classifier')
```

```
INFO:tensorflow:Assets written to: saved_classifier\assets
```

```python
import tensorflow
loaded_model = tensorflow.keras.models.load_model(
'saved_classifier')
```

References

Abbe, E.(2017). Community detection and stochastic block models: recent developments. *The Journal of Machine Learning Research 18*(1), 6446–6531.

Bansal, N., Blum, A. ., & Chawla, S. (2004). Correlation clustering. *Machine learning 56*(1), 89–113.

BAYES. (1958). An essay towards solving a problem in the doctrine of chances. *Biometrika 45*(3-4), 296–315.

Bernoulli, J. (1713). Ars conjectandi: Usum & applicationem praecedentis doctrinae in civilibus. *Moralibus & Oeconomicis 4*(1713), 1713.

Bertsekas, D., & Tsitsiklis, J. N. (2008). *Introduction to probability* (Vol. 1). Athena Scientific.

Blitzstein, J. K. , & Hwang, J. (2019). *Introduction to probability*. Crc Press.

Boyd, S. P. , & Vandenberghe, L. (2004). *Convex optimization*. Cambridge university press.

Browning, S. R., & Browning, B. L. (2011). Haplotype phasing: existing methods and new developments. *Nature Reviews Genetics 12*(10), 703–714.

Chartrand, G. (1977). *Introductory graph theory*. Courier Corporation.

Chen, J. , & Yuan, B. (2006). Detecting functional modules in the yeast protein–protein interaction network. *Bioinformatics 22*(18), 2283–2290.

Chen, Y., Kamath, G., Suh, C., & Tse, D. (2016). Sdhap: haplotype assembly for diploids and polyploids via semi-definite programming. *BMC genomics 16*(1), 1–16.

Cover, T., & Joy, A. T.(2006). *Elements of information theory*. Wiley-Interscience.

Cover, T. M. (1999). Elements of information theory. John Wiley & Sons.

Das, S., & Vikalo, H. (2015). Sdhap: haplotype assembly for diploids and polyploids via semi-definite programming. *BMC genomics*, 16 (1), 1–16.

DasGupta, A. (2011). *Probability for statistics and machine learning: fundamentals and advanced topics*. Springer.

Erdős, P., Rényi, A., et al. 1960. On the evolution of random graphs. *Publ. Math. Inst. Hung. Acad. Sci 5*(1), 17–60.

Feller, W. (1991). *An introduction to probability theory and its applications, volume 2* (Vol. 81). John Wiley & Sons.

Flajolet, P., Gardy, D., & Thimonier, L. 1992. Birthday paradox, coupon collectors, caching algorithms and self-organizing search. *Discrete Applied Mathematics 39*(3), 207–229.

Forney, G. D. (1973). The viterbi algorithm. *Proceedings of the IEEE 61*(3), 268–278.

Fortunato, S. (2010). Community detection in graphs. *Physics reports 486*(3-5), 75–174.

Freedman, D., Pisani, R., & Purves, R. (2007). *Statistics*. W.W. Norton & Co.

© The Editor(s) (if applicable) and The Author(s), under exclusive license to Springer Nature Singapore Pte Ltd. 2025
C. Suh, *Probability for Information Technology*,
https://doi.org/10.1007/978-981-97-4032-1

Gagniuc, P. A. (2017). *Markov chains: from theory to implementation and experimentation*. John Wiley & Sons.

Gallager, R. G. (2013). *Stochastic processes: theory for applications*. Cambridge University Press.

Garnier, J.-G. , & Quetelet, A. (1838). *Correspondance mathématique et physique* (Vol. 10). Impr. d'H. Vandekerckhove.

Gauß, C. F. (1809). *Theoria motus corporvm coelestivm in sectionibvs conicis solem ambientivm*. Perthes et Besser.

Girvan, M., & Newman, M. E. (2002). Proceedings of the fourteenth international conference on artificial intelligence and statistics (pp. 315–323).

Glorot, X., Bordes, A., & Bengio, Y. (2011). Deep sparse rectifier neural networks. In *Proceedings of the fourteenth international conference on artificial intelligence and statistics* (pp. 315–323).

Golub, G. H., & Van Loan, C. F. (2013). *Matrix computations*. JHU press.

Hinton, G., Srivastava, N., & Swersky, K. (2012). Neural networks for machine learning lecture 6a overview of mini-batch gradient descent. *Cited on 14*(8), 2.

Ivakhnenko, A. G. (1971). Clustering partially observed graphs via convex optimization. In *Icml*.

Jalali, A., Chen, Y., Sanghavi, S., & Xu, H. (2011). Clustering partially observed graphs via convex optimization. In *Icml*.

Jaynes, E. T. (2003). *Probability theory: The logic of science*. Cambridge university press.

Johnson, J. B. (1928). Thermal agitation of electricity in conductors. *Physical review 32*(1), 97.

Kay, S. (2006). *Intuitive probability and random processes using matlab®*. Springer Science & Business Media.

Kingma, D. P., & Ba, J. (2014). Adam: A method for stochastic optimization. arXiv preprint arXiv:1412.6980 .

Koller, D., & Friedman, N. (2009). Probabilistic graphical models: principles and techniques. MIT press.

Krizhevsky, A., Sutskever, I., & Hinton, G. E. (2012). Imagenet classification with deep convolutional neural networks. *Advances in neural information processing systems 25*.

Kroese, D. P., Brereton, T., Taimre, T., & Botev, Z. I. (2014). Why the monte carlo method is so important today. *Wiley Interdisciplinary Reviews: Computational Statistics 6*(6), 386–392.

LeCun, Y., Bottou, L., Bengio, Y., & Haffner, P. (1998). Gradient-based learning applied to document recognition. *Proceedings of the IEEE, 86* (11), 2278–2324.

Lemaréechal, C. (2012). Cauchy and the gradient method. Doc Math Extra, 251 (254), 10.

Markov, A. A. (1906). Rasprostranenie zakona bol'shih chisel na velichiny, zavisyaschie drug ot druga. *Izvestiya Fiziko-matematicheskogo obschestva pri Kazanskom universitete, 15* (135-156), 18.

Marsden, J. E., & Tromba, A. (2003). *Vector calculus*. Macmillan.

Meta. (2023). Investor earnings report for 3q 2023.

News, B. (2016). Artificial intelligence: Google's alphago beats go master lee se-dol.

Nyquist, H. (1928). Certain topics in telegraph transmission theory. *Transactions of the American Institute of Electrical Engineers, 47* (2), 617–644.

Oppenheim, A. V., Willsky, A. S., Nawab, S. H., & Ding, J.-J. (1997). *Signals and systems* (Vol. 2). Prentice hall Upper Saddle River, NJ.

Patterson, D. A., Gibson, G., & Katz, R. H. (1988). A case for redundant arrays of inexpensive disks (raid). In *Proceedings of the 1988 acm sigmod international conference on management of data* (pp. 109–116).

Pitman, J. (2012). *Probability*. Springer Science & Business Media.

Pólya, G. (1920). Über den zentralen grenzwertsatz der wahrscheinlichkeitsrechnung und das momentenproblem. *Mathematische Zeitschrift 8*(3-4), 171–181.

Polyak, B. T. (1964). Some methods of speeding up the convergence of iteration methods. *Ussr computational mathematics and mathematical physics 4*(5), 1–17.

Pouwelse, J., Garbacki, P., Epema, D., & Sips, H. (2005). The bittorrent p2p file-sharing system: Measurements and analysis. In *Peer-to-peer systems iv: 4th international workshop, iptps 2005, ithaca, ny, usa, february 24-25, 2005. revised selected papers 4* (pp. 205–216).

Rabiner, L., & Juang, B. (1986). An introduction to hidden markov models. *ieee assp magazine, 3* (1), 4–16.

Rosenblatt, F. (1958). The perceptron: a probabilistic model for information storage and organization in the brain. *Psychological review, 65* (6), 386.

Ross, S. M. (2014). A first course in probability.

Samuel, A. L. (1967). Some studies in machine learning using the game of checkers. ii–recent progress. *IBM Journal of research and development 11*(6), 601–617.

Shannon, C. E. (2001). A mathematical theory of communication. *ACM SIGMOBILE mobile computing and communications review 5*(1), 3–55.

Shen, J., Tang, T., & Wang, L.-L. (2011). *Spectral methods: algorithms, analysis and applications* (Vol. 41). Springer Science & Business Media.

Si, H., Vikalo, H., & Vishwanath, S. (2014). Haplotype assembly: An information theoretic view. In *2014 ieee information theory workshop (itw 2014)* (pp. 182–186).

Silver, D., Huang, A., Maddison, C. J., Guez, A., Sifre, L., Van Den Driessche, G., ... others (2016). Mastering the game of go with deep neural networks and tree search. *nature 529*(7587), 484–489.

Stewart, J. (2015). *Calculus*. Cengage Learning.

Strang, G. (2022). *Introduction to linear algebra*. SIAM.

Suh, C. (2022). *Convex optimization for machine learning*. Now Publishers.

Suh, C. (2023a). *Communication principles for data science*. Springer.

Suh, C. (2023b). *Information theory for data science*. Now Publishers.

Unpingco, J. (2016). *Python for probability, statistics, and machine learning* (Vol. 1). Springer.

Vaseghi, S. V. (2008). *Advanced digital signal processing and noise reduction*. John Wiley & Sons.

Walrand, J. (2021). *Probability in electrical engineering and computer science: An application-driven course*. Springer Nature.

Yates, R. D., & Goodman, D. J. (2014). *Probability and stochastic processes: a friendly introduction for electrical and computer engineers*. John Wiley & Sons.

Yerushalmy, J. 1947. Statistical problems in assessing methods of medical diagnosis, with special reference to x-ray techniques. *Public Health Reports (1896-1970)* 1432–1449.

Index

© The Editor(s) (if applicable) and The Author(s), under exclusive license to Springer
Nature Singapore Pte Ltd. 2025
C. Suh, *Probability for Information Technology*,
https://doi.org/10.1007/978-981-97-4032-1

GPSR Compliance

The European Union's (EU) General Product Safety Regulation (GPSR) is a set of rules that requires consumer products to be safe and our obligations to ensure this.

If you have any concerns about our products, you can contact us on ProductSafety@springernature.com

In case Publisher is established outside the EU, the EU authorized representative is:

Springer Nature Customer Service Center GmbH
Europaplatz 3
69115 Heidelberg, Germany

The manufacturer's authorised representative in the EU is Springer
Nature Customer Service Centre GmbH, Europaplatz 3, 69115 Heidelberg,
Germany. If you have any concerns regarding our products, please
contact ProductSafety@springernature.com

Printed and bound by CPI Group (UK) Ltd, Croydon, CR0 4YY
29/04/2026
02099466-0005